Second Edition

BEBOP TO THE BOOLEAN BOOGIE

An unconventional guide to electronics fundamentals, components, and processes

by Clive (call me "Max") Maxfield

Foreword by Pete Waddell,
Publisher of *Printed Circuit Design*

 Newnes

Amsterdam   Boston   London   New York   Oxford   Paris
San Diego   San Francisco   Singapore   Sydney   Toyko

Newnes is an imprint of Elsevier Science.

Recognizing the importance of preserving what has been written, Elsevier Science prints its books on acid-free paper whenever possible.

Library of Congress Cataloging-in-Publication Data
Maxfield, Clive, 1957–
    Bebop to the boolean boogie : an unconventional guide to electronics fundamentals, components, and processes / by Clive (call me "Max") Maxfield ; foreword by Pete Waddell.—2nd ed.
        p.   cm
    Includes bibliographical references and index.
    ISBN 0-7506-7543-8 (alk. paper)
    1. Digital electronics—Popular works.   I. Title.
TK7868.D5 M323 2002
621.381—dc21   2002038930

British Library Cataloguing-in-Publication Data
A catalogue record for this book is available from the British Library.

The publisher offers special discounts on bulk orders of this book. For information, please contact:

Manager of Special Sales
Elsevier Science
200 Wheeler Road
Burlington, MA 01803
Tel: 781-313-4700
Fax: 781-313-4882

For information on all Newnes publications available, contact our World Wide Web home page at: http://www.newnespress.com

Printed in the United States of America

10 9 8 7 6 5 4 3 2 1

Foreword

My first exposure to the unique writing style of Clive (call me "Max") Maxfield was a magazine article that he co-wrote with an associate. The article was technically brilliant (he paid me to say that) and very informative, but it was the short biography at the end of the piece that I enjoyed the most. I say enjoyed the most because, as you will soon learn, Max does not necessarily follow the herd or dance to the same drummer as the masses. Trade journals have a reputation for being informative and educational but also as dry as West Texas real estate.

Anyway, Max's personally submitted biography not only included a message from his mom, but also made mention of the fact that he (Max) is taller than his co-author, who just happened to be his boss at the time. Now to some people this may seem irrelevant, but to our readers (and Max's boss), these kind of things—trivial as they may seem to the uninitiated— are what helps us to maintain our off-grid sense of the world. Max has become, for better or worse, a part of that alternate life experience.

So now it's a couple of years later, and Max has asked me to write a few words by way of introduction. Personally, I think that the title of this tome alone (hmmm, a movie?) should provide some input as to what you can expect. But, for those who require a bit more: be forewarned, dear reader, you will probably learn far more than you could hope to expect from *Bebop to the Boolean Boogie*, just because of the unique approach Max has to technical material. The author will guide you from the basics through a minefield of potentially boring theoretical mish-mash, to a Nirvana of understanding. You will not suffer that fate familiar to every reader: rereading paragraphs over and over wondering what in the world the author was trying to say. For a limey, Max shoots amazingly well and from the hip, but in a way that will keep you interested and amused. If you are not vigilant, you may not only learn something, but you may even enjoy the process. The only further advice I can give is to "expect the unexpected."

– Pete Waddell, Publisher, *Printed Circuit Design*
Literary genius (so says his mom), and taller than Max by ⅛"

Contents

Section 2 Components & Processes

*This book is dedicated to my Auntie Barbara,
whose assiduous scrubbing in my younger years
has left me the proud owner of the cleanest pair
of knees in the known universe!*

About this Book

Note from the author with regard to this second edition.

I awoke one Saturday morning in July 1992 with the idea that it would be "sort of cool" to stroll into a bookshop and see something I'd written on the shelves. So with no clue as to what this would actually entail, I started penning the first edition of *Bebop to the Boolean Boogie*, which eventually hit the streets in 1995.

Much to my surprise, *Bebop* quickly found use at Yale University as part of an introductory electronics course (it was subsequently adopted by a number of other universities around the world), and it soon became required reading for sales and marketing groups at a number of high-tech companies in Silicon Valley and across the USA.

Time passed by (as is its wont), and suddenly it was seven years later and we were in a new millennium! Over these last few years, electronics and computing technology has progressed in leaps and bounds. For example, in 1995, an integrated circuit containing around 14 million transistors was considered to be relatively state-of the art. By the summer of 2002, however, Intel had announced a test chip containing 330 million transistors!

And it's not just improvements to existing technologies, because over the last few years entirely new ideas like carbon nanotubes have made their appearance on the scene. Therefore, by popular demand, I've completely revamped *Bebop* from cover to cover, revising the nitty-gritty details to reflect the latest in technology, and adding a myriad of new facts, topics, and nuggets of trivia (see especially the bonus Chapter 22 on the CD ROM accompanying the book). Enjoy!

This outrageously interesting book has two namesakes, Bebop, a jazz style known for its fast tempos and agitated rhythms, and Boolean algebra, a branch of mathematics that is the mainstay of the electronics designer's tool chest. *Bebop to the Boolean Boogie* meets the expectations set by both, because it leaps from topic to topic with the agility of a mountain goat, and it will become your key reference guide to understanding the weird and wonderful world of electronics.

Bebop to the Boolean Boogie provides a wide-ranging but comprehensive introduction to the electronics arena, roaming through the fundamental concepts, and rampaging through electronic components and the processes used to create them. As a bonus, nuggets of trivia are included with which you can amaze your family and friends; for example, Greenland Eskimos have a base twenty number system because they count using both fingers and toes.

Section 1: Fundamental Concepts starts by considering the differences between analog and digital views of the world. We then proceed rapidly through atomic theory and semiconductor switches to primitive logic functions and their electronic implementations. The concepts of alternative numbering systems are presented, along with binary arithmetic, Boolean algebra, and Karnaugh map representations. Finally, the construction of more complex logical functions is considered along with their applications.

Section 2: Components and Processes is where we consider the components from which electronic systems are formed and the processes required to construct them. The construction of integrated circuits is examined in some detail, followed by introductions to memory devices, programmable devices, and application-specific devices. The discussion continues with hybrids, printed circuit boards, and multichip modules. We close with an overview of some alternative and future technologies along with a history of where everything came from. Also, there's a bonus chapter (Chapter 22), *An Illustrated History of Electronics and Computing,* on the CD-ROM accompanying this book, that will answer questions you didn't even think to ask!

This book is of particular interest to electronics students. Additionally, by clarifying the techno-speech used by engineers, the book is of value to anyone who is interested in understanding more about electronics but lacks a strong technical background.

Except where such interpretation is inconsistent with the context, the singular shall be deemed to include the plural, the masculine shall be deemed to include the feminine, and the spelling (and the punctuation) shall be deemed to be correct!

About the Author

Clive "Max" Maxfield is 6'1" tall, outrageously handsome, English and proud of it. In addition to being a hero, trendsetter, and leader of fashion, he is widely regarded as an expert in all aspects of electronics (at least by his mother).

After receiving his B.Sc. in Control Engineering in 1980 from Sheffield Polytechnic (now Sheffield Hallam University), England, Max began his career as a designer of central processing units for mainframe computers. To cut a long story short, Max now finds himself President of TechBites Interactive (www.techbites.com). A marketing consultancy, TechBites specializes in communicating the value of technical products and services to non-technical audiences through such mediums as websites, advertising, technical documents, brochures, collaterals, and multimedia.

In his spare time (Ha!), Max is co-editor and co-publisher of the web-delivered electronics and computing hobbyist magazine *EPE Online* (www.epemag.com) and a contributing editor to www.eedesign.com. In addition to numerous technical articles and papers appearing in magazines and at conferences around the world, Max is also the author of *Designus Maximus Unleashed (Banned in Alabama)* and co-author of *Bebop BYTES Back (An Unconventional Guide to Computers)*.

On the off-chance that you're still not impressed, Max was once referred to as an *"industry notable"* and a *"semiconductor design expert"* by someone famous who wasn't prompted, coerced, or remunerated in any way!

Acknowledgments

Special thanks for technical advice go to Alvin Brown, Alon Kfir, Don Kuk, and Preston Jett, the closest thing to living encyclopedic reference manuals one could hope to meet. (The reason that the text contains so few bibliographic references is due to the fact that I never had to look anything up—I simply asked the relevant expert for the definitive answer.)

I would also like to thank Dave Thompson from Mentor Graphics, Tamara Snowden and Robert Bielby from Xilinx, Stuart Hamilton from NEC, Richard Gordon and Gary Smith from Gartner Dataquest, Richard Goering from EE Times, high-speed design expert Lee Ritchey from Speeding Edge, and circuit board technologist Happy Holden from Westwood Associates, all of whom helped out with critical nuggets of information just when I needed them the most.

Thanks also to Joan Doggrell, who labored furiously to meet my ridiculous deadlines. An old friend and expert copy editor, Joan not only corrected my syntax and grammar, but also offered numerous suggestions that greatly improved the final result. (In the unlikely event that any errors did creep in, they can only be attributed to cosmic rays and have nothing whatsoever to do with me.)

Last but not least, I should also like to mention my daughters—Abby and Lucie—without whom this book would never have materialized (they so depleted my financial resources that I was obliged to look for a supplemental source of income).

– Clive (Max) Maxfield, June 2002

Analog Versus Digital

It was a dark and stormy night . . . always wanted to start a book this way, and this is as good a time as any, but we digress . . .

Now sit up and pay attention because this bit is important. Electronic engineers split their world into two views called *analog* and *digital*, and it's necessary to understand the difference between these views to make much sense out of the rest of this book.[1]

A digital quantity is one that can be represented as being in one of a finite number of states, such as 0 and 1, ON and OFF, UP and DOWN, and so on. As an example of a simple digital system, consider a light switch in a house. When the switch is UP, the light is ON, and when the switch is DOWN, the light is OFF.[2] By comparison, a light controlled by a dimmer switch provides an example of a simple analog system.

Made famous by the Peanuts cartoon character, Snoopy, the phrase "*It was a dark and stormy night . . .*" is actually the opening sentence to an 1830 book by the British author Edward George Bulwer-Lytton. A legend in his own lunchtime, Bulwer-Lytton became renowned for penning exceptionally bad prose, of which the opening to his book *Paul Clifford* set the standard for others to follow. For your delectation and delight, the complete opening sentence was: "*It was a dark and stormy night; the rain fell in torrents— except at occasional intervals, when it was checked by a violent gust of wind which swept up the streets (for it is in London that our scene lies), rattling along the housetops, and fiercely agitating the scanty flame of the lamps that struggled against the darkness.*" Where else are you going to go to learn nuggets of trivia like this? As you can see, this isn't your mother's electronics book!

---

[1] In England, "analog" is spelled "analogue" (and pronounced with a really cool accent).

[2] At least, that's the way they work in America. It's the opposite way round in England, and you take your chances in the rest of the world.

We can illustrate the differences in the way these two systems work by means of a graph-like diagram (Figure 1-1). Time is considered to progress from left to right, and the solid lines, which engineers often refer to as *waveforms*, indicate what is happening.

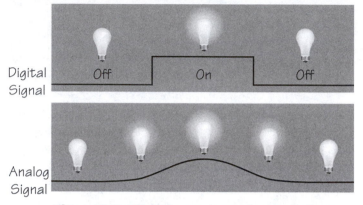

Figure 1-1. Digital versus analog waveforms

In this figure, the digital waveform commences in its OFF state, changes to its ON state, and then returns to its OFF state. In the case of the analog waveform, however, we typically don't think in terms of ON and OFF. Rather, we tend to regard things as being *more* OFF or *more* ON with an infinite number of values between the two extremes.

One interesting point about digital systems is that they can have more than two states. For example, consider a fun-loving fool sliding down a ramp mounted alongside a staircase (Figure 1-2).

In order to accurately determine this person's position on the ramp, an independent observer would require the use of a tape measure. Alternatively, the observer could estimate the ramp-slider's approximate location in relation to the nearest stair. The exact position on the ramp, as measured using the tape measure, would be considered to be an analog value. In this case, the analog value most closely represents the real world and can be as precise as the measuring

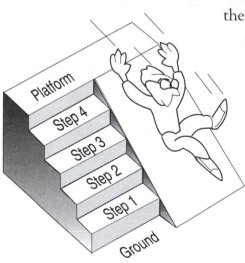

Figure 1-2. Staircase and ramp

technique allows. By comparison, an estimation based on the nearest stair would be considered to be a digital value. As was previously noted, a digital value is represented as being in one of a finite number of discrete states. These states are called *quanta* (from the Latin neuter of *quantus*, meaning "how great") and the accuracy, or resolution, of a digital value is dependent on the number of quanta employed to represent it.

Assume that at some starting time we'll call T_0 ("time zero"), our thrill-seeker is balanced at the top of the ramp preparing to take the plunge. He commences sliding at time T_1 and reaches the bottom of the ramp at time T_2. Analog and digital waveforms can be plotted representing the location of the person on the ramp as a function of time (Figure 1-3).

Once again, the horizontal axis in both waveforms represents the passage of time, which is considered to progress from left to right. In the case of the analog waveform, the vertical axis is used to represent the thrill-seeker's exact location in terms of height above the ground, and is therefore annotated with real, physical units. By comparison, the vertical axis for the digital waveform is annotated with abstract labels, which do not have any units associated with them.

To examine the differences between analog and digital views in more detail, let's consider a brick suspended from a wooden beam by a piece of elastic. If the brick is left to its own devices, it will eventually reach a stable state in which the pull of gravity is balanced by the tension in the elastic (Figure 1-4).

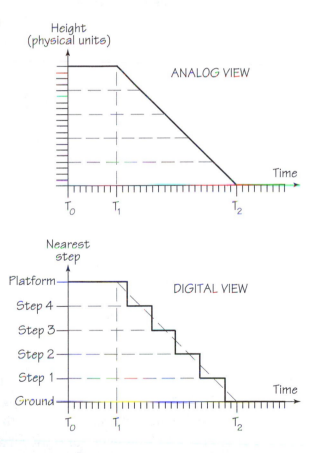

Figure 1-3. Analog and digital waveforms showing the position of the person sliding down the ramp

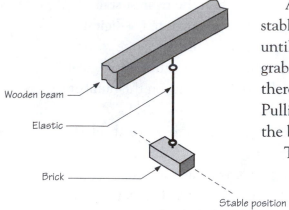

Wooden beam

Elastic

Brick

Stable position
(tension in elastic balances pull of gravity)

Figure 1-4. Brick suspended by elastic

Assume that at time T_0 the system is in its stable state. The system remains in this state until time T_1, when an inquisitive passerby grabs hold of the brick and pulls it down, thereby increasing the tension on the elastic. Pulling the brick down takes some time, and the brick reaches its lowest point at time T_2. The passerby hangs around for a while looking somewhat foolish, releases the brick at time T_3, and thereafter exits from our story. The brick's resulting motion may be illustrated using an analog waveform (Figure 1-5).

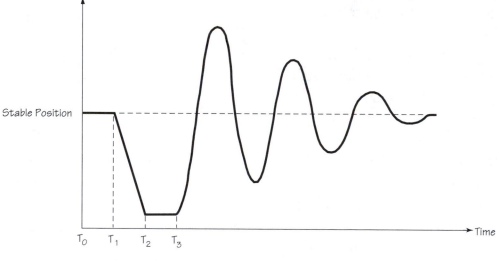

Height

Stable Position

T_0 T_1 T_2 T_3 Time

Figure 1-5. Brick on elastic: analog waveform

Now consider a digital view of the brick's motion represented by two quanta labeled LOW and HIGH. The LOW quanta may be taken to represent any height less than or equal to the system's stable position, and the HIGH quanta therefore represents any height greater than the stable position (Figure 1-6). (Our original analog waveform is shown as a dotted line.)

Although it is apparent that the digital view is a subset of the analog view, digital representations often provide extremely useful approximations to

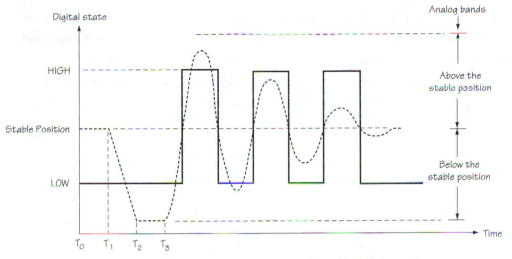

Figure 1-6. Brick on elastic: two-quanta digital waveform

the real world. If the only requirement in the above example is to determine whether the brick is above or below the stable position, then the digital view is the most appropriate.

The accuracy of a digital view can be improved by adding more quanta. For example, consider a digital view with five quanta: LOW, LOW-MIDDLE, MIDDLE, HIGH-MIDDLE, and HIGH. As the number of quanta increases, the digital view more closely approximates the analog view (Figure 1-7).

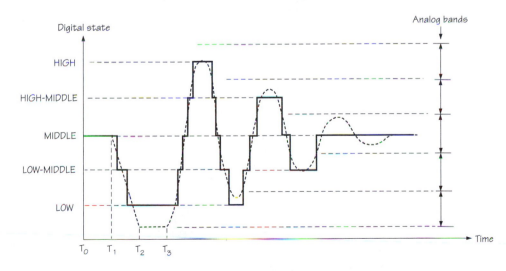

Figure 1-7. Brick on elastic: five-quanta digital waveform

In the real world, every electronic component behaves in an analog fashion. However, these components can be connected together so as to form functions whose behavior is amenable to digital approximations. This book concentrates on the digital view of electronics, although certain aspects of analog designs and the effects associated with them are discussed where appropriate.

Atoms, Molecules, and Crystals

Matter, the stuff that everything is made of, is formed from *atoms*. The heart of an atom, the *nucleus*, is composed of *protons* and *neutrons* and is surrounded by a "cloud" of *electrons*.[1] For example, consider an atom of the gas helium, which consists of two protons, two neutrons, and two electrons (Figure 2-1).

It may help to visualize the electrons as orbiting the nucleus in the same way that the moon orbits the earth. In the real world things aren't this simple, but the concept of orbiting electrons serves our purpose here.

Each proton carries a single positive (+ve) charge, and each electron carries a single negative (−ve) charge. The neutrons are neutral and act like glue, holding the nucleus together and resisting the natural tendency of the protons to repel each other. Protons and neutrons are approximately the same size, while electrons are very much smaller. If a basketball were used to represent the nucleus of a helium atom, then, on the same scale, softballs could represent the individual protons and neutrons, while large garden peas could represent the electrons. In this case, the diameter of an electron's orbit would be approximately equal to the length of 250 American

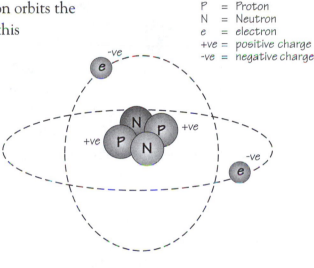

P = Proton
N = Neutron
e = electron
+ve = positive charge
−ve = negative charge

Figure 2-1. Helium atom

---

[1] We now know that protons and neutrons are formed from fundamental particles called *quarks*, of which there are six flavors: *up*, *down*, *charm*, *strange*, *top* (or *truth*), and *bottom* (or *beauty*). Quarks are so weird that they have been referred to as *"The dreams that stuff is made from,"* and they are way beyond the scope of this book.

football fields (excluding the end zones)! Thus, the majority of an atom consists of empty space. If all the empty space were removed from the atoms that form a camel, it *would* be possible for the little rascal to pass through the eye of a needle![2,3,4]

The number of protons determines the type of the element; for example, hydrogen has one proton, helium two, lithium three, etc. Atoms vary greatly in size, from hydrogen with its single proton to those containing hundreds of protons. The number of neutrons does not necessarily equal the number of protons. There may be several different flavors, or *isotopes*, of the same element differing only in their number of neutrons; for example, hydrogen has three isotopes with zero, one, and two neutrons, respectively.

Left to its own devices, each proton in the nucleus will have a complementary electron. If additional electrons are forcibly added to an atom, the result is a *negative ion* of that atom; if electrons are forcibly removed from an atom, the result is a *positive ion*.

In an atom where each proton is balanced by a complementary electron, one would assume that the atom would be stable and content with its own company, but things are not always as they seem. Although every electron contains the same amount of negative charge, electrons orbit the nucleus at different levels known as *quantum levels* or *electron shells*. Each electron shell requires a specific number of electrons to fill it; the first shell requires two electrons, the second requires eight, etc. Thus, although a hydrogen atom contains both a proton and an electron and is therefore electrically balanced, it is still not completely happy. Given a choice, hydrogen would prefer to have a second electron to fill its first electron shell. However, simply adding a second electron is not the solution; although the first electron shell would now be filled, the extra electron would result in an electrically unbalanced negative ion.

Obviously this is a bit of a poser, but the maker of the universe came up with a solution; atoms can use the electrons in their outermost shell to form

---

[2] I am of course referring to the Bible verse: *"It is easier for a camel to go through the eye of a needle, than for a rich man to enter the Kingdom of God."* (Mark 10:25).

[3] In fact, the "needle" was a small, man-sized gate located next to the main entrance to Jerusalem.

[4] The author has discovered to his cost that if you call a zoo to ask the cubic volume of the average adult camel, they treat you as if you are a complete idiot . . . go figure!

bonds with other atoms. The atoms share each other's electrons, thereby forming more complex structures. One such structure is called a *molecule*; for example, two hydrogen atoms (chemical symbol H), each comprising a single proton and electron, can bond together and share their electrons to form a hydrogen molecule (chemical symbol H_2) (Figure 2-2).

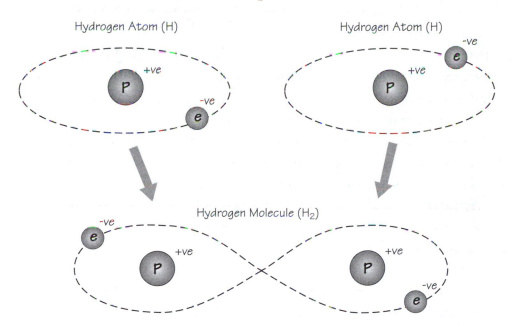

Figure 2-2. Two hydrogen atoms bonding to form a hydrogen molecule

These types of bonds are called *valence bonds*. The resulting hydrogen molecule contains two protons and two electrons from its constituent atoms and so remains electrically balanced. However, each atom lends its electron to its partner and, at the same time, borrows an electron from its partner. This can be compared to two circus jugglers passing objects between each other—the quickness of the hand deceives the eye. The electrons are passing backwards and forwards between the atoms so quickly that each atom is fooled into believing it has two electrons. The first electron shell of each atom appears to be completely filled and the hydrogen molecule is therefore stable.

Even though the hydrogen molecule is the simplest molecule of all, the previous illustration demanded a significant amount of time, space, and effort. Molecules formed from atoms containing more than a few protons and electrons would be well nigh impossible to represent in this manner. A simpler form of representation is therefore employed, with two dashed lines indicating the sharing of two electrons (Figure 2-3).

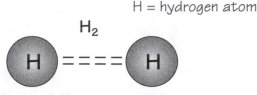

H = hydrogen atom

H_2

Figure 2-3. Alternative representation of a hydrogen molecule

Now contrast the case of hydrogen with helium. Helium atoms each have two protons and two electrons and are therefore electrically balanced. Additionally, as helium's two electrons completely fill its first electron shell, this atom is very stable.[5] This means that, under normal circumstances, helium atoms do not go around casually making molecules with every other atom they meet.

Molecules can also be formed by combining different types of atoms. An oxygen atom (chemical symbol O) contains eight protons and eight electrons. Two of the electrons are used to fill the first electron shell, which leaves six left over for the second shell. Unfortunately for oxygen, its second shell would ideally prefer eight electrons to fill it. Each oxygen atom can therefore form two bonds with other atoms—for example, with two hydrogen atoms to form a water molecule (chemical symbol H_2O) (Figure 2-4). (The reason the three atoms in the water molecule are not shown as forming a straight line is discussed in the section on nanotechnology in Chapter 21.)

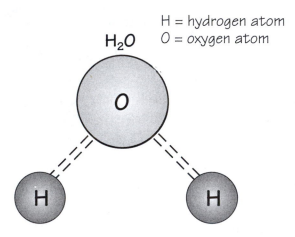

H = hydrogen atom
O = oxygen atom

H_2O

Figure 2-4. Water molecule

Each hydrogen atom lends its electron to the oxygen atom and at the same time borrows an electron from the oxygen atom. This leads both of the hydrogen atoms to believe they have two electrons in their first electron shell. Similarly, the oxygen atom lends two electrons (one to each hydrogen atom) and at the same time borrows two electrons (one from each hydrogen atom).

---

[5] Because helium is so stable, it is known as an *inert*, or *noble*, gas (the latter appellation presumably comes from the fact that helium doesn't mingle with the commoners <grin>).

When the two borrowed electrons are added to the original six in the oxygen atom's second shell, this shell appears to contain the eight electrons necessary to fill it. Thus, all the atoms in the water molecule are satisfied with their lot and the molecule is stable.

Structures other than molecules may be formed when atoms bond; for example, *crystals.* Carbon, silicon, and germanium all belong to the same family of elements; each has only four electrons in its outermost electron shell. Silicon has 14 protons and 14 electrons; two electrons are required to fill the first electron shell and eight to fill the second shell; thus, only four remain for the third shell, which would ideally prefer eight. Under the appropriate conditions, each silicon atom will form bonds with four other silicon atoms, resulting in a three-dimensional silicon crystal[6] (Figure 2-5).

The electrons used to form the bonds in crystalline structures such as silicon are tightly bound to their respective atoms. Yet another structure is presented by metals such as copper, silver, and gold. Metals have an *amorphous crystalline structure* in which their shared electrons have relatively weak bonds and may easily migrate from one atom to another.

Apart from the fact that atoms are the basis of life, the universe, and everything as we know it, they are also fundamental to the operation of the components used in electronic designs. Electricity may be considered to be vast herds of electrons migrating from one place to another, while electronics is the science of controlling these herds: starting them, stopping them, deciding where they can roam, and determining what they are going to do when they get there.

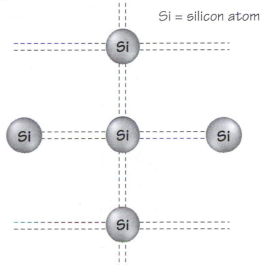

Si = silicon atom

Figure 2-5. Simplified (two-dimensional) representation of the three-dimensional structure of crystalline silicon

---

[6] An equivalent structure formed from carbon atoms is known as diamond.

Conductors and Insulators; Voltage, Current, Resistance, Capacitance, and Inductance

A substance that conducts electricity easily is called a *conductor*. Metals such as copper are very good conductors because the bonds in their amorphous crystalline structures are relatively weak, and the bonding electrons can easily migrate from one atom to another. If a piece of copper wire is used to connect a source with an excess of electrons to a target with too few electrons, the wire will conduct electrons between them (Figure 3-1).

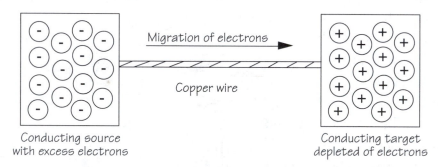

Figure 3-1. Electrons flowing through a copper wire

If we consider electricity to be the migration of electrons from one place to another, then we may also say that it flows from the more negative source to the more positive target. As an electron *jumps* from the negative source into the wire, it *pushes* the nearest electron in the wire out of the way. This electron pushes another in turn, and the effect ripples down the wire until an electron at the far end of the wire is ejected into the more positive target. When an electron arrives in the positive target, it neutralizes one of the positive charges.

An individual electron will take a surprisingly long time to migrate from one end of the wire to the other; however, the time between an electron entering one end of the wire and causing an equivalent electron to be ejected from the other end is extremely fast.[1]

As opposed to a conductor, a substance which does not conduct electricity easily is called an *insulator*. Materials such as rubber are very good insulators because the electrons used to form bonds are tightly bound to their respective atoms.[2]

Voltage, Current, and Resistance

One measure of whether a substance is a conductor or an insulator is how much it resists the flow of electricity. Imagine a tank of water to which two pipes are connected at different heights; the water ejected from the pipes is caught in two buckets A and B (Figure 3-2).

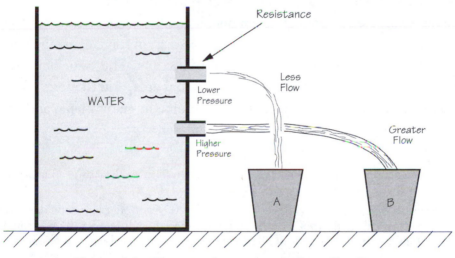

**Figure 3-2. Water tank representation of voltage,
current, and resistance**

---

[1] For a copper wire isolated in a vacuum, the speed of a signal propagating through the wire is only fractionally less than the speed of light. However, the speed of a signal is modified by a variety of factors, including any materials surrounding or near the conductor. Signal speeds in electronic circuits vary, but are typically in the vicinity of half the speed of light.

[2] In reality, everything conducts if presented with a sufficiently powerful electric potential. For example, if you don a pair of rubber boots and then fly a kite in a thunderstorm, your rubber boots won't save you when the lightning comes racing down the kite string! (Bearing in mind that litigation is a national sport in America, do *NOT* try this at home unless you are a professional.)

Let's assume that the contents of the tank are magically maintained at a constant level. The water pressure at the end of a pipe inside the tank depends on the depth of the pipe with respect to the surface level. The difference in pressure between the ends of a pipe inside and outside the tank causes water to flow. The amount of water flowing through a pipe depends on the water pressure and on the resistance to that pressure determined by the pipe's cross-sectional area. A thin pipe with a smaller cross-sectional area will present more resistance to the water than a thicker pipe with a larger cross-sectional area. Thus, if both pipes have the same cross-sectional area, bucket B will fill faster than bucket A.

In electronic systems, the flow of electricity is called *current* measured in units of *amperes* or *amps*;[3,4] the resistance to electrical flow is simply called *resistance* measured in units of *ohms*.[5] and the electrical equivalent to pressure is called *voltage*, or electric potential, measured in units of *volts*.[6]

The materials used to connect components in electronic circuits are typically selected to have low resistance values; however, in some cases engineers make use of special resistive components called *resistors*. The value of resistance (R) depends on the resistor's length, cross-sectional area, and the resistivity of the material from which it is formed. Resistors come in many shapes and sizes; a common example could be as shown in Figure 3-3.[7,8]

---

[3] The term *amp* is named after the French mathematician and physicist André-Marie Ampère, who formulated one of the basic laws of electromagnetism in 1820.

[4] An amp corresponds to approximately 6,250,000,000,000,000,000 electrons per second flowing past a given point in an electrical circuit (not that the author counted them himself, you understand; this little nugget of information is courtesy of Microsoft's multimedia encyclopedia, *Encarta*).

[5] The term *ohm* is named after the German physicist Georg Simon Ohm, who defined the relationship between voltage, current, and resistance in 1827 (we now call this *Ohm's Law*).

[6] The term *volt* is named after the Italian physicist Count Alessandro Giuseppe Antonio Anastastio Volta, who invented the electric battery in 1800. (Having said this, some people believe that an ancient copper-lined jar found in an Egyptian pyramid was in fact a primitive battery . . . there again, some people will believe anything. Who knows for sure?)

[7] In addition to the simple resistor shown here, there are also variable resistors (sometimes called potentiometers), in which a third "center" connection is made via a conducting slider. Changing the position of the slider (perhaps by turning a knob) alters the relative resistance between the center connection and the two ends.

[8] There are also a variety of sensor resistors, including light-dependent resistors (LDRs) whose value depends on the amount of light falling on them, heat-dependent resistors called *thermistors*, and voltage-dependent resistors called VDRs or *varistors*.

In a steady-state system where everything is constant, the voltage, current, and resistance are related by a rule called *Ohm's Law*, which states that voltage (V) equals current (I) multiplied by resistance (R). An easy method for remembering

Approx. actual size

(a) Discrete Component
(b) Symbol

Figure 3-3. Resistor: component and symbol

Ohm's Law is by means of a diagram known as *Ohm's Triangle* (Figure 3-4).

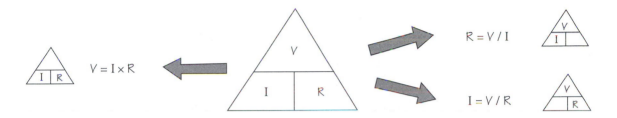

Figure 3-4. Ohm's Triangle

Consider a simple electrical circuit comprising two wires with electrical potentials of 5 volts and 0 volts connected by a resistor of 10 ohms (Figure 3-5).[9,10]

This illustration shows the direction of current flow as being from the more positive (+5 volts) to the more negative (0 volts). This may seem strange, as we know that current actually consists of electrons migrating from a negative source to a positive target.

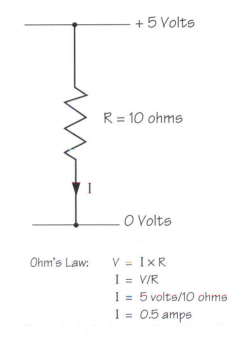

Ohm's Law:
$$V = I \times R$$
$$I = V/R$$
$$I = 5 \text{ volts}/10 \text{ ohms}$$
$$I = 0.5 \text{ amps}$$

Figure 3-5. Current flowing through a resistor

---

[9] Instead of writing "5 volts," engineers would simply use "5V" (a useful rule to remember is that no space is used for a single-letter qualifier like "5V," but spaces are used for multi-letter qualifiers like "5 Hz").

[10] The Greek letter omega "Ω" is used to represent resistance, so instead of writing "10 ohms," engineers would typically use "10Ω."

The reason for this inconsistency is that the existence of electricity was discovered long before it was fully understood. Electricity as a phenomenon was known for quite some time, but it wasn't until the early part of the 20th century that George Thomson proved the existence of the electron at the University of Aberdeen, Scotland. The men who established the original electrical theories had to make decisions about things they didn't fully understand. The direction of current flow is one such example; for a variety of reasons, it was originally believed that current flowed from positive to negative. As you may imagine, this inconsistency can, and does, cause endless problems.

Capacitance

Now imagine a full water tank A connected by a blocked pipe to an empty water tank B (Figure 3-6a). Assume that the contents of tank A are magically maintained at the same level regardless of the amount of water that is removed. At some time T_O ("time zero"), the pipe is unblocked and tank B starts to fill. By the time we'll call T_{FULL}, tank B will have reached the same level as tank A (Figure 3-6b).

The speed with which tank B fills depends on the rate with which water flows between the two tanks. The rate of water flow depends on the difference in pressure between the two ends of the pipe and any resistance to the flow

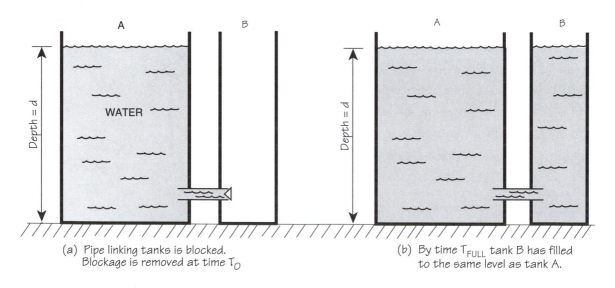

(a) Pipe linking tanks is blocked. Blockage is removed at time T_O

(b) By time T_{FULL} tank B has filled to the same level as tank A.

Figure 3-6. Water tank representation of capacitance

caused by the pipe. When the water starts to flow between the tanks at time T_O, there is a large pressure differential between the end of the pipe in tank A and the end in tank B; however, as tank B fills, the pressure differential between the tanks is correspondingly reduced. This means that tank B fills faster at the beginning of the process than it does

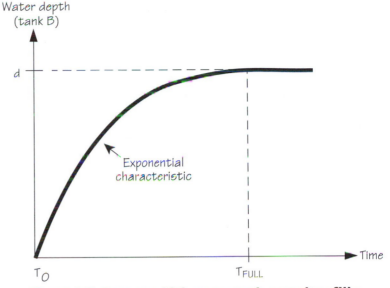

Figure 3-7. Rate at which water tank capacitor fills

at the end. The rate at which tank B fills has an *exponential* characteristic best illustrated by a graph (Figure 3-7).

The electronic equivalent of tank B stores electrical charge. This ability to store charge is called *capacitance* measured in units of *farads*.[11] Capacitances occur naturally in electronic circuits, and engineers generally try to ensure that their values are as low as possible; however, in some cases, designers may make use of special capacitive components called *capacitors*. One type of capacitor is formed from two metal plates separated by a layer of insulating material; the resulting capacitance (C) depends on the surface area of the plates, the size of the gap between them, and the material used to fill the gap. Capacitors come in many shapes and sizes; a common example could be as shown in Figure 3-8.

Approximate actual size

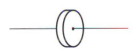

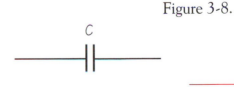

(a) Discrete component

(b) Symbol

Figure 3-8. Capacitor: component and symbol

[11] The term *farad* is named after the British scientist Michael Faraday, who constructed the first electric motor in 1821.

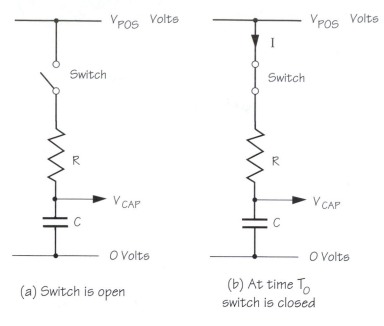

(a) Switch is open

(b) At time T_O switch is closed

Figure 3-9. Resistor-capacitor-switch circuit

Now consider a simple circuit consisting of a resistor, a capacitor, and a switch. Initially the switch is in the OPEN (OFF) position, the capacitor voltage V_{CAP} is 0 volts, and no current is flowing (Figure 3-9).

When the switch is CLOSED (turned ON), any difference in potential between V_{POS} and V_{CAP} will cause current to flow through the resistor. As usual, the direction of current flow is illustrated in the classical rather than the actual sense. The current flowing through the resistor causes the capacitor to charge towards V_{POS}. But as the capacitor charges, the difference in voltage between V_{POS} and V_{CAP} decreases, and consequently so does the current (Figure 3-10).

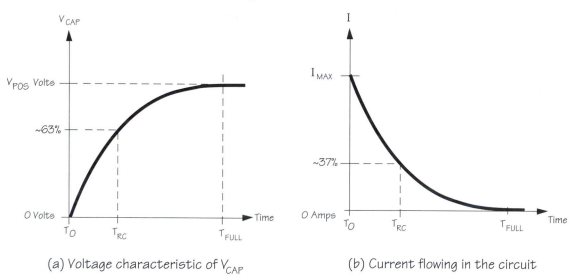

(a) Voltage characteristic of V_{CAP}

(b) Current flowing in the circuit

Figure 3-10. Voltage and current characteristics of resistor-capacitor-switch circuit

The maximum current I_{MAX} occurs at time T_O when there is the greatest difference between V_{POS} and V_{CAP}; from Ohm's Law, $I_{MAX} = V_{POS}/R$. The capacitor is considered to be completely charged by time T_{FULL}, at which point the flow of current falls to zero.

The time T_{RC} equals $R \times C$ and is known as the *RC time constant*. With R in ohms and C in farads, the resulting T_{RC} is in units of seconds. The RC time constant is approximately equal to the time taken for V_{CAP} to achieve 63% of its final value and for the current to fall to 37% of its initial value.[12]

Inductance

This is the tricky one. The author has yet to see a water-based analogy for inductance that didn't leak like a sieve <grin>. Consider two electric fans facing each other on a desk. If you turn one of the fans on, the other will start to spin in sympathy. That is, the first fan *induces* an effect in the second. Well, electrical inductance is just like this, but different.

What, you want more? Oh well, how about this then . . . a difference in electrical potential between two ends of a conducting wire causes current to flow, and current flowing through a wire causes an electromagnetic field to be generated around that wire (Figure 3-11).

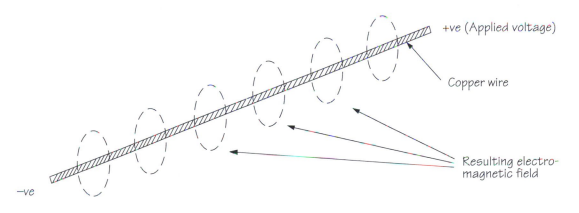

+ve (Applied voltage)

Copper wire

Resulting electro-magnetic field

−ve

Figure 3-11. Current flowing through a wire generates an electromagnetic field

---

[12] During each successive T_{RC} time constant, the capacitor will charge 63% of the *remaining* distance to the maximum voltage level. A capacitor is generally considered to be fully charged after five time constants.

Correspondingly, if a piece of wire is moved through an externally generated electromagnetic field, it cuts the lines of electromagnetic flux, resulting in an electrical potential being generated between the two ends of the wire (Figure 3-12).

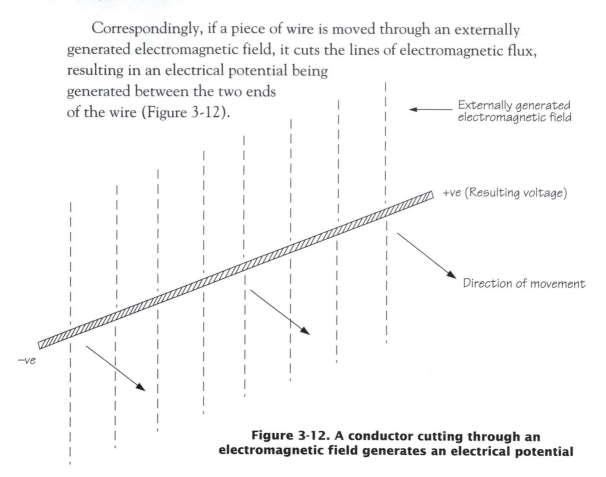

Figure 3-12. A conductor cutting through an electromagnetic field generates an electrical potential

Engineers sometimes make use of components called *inductors*, which may be formed by winding a wire into a coil. When a current is passed through the coil, the result is an intense electromagnetic field (Figure 3-13).

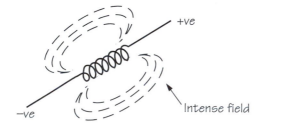

(a) Inductor Component

(b) Symbol

Figure 3-13. Inductor: component and symbol

Now consider a simple circuit consisting of a resistor, an inductor, and a switch. Initially the switch is in the OPEN (OFF) position, the inductor voltage V_{IND} is at V_{POS} volts, and no current is flowing (Figure 3-14).

As the inductor is formed from a piece of conducting wire, one might expect that closing the switch at time T_0 would immediately cause V_{IND} to drop to 0 volts; however, when the switch is CLOSED (turned ON) and

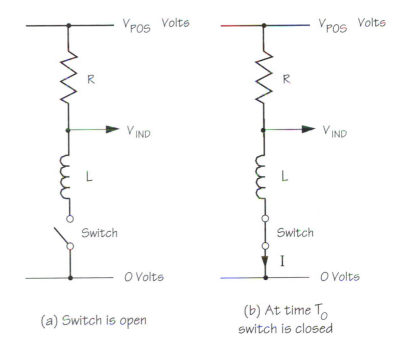

(a) Switch is open

(b) At time T_0 switch is closed

Figure 3-14. Resistor-inductor-switch circuit

current begins to flow, the inductor's electromagnetic field starts to form. As the field grows in strength, the lines of flux are pushed out from the center, and in the process they cut through the loops of wire forming the coil. This has the same effect as moving a conductor through an electromagnetic field and a voltage differential is created between the ends of the coil. This generated voltage is such that it attempts to oppose the changes causing it (Figure 3-15).

This effect is called *inductance*, the official unit of which is the *henry*.[13] As time progresses, the coil's inductance is overcome and its electromagnetic field is fully established. Thus, by the time we'll call T_{STABLE}, the inductor appears little different from any other piece of wire in the circuit (except for its concentrated electromagnetic field). This will remain the case until some new event occurs to disturb the circuit—for example, opening the switch again.

Inductors are typically formed by coiling a wire around a ferromagnetic rod. When you strike a musical tuning fork, it rings with a certain frequency

---

[13] The term *henry* is named after the American inventor Joseph Henry, who discovered inductance in 1832.

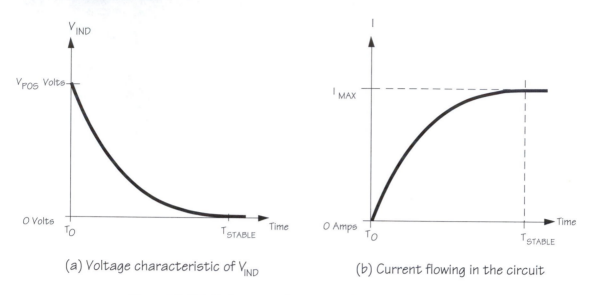

(a) Voltage characteristic of V_{IND} (b) Current flowing in the circuit

Figure 3-15. Voltage and current characteristics of resistor-inductor-switch circuit

depending on its physical characteristics (size, material, etc.). Similarly, an inductor has a natural resonant frequency depending on the diameter of the rod, the material used to form the rod, the number of coils, etc. For this reason, inductors may form part of the tuning circuits used in radios (they also have many other uses that we won't go into here).

Once a circuit has reached a steady-state condition where nothing is changing, any capacitors act like OPEN (OFF) switches and any inductors act like simple pieces of wire. The effects of these components only come into play when signals are transitioning between different values.

Unit Qualifiers

Engineers often work with very large or very small values of voltage, current, resistance, capacitance, and inductance. As an alternative to writing endless zeros, electrical quantities can be annotated with the qualifiers given in Table 3-1. For example, 15 MΩ[14] (15 megaohms) means fifteen million ohms, 4 mA (4 milliamps) means four thousandths of an amp, and 20 fF (20 femtofarads) means a very small capacitance indeed.

[14] Remember that the Greek letter omega "Ω" is used to represent resistance.

Table 3-1. Unit Qualifiers

Qualifier	Symbol	Factor	Name (U.S.)
yotta	Y	10^{24}	septillion (one million million million million)
zetta	Z	10^{21}	sextillion (one thousand million million million)
exa	E	10^{18}	quintillion (one million million million)
peta	P	10^{15}	quadrillion (one thousand million million)
tera	T	10^{12}	trillion (one million million)
giga	G	10^{9}	billion (one thousand million)[15]
mega	M	10^{6}	million
kilo	k	10^{3}	thousand
milli	m	10^{-3}	thousandth
micro	μ	10^{-6}	millionth
nano	n	10^{-9}	billionth (one thousandth of one millionth)
pico	p	10^{-12}	trillionth (one millionth of one millionth)
femto	f	10^{-15}	quadrillionth (one thousandth of one millionth of one millionth)
atto	a	10^{-18}	quintillionth (one millionth of one millionth of one millionth)
zepto	z	10^{-21}	sextillionth (one thousandth of one millionth of one millionth of one millionth)
yocto	y	10^{-24}	septillionth (one millionth of one millionth of one millionth of one millionth)

One last point that's worth noting is that the qualifiers *kilo*, *mega*, *giga*, and so forth mean slightly different things when we use them to describe the size of computer memory. The reasoning behind this (and many other mysteries) is revealed in Chapter 15.

---

[15] In Britain, "billion" traditionally used to mean "*a million million*" (10^{12}). However, for reasons unknown, the Americans decided that "billion" should mean "*a thousand million*" (10^9). In order to avoid the confusion that would otherwise ensue, most countries in the world (including Britain) have decided to go along with the Americans.

Semiconductors: Diodes and Transistors

As we noted earlier, electricity may be considered to be vast herds of electrons migrating from one place to another, while electronics is the science of controlling these herds. Ever since humans discovered electricity (as opposed to electricity—in the form of lightning—discovering us), taming the little rascal and bending it to our will has occupied a lot of thought and ingenuity.

The first, and certainly the simplest, form of control is the humble mechanical switch. Consider a circuit consisting of a switch, a power supply (say a battery), and a light bulb (Figure 4-1).

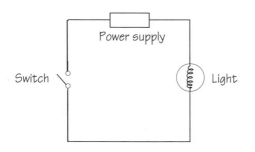

Figure 4-1. The simplest control device is a switch

When the switch is CLOSED, the light is ON, and when the switch is OPEN, the light is OFF. As we'll see in Chapter 5, we can actually realize interesting logical functions by connecting switches together in different ways. However, if mechanical switches were all we had to play with, the life of an electronics engineer would be fairly boring, so something with a bit more "zing" was required . . .

The Electromechanical Relay

By the end of the nineteenth century, when Queen Victoria still held sway over all she surveyed, the most sophisticated form of control for electrical systems was the electromechanical relay. This device consisted of a rod of iron (or some other ferromagnetic material) wrapped in a coil of wire. Applying an electrical potential across the ends of the coil caused the iron rod to act like a magnet. The magnetic field could be used to attract another piece of iron acting as a switch. Removing the potential from the coil caused the iron bar to lose its magnetism, and a small spring would return the switch to its inactive state.

The relay is a digital component, because it is either ON or OFF. By connecting relays together in different ways, it's possible to create all sorts of things. Perhaps the most ambitious use of relays was to build gigantic electro-mechanical computers, such as the Harvard Mark 1. Constructed between

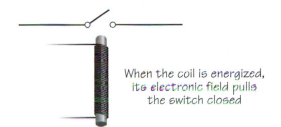

When the coil is energized, its electronic field pulls the switch closed

Figure 4-2. The electromechanical relay

1939 and 1944, the Harvard Mark 1 was 50 feet long, 8 feet tall, and contained over 750,000 individual components.

The problem with relays (especially the types that were around in the early days) is that they can only switch a limited number of times a second. This severely limits the performance of a relay-based computer. For example, the Harvard Mark 1 took approximately six seconds to multiply two numbers together, so engineers were interested in anything that could switch faster...

The First Vacuum Tubes

In 1879, the legendary American inventor Thomas Alva Edison publicly demonstrated his incandescent electric light bulb for the first time. This is the way it worked. A filament was mounted inside a glass bulb. Then all the air was sucked out, leaving a vacuum. When electricity was passed through the filament, it began to glow brightly (the vacuum stopped it from bursting into flames).

A few years later in 1883, one of Edison's assistants discovered that he could detect electrons flowing through the vacuum from the lighted filament to a metal plate mounted inside the bulb. Unfortunately, Edison didn't develop this so-called *Edison Effect* any further. In fact, it wasn't until 1904 that the English physicist Sir John Ambrose Fleming used this phenomenon to create the first vacuum tube.[1] This device, known as a *diode*, had two terminals and conducted electricity in only one direction (a feat that isn't as easy to achieve as you might think).

---

[1] Vacuum tubes are known as *valves* in England. This is based on the fact that they can be used to control the flow of electricity, similar in concept to the way in which their mechanical namesakes are used to control the flow of fluids.

In 1906, the American inventor Lee de Forest introduced a third electrode into his version of a vacuum tube. The resulting *triode* could be used as both an amplifier and a switch. De Forest's triodes revolutionized the broadcasting industry (he presented the first live opera broadcast and the first news report on radio). Furthermore, their ability to act as switches was to have a tremendous impact on digital computing.

One of the most famous early electronic digital computers is the *electronic numerical integrator and calculator (ENIAC)*, which was constructed at the University of Pennsylvania between 1943 and 1946. Occupying 1,000 square feet, weighing in at 30 tons, and employing 18,000 vacuum tubes, ENIAC was a monster . . . but it was a monster that could perform fourteen multiplications or 5,000 additions a second, which was way faster than the relay-based Harvard Mark 1.

However, in addition to requiring enough power to light a small town, ENIAC's vacuum tubes were horrendously unreliable, so researchers started looking for a smaller, faster, and more dependable alternative that didn't demand as much power . . .

Semiconductors

Most materials are conductors, insulators, or something in-between, but a special class of materials known as *semiconductors* can be persuaded to exhibit both conducting and insulating properties. The first semiconductor to undergo evaluation was the element germanium (chemical symbol Ge). However, for a variety of reasons, silicon (chemical symbol Si) replaced germanium as the semiconductor of choice. As silicon is the main constituent of sand and one of the most common elements on earth (silicon accounts for approximately 28% of the earth's crust), we aren't in any danger of running out of it in the foreseeable future.

Pure crystalline silicon acts as an insulator; however, scientists at Bell Laboratories in the United States found that, by inserting certain impurities into the crystal lattice, they could make silicon act as a conductor. The process of inserting the impurities is known as *doping*, and the most commonly used *dopants* are boron atoms with three electrons in their outermost electron shells and phosphorus atoms with five.

If a pure piece of silicon is surrounded by a gas containing boron or phosphorus and heated in a high-temperature oven, the boron or phosphorus atoms will permeate the crystal lattice and displace some silicon atoms without disturbing other atoms in the vicinity. This process is known as *diffusion*. Boron-doped silicon is called *P-type* silicon and phosphorus-doped silicon is called *N-type* (Figure 4-3).[2]

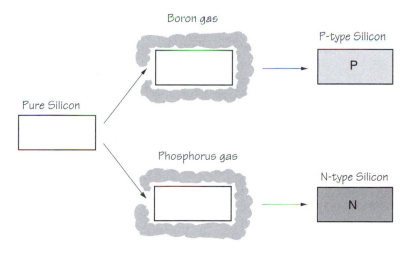

Figure 4-3. Creating P-type and N-type silicon

Because boron atoms have only three electrons in their outermost electron shells, they can only make bonds with three of the silicon atoms surrounding them. Thus, the site (location) occupied by a boron atom in the silicon crystal will accept a free electron with relative ease and is therefore known as an *acceptor*. Similarly, because phosphorous atoms have five electrons in their outermost electron shells, the site of a phosphorous atom in the silicon crystal will donate an electron with relative ease and is therefore known as a *donor*.

---

[2] If you ever see an illustration of an integrated circuit, you may see symbols like *n*, *n*+, *n*++, *p*, *p*+, and *p*++. In this case, the n and p stand for standard N-Type and P-type material (which we might compare to an average guy), the *n*+ and *p*+ indicate a heavier level of doping (say a bodybuilder or the author flexing his rippling muscles on the beach), and the *n*++ and *p*++ indicate a really high level of doping (like a weightlifter on steroids).

Semiconductor Diodes

As was noted above, pure crystalline silicon acts as an insulator. By comparison, both P-type and N-type silicon are reasonably good conductors (Figure 4-4).

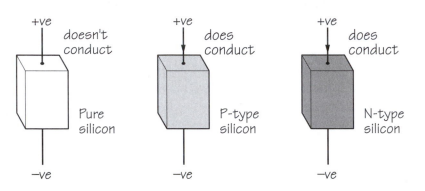

+ve and −ve indicate positive and negative voltage sources, respectively (for example, they could be wires connected to the terminals of a battery)

Figure 4-4. Pure P-type and N-type silicon

When things start to become really interesting, however, is when a piece of silicon is doped such that part is P-type and part is N-type (Figure 4-5).

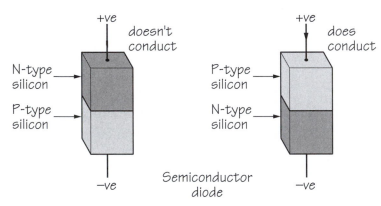

Figure 4-5. Mixing P-type and N-type silicon

The silicon with both P-type and N-type material conducts electricity in only one direction; in the other direction it behaves like an OPEN (OFF) switch. These structures, known as *semiconductor diodes*,[3] come in many shapes and sizes; an example could be as shown in Figure 4-6.

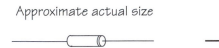
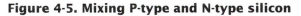

(a) Diode component   (b) Symbol

Figure 4-6. Diode: Component and symbol

[3] The "semiconductor" portion of *semiconductor diode* was initially used to distinguish these components from their vacuum tube-based cousins. As semiconductors took over, everyone started to just refer to them as *diodes*.

If the triangular body of the symbol is pointing in the classical direction of current flow (more positive to more negative), the diode will conduct. An individually packaged diode consists of a piece of silicon with connections to external leads, all encapsulated in a protective package (the silicon is typically smaller than a grain of sand). The package protects the silicon from moisture and other impurities and, when the diode is operating, helps to conduct heat away from the silicon.

Due to the fact that diodes (and transistors as discussed below) are formed from solids—as opposed to vacuum tubes, which are largely formed from empty space—people started to refer to them as *solid-state* electronics.

Bipolar Junction Transistors

More complex components called *transistors* can be created by forming a sandwich out of three regions of doped silicon. One family of transistors is known as bipolar junction transistors (BJTs)[4] of which there are two basic types called *NPN* and *PNP*; these names relate to the way in which the silicon is doped (Figure 4-7).

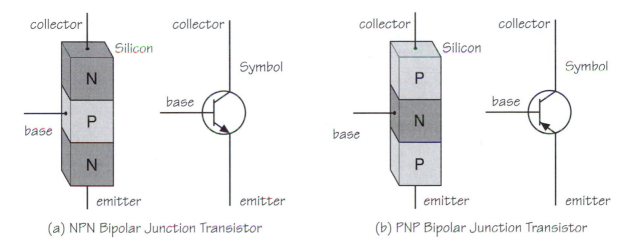

(a) NPN Bipolar Junction Transistor   (b) PNP Bipolar Junction Transistor

Figure 4-7. Bipolar junction transistors (BJTs)

---

[4] In conversation, the term BJT is spelled out as "B-J-T".

In the analog world, a transistor can be used as a voltage amplifier, a current amplifier, or a switch; in the digital world, a transistor is primarily considered to be a switch. The structure of a transistor between the *collector* and *emitter* terminals is similar to that of two diodes connected back-to-back. Two diodes connected in this way would typically not conduct; however, when signals are applied to the *base* terminal, the transistor can be turned ON or OFF. If the transistor is turned ON, it acts like a CLOSED switch and allows current to flow between the collector and the emitter; if the transistor is turned OFF, it acts like an OPEN switch and no current flows. We may think of the collector and emitter as *data* terminals, and the base as the *control* terminal.

As for a diode, an individually packaged transistor consists of the silicon,

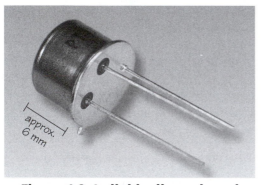

approx. 6 mm

Figure 4-8. Individually packaged transistor
(*photo courtesy of Alan Winstanley*)

with connections to external leads, all encapsulated in a protective package (the silicon is typically smaller than a grain of sand). The package protects the silicon from moisture and other impurities and helps to conduct heat away from the silicon when the transistor is operating. Transistors may be packaged in plastic or in little metal cans about a quarter of an inch in diameter with three leads sticking out of the bottom (Figure 4-8).

Metal-Oxide Semiconductor Field-Effect Transistors

Another family of transistors is known as metal-oxide semiconductor field-effect transistors (MOSFETs)[5] of which there are two basic types called *n-channel* and *p-channel*; once again these names relate to the way in which the silicon is doped (Figure 4-9).[6]

In the case of these devices, the *drain* and *source* form the *data* terminals and the *gate* acts as the *control* terminal. Unlike bipolar devices, the control

---

[5] In conversation, the term MOSFET is pronounced as a single word, where "MOS" rhymes with "boss" and "FET" rhymes with "bet".

[6] Nothing is simple. In fact the MOSFETs discussed in this book are known as *enhancement-type* devices, which are OFF unless a control signal is applied to turn them ON. There are also *depletion-type* devices, which are ON unless a control signal is applied to turn them OFF.

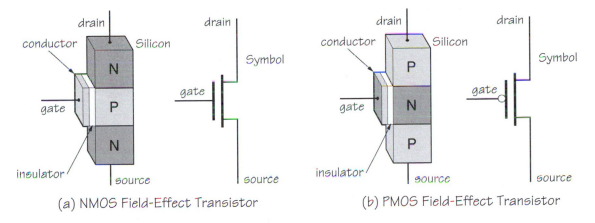

(a) NMOS Field-Effect Transistor   (b) PMOS Field-Effect Transistor

Figure 4-9. Metal-oxide semiconductor field-effect transistors (MOSFETs)

terminal is connected to a conducting plate, which is insulated from the silicon by a layer of non-conducting oxide. In the original devices the conducting plate was metal—hence the term *metal-oxide*. When a signal is applied to the gate terminal, the plate, insulated by the oxide, creates an electromagnetic field, which turns the transistor ON or OFF—hence the term *field-effect*.

Now this is the bit that always confuses the unwary, because the term "channel" refers to the piece of silicon under the gate terminal, that is, the piece linking the drain and source regions. But the channel in the n-channel device is formed from P-type material, while the channel in the p-channel device is formed from N-type material.

At first glance, this would appear to be totally counter-intuitive, but there is reason behind the madness. Let's consider the n-channel device. In order to turn this ON, a positive voltage is applied to the gate. This positive voltage attracts negative electrons in the P-type material and causes them to accumulate beneath the oxide layer where they form a negative channel—hence the term *n-channel*. In fact, saying "n-channel" and "p-channel" is a bit of a mouthful, so instead we typically just refer to these as NMOS and PMOS transistors, respectively.[7]

This book concentrates on MOSFETs, because their symbols, construction, and operation are easier to understand than those of bipolar junction transistors.

---

[7] In conversation, NMOS and PMOS are pronounced "N-MOS" and "P-MOS", respectively. That is, by spelling out the first letter followed by "MOS" to rhyme with "boss".

The Transistor as a Switch

To illustrate the application of a transistor as a switch, first consider a simple circuit comprising a resistor and a real switch (Figure 4-10).

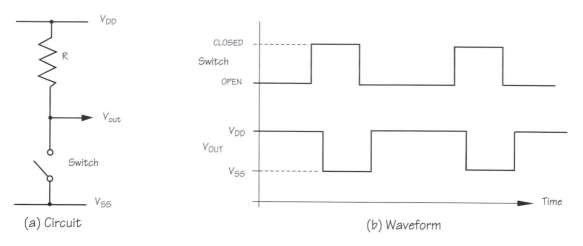

(a) Circuit   (b) Waveform

Figure 4-10. Resistor-switch circuit

The labels V_{DD} and V_{SS} are commonly used in circuits employing MOSFETs. At this point we have little interest in their actual values and, for the purpose of these examples, need only assume that V_{DD} is more positive than V_{SS}.

When the switch is OPEN (OFF), V_{OUT} is connected via the resistor to V_{DD}; when the switch is CLOSED (ON), V_{OUT} is connected via the switch directly to V_{SS}. In this latter case, V_{OUT} takes the value V_{SS} because, like people, electricity takes the path of least resistance, and the resistance to V_{SS} through the closed switch is far less than the resistance to V_{DD} through the resistor. The waveforms in the illustration above show a delay between the switch operating and V_{OUT} responding. Although this delay is extremely small, it is important to note that there will always be some element of delay in any physical system.

Now consider the case where the switch is replaced with an NMOS transistor whose control input can be switched between V_{DD} and V_{SS} (Figure 4-11).

When the control input to an NMOS transistor is connected to V_{SS}, the transistor is turned OFF and acts like an OPEN switch; when the control input is connected to V_{DD}, the transistor is turned ON and acts like a closed switch. Thus, the transistor functions in a similar manner to the switch. However, a switch is controlled by hand and can only be operated a few times a second,

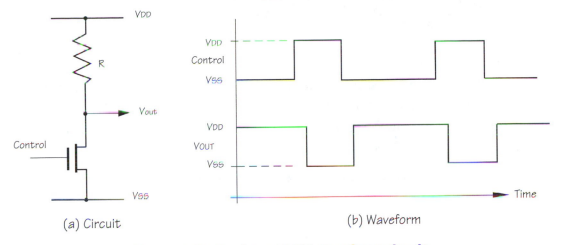

Figure 4-11. Resistor-NMOS transistor circuit

but a transistor's control input can be driven by other transistors, allowing it to be operated millions of times a second.

Gallium Arsenide Semiconductors

Silicon is known as a *four-valence semiconductor* because it has four electrons available to make bonds in its outermost electron shell. Although silicon is the most commonly used semiconductor, there is another that requires some mention. The element gallium (chemical symbol Ga) has three electrons available in its outermost shell and the element arsenic (chemical symbol As) has five. A crystalline structure of gallium arsenide (GaAs) is known as a III-V valence semiconductor[8] and can be doped with impurities in a similar manner to silicon.

In a number of respects, GaAs is preferable to silicon, not the least of which is that GaAs transistors can switch approximately eight times faster than their silicon equivalents. However, GaAs is hard to work with, which results in GaAs transistors being more expensive than their silicon cousins.

Light-Emitting Diodes

On February 9, 1907, one of Marconi's engineers, Mr. H.J. Round of New York, NY, had a letter published in Electrical World magazine as follows:

---

[8] In conversation, the Roman Numerals III-V are pronounced "three-five."

> *A Note on Carborundum*
>
> To the editors of Electrical World:
> Sirs: During an investigation of the unsymmetrical passage of current through
> a contact of carborundum and other substances a curious phenomenon was
> noted. On applying a potential of 10 volts between two points on a crystal of
> carborundum, the crystal gave out a yellowish light.

Mr. Round went on to note that some crystals gave out green, orange, or
blue light. This is quite possibly the first documented reference to the effect
upon which special components called *light-emitting diodes (LEDs)* are based.[9]

Sad to relate, no one seemed particularly interested in Mr. Round's
discovery, and nothing really happened until 1922, when the same phenomenon
was observed by O.V Losov in Leningrad. Losov took out four patents between
1927 and 1942, but he was killed during the Second World War and the details
of his work were never discovered.

In fact, it wasn't until 1951, following the discovery of the bipolar transistor,
that researchers really started to investigate this effect in earnest. They found
that by creating a semiconductor diode from a compound semiconductor
formed from two or more elements—such as gallium arsenide (GaAs)—light
is emitted from the PN junction, that is, the junction between the P-Type and
N-type doped materials.

As for a standard diode, a LED conducts electricity in only one direction
(and it emits light only when it's conducting). Thus, the symbol for an LED is
similar to that for a normal diode, but with
two arrows to indicate light being emitted
(Figure 4-12).

A LED formed from pure gallium arsenide
emits infrared light, which is useful for sensors,
but which is invisible to the human eye. It was

Figure 4-12. Symbol for a LED

discovered that adding aluminum to the semiconductor to give aluminum
gallium arsenide (AlGaAs) resulted in red light humans could see. Thus, after

---

[9] In conversation, the term LED may be spelled out as "L-E-D" or pronounced as a single word
to rhyme with "bed".

much experimentation and refinement, the first red LEDs started to hit the streets in the late 1960s.

LEDs are interesting for a number of reasons, not the least of which is that they are extremely reliable, they have a very long life (typically 100,000 hours as compared to 1,000 hours for an incandescent light bulb), they generate very pure, saturated colors, and they are extremely energy efficient (LEDs use up to 90% less energy than an equivalent incandescent bulb).

Over time, more materials were discovered that could generate different colors. For example, gallium phosphide gives green light, and aluminum indium gallium phosphite can be used to generate yellow and orange light. For a long time, the only color missing was blue. This was important because blue light has the shortest wavelength of visible light, and engineers realized that if they could build a blue laser diode, they could quadruple the amount of data that could be stored on, and read from, a CD-ROM or DVD.

However, although semiconductor companies spent hundreds of millions of dollars desperately trying to create a blue LED, the little rapscallion remained elusive for more than three decades. In fact, it wasn't until 1996 that the Japanese Electrical Engineer Shuji Nakamura demonstrated a blue LED based on gallium nitride. Quite apart from its data storage applications, this discovery also makes it possible to combine the output from a blue LED with its red and green cousins to generate white light. Many observers believe that this may ultimately relegate the incandescent light bulb to the museum shelf.

Chapter

5

Primitive Logic Functions

Consider an electrical circuit consisting of a power supply, a light, and two switches connected in *series* (one after the other). The switches are the inputs to the circuit and the light is the output. A truth table provides a convenient way to represent the operation of the circuit (Figure 5-1).

As the light is only ON when both the *a* **and** *b* switches are CLOSED (ON), this circuit could be said to perform a 2-input **AND** function.[1] In fact, the results depend on the way in which the switches are connected; consider another circuit in which two switches are connected in *parallel* (side by side) (Figure 5-2).

In this case, as the light is ON when either *a* **or** *b* are CLOSED (ON), this circuit could be said to provide a 2-input **OR** function.[2] In a limited respect, we might consider that these circuits are making simple logical decisions; two switches offer four combinations of OPEN (OFF) and CLOSED (ON), but only certain combinations cause the light to be turned ON.

Logic functions such as AND and OR are generic concepts that can be implemented in a variety of ways, including switches as illustrated above, transistors for use in computers, and even pneumatic devices for use in hostile environments such as steel works or nuclear reactors. Thus, instead of drawing circuits using light switches, it is preferable to make use of more abstract forms of representation. This permits designers to specify the function of systems with minimal consideration as to their final physical realization. To facilitate this, special symbols are employed to represent logic functions, and truth table assignments are specified using the abstract terms FALSE and TRUE. This is because assignments such as OPEN, CLOSED, ON, and OFF may imply a particular implementation.

---

[1] A 3-input version could be constructed by adding a third switch in series with the first two.

[2] A 3-input version could be constructed by adding a third switch in parallel with the first two.

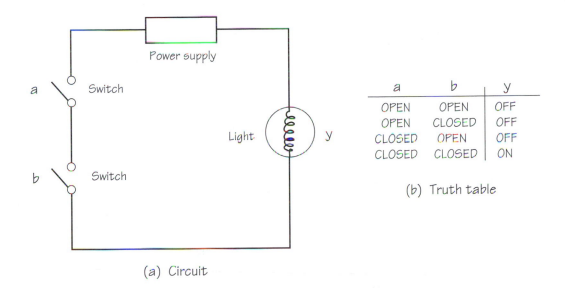

a	b	y
OPEN	OPEN	OFF
OPEN	CLOSED	OFF
CLOSED	OPEN	OFF
CLOSED	CLOSED	ON

(b) Truth table

(a) Circuit

Figure 5-1 Switch representation of a 2-input AND function

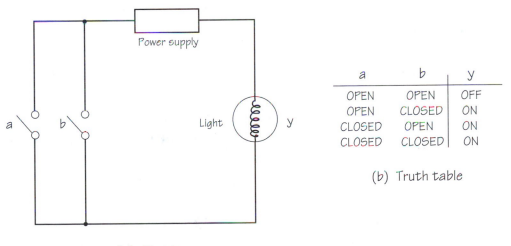

a	b	y
OPEN	OPEN	OFF
OPEN	CLOSED	ON
CLOSED	OPEN	ON
CLOSED	CLOSED	ON

(b) Truth table

(a) Circuit

Figure 5.2: Switch representation of a 2-input OR function

BUF and NOT Functions

The simplest of all the logic functions are known as BUF and NOT (Figure 5-3).

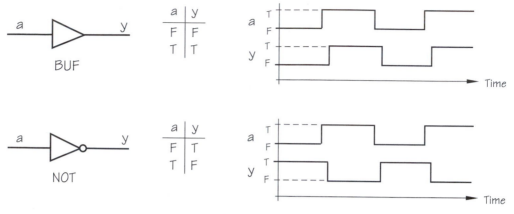

Figure 5-3. BUF and NOT functions

The F and T values in the truth tables are shorthand for FALSE and TRUE, respectively. The output of the BUF function has the same value as the input to the function; if the input is FALSE the output is FALSE, and if the input is TRUE the output is TRUE. By comparison, the small circle, or bobble,[3] on the output of the NOT symbol indicates an inverting function; if the input is FALSE the output is TRUE, and if the input is TRUE the output is FALSE.[4]

As a reminder that these abstract functions will eventually have physical realizations, the waveforms show delays between transitions on the inputs and corresponding responses at the outputs. The actual values of these delays depend on the technology used to implement the functions, but it is important to note that in any physical implementation there will always be some element of delay.

Now consider the effect of connecting two NOT functions in *series* (one after the other) as shown in Figure 5-4.

The first NOT gate inverts the value from the input, and the second NOT gate inverts it back again. Thus, the two negations cancel each other out (sort of like "*two wrongs* do *make a right*"). The end result is equivalent to that of a BUF function, except that each NOT contributes an element of delay.

---

[3] Some engineers use the term *bubble*, others say *bobble*, and the rest try to avoid mentioning it at all.

[4] A commonly-used alternative name for a NOT function is INV (short for inverter).

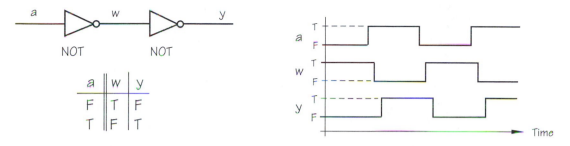

Figure 5-4. Two NOT functions connected together in series

AND, OR, and XOR Functions

Three slightly more complex functions are known as AND, OR, and XOR (Figure 5-5).[5]

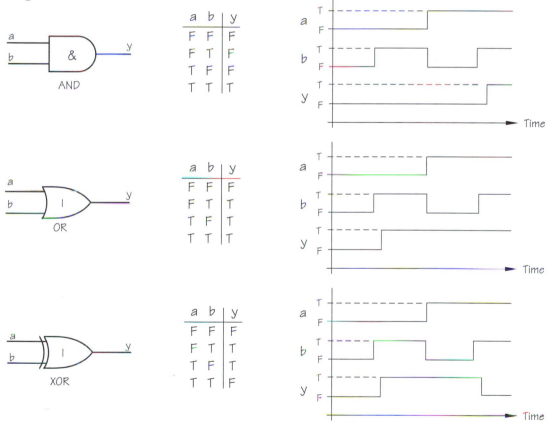

Figure 5-5. AND, OR, and XOR functions

---

[5] In conversation, the term XOR is pronounced "X-OR"; that is, spelling the letter "X" followed by "OR" to rhyme with "door".

The AND and OR representations shown here are the abstract equivalents of our original switch examples. In the case of the AND, the output is only TRUE if *both* a **and** b are TRUE; in the case of the OR, the output is TRUE if *either* a **or** b are TRUE. In fact, the OR should more properly be called an *inclusive-OR*, because the TRUE output cases *include* the case when both inputs are TRUE. Contrast this with the *exclusive-OR*, or XOR, where the TRUE output cases *exclude* the case when both inputs are TRUE.

NAND, NOR, and XNOR Functions

Now consider the effect of appending a NOT function to the output of the AND function (Figure 5-6).

This combination of functions occurs frequently in designs. Similarly, the outputs of the OR and XOR functions are often inverted with NOT functions. This leads to three more primitive functions called NAND (NOT-AND), NOR (NOT-OR) and NXOR (NOT-XOR).[6] In practice, however, the NXOR is almost always referred to as an XNOR (exclusive-NOR) (Figure 5-7).[7]

The bobbles on their outputs indicate that these are inverting functions. One way to visualize this is that the symbol for the NOT function has been forced back into the preceding symbol until only the bobble remains visible.

Of course, if we appended a NOT function to the output of a NAND, we'd end up back with our original AND function again. Similarly, appending a NOT to a NOR or an XNOR results in an OR and XOR, respectively.

Not a Lot

And that's about it. In reality there are only eight simple functions (BUF, NOT, AND, NAND, OR, NOR, XOR, and XNOR) from which everything else is constructed. In fact, some might argue that there are only seven core functions because you can construct a BUF out of two NOTs, as was discussed earlier.

---

[6] In conversation, the terms NAND and NOR are pronounced as single words to rhyme with "band" and "door", respectively.

[7] In conversation, the term XNOR is pronounced "X-NOR," that is, spelling the letter "X" followed by "NOR" to rhyme with "door."

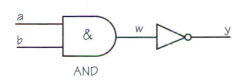

a	b	w	y
F	F	F	T
F	T	F	T
T	F	F	T
T	T	T	F

Figure 5-6. AND function followed by a NOT function

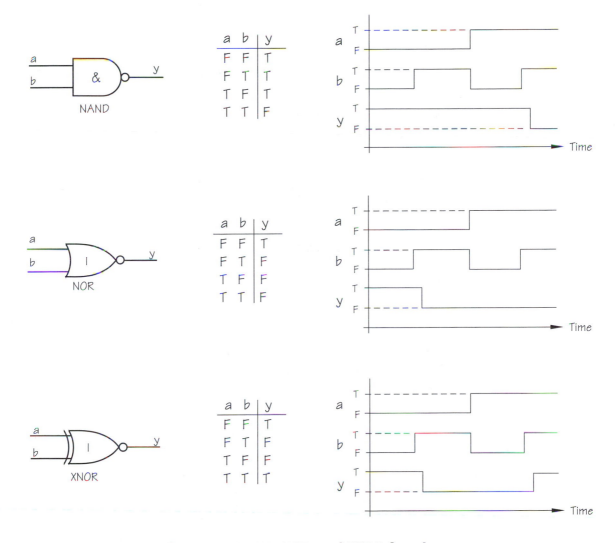

Figure 5-7. NAND, NOR, and XNOR functions

Actually, if you want to go down this path, you can construct all of the above functions using one or more NAND gates (or one or more NOR gates). For example, if you connect the two inputs of a NAND gate together, you end up with a NOT as shown in Figure 5-8 (you can achieve the same effect by connecting the two inputs of a NOR gate together).

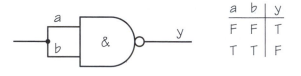

a	b	y
F	F	T
T	T	F

NAND acting as NOT

Figure 5-8. Forming a NOT from a NAND

As the inputs a and b are connected together, they have to carry identical values, so we end up showing only two rows in the truth table. We also know that if we invert the output from a NAND, we end up with an AND. So we could append a NAND configured as a NOT to the output of another NAND to generate an AND (Figure 5-9).

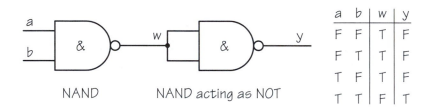

a	b	w	y
F	F	T	F
F	T	T	F
T	F	T	F
T	T	F	T

NAND   NAND acting as NOT

Figure 5-9. Forming an AND from two NANDs

Later on in Chapter 9, we'll discover how to transform functions formed from ANDs into equivalent functions formed from ORs and vice versa. Coupled with what we've just seen here, this would allow us to build anything we wanted out of a bunch of 2-input NAND (or NOR) functions.

Functions versus Gates

Simple functions such as BUF, NOT, AND, NAND, OR, NOR, XOR, and XNOR are often known as *primitive gates*, *primitives*, *logic gates*, or simply *gates*.[8] Strictly speaking, the term *logic function* implies an abstract mathematical relationship, while *logic gate* implies an underlying physical implementation. In practice, however, these terms are often used interchangeably.

More complex functions can be constructed by combining primitive gates in different ways. A complete design—say a computer—employs a great many gates connected together to achieve the required result. When the time arrives to translate the abstract representation into a particular physical implementation, the logic symbols are converted into appropriate equivalents such as switches, transistors, or pneumatic valves. Similarly, the FALSE and TRUE logic values are mapped into appropriate equivalents such as switch positions, voltage levels, or air pressures. The majority of designs are translated into a single technology. However, one of the advantages of abstract representations is that they allow designers to implement different portions of a single design in dissimilar technologies with relative ease. Throughout the remainder of this book we will be concentrating on electronic implementations.

Finally, if some of the above seems to be a little esoteric, consider a real-world example from your home, such as two light switches mounted at opposite ends of a hallway controlling the same light. If both of the switches are UP or DOWN the light will be ON; for any other combination the light will be OFF. Constructing a truth table reveals a classic example of an XNOR function.

---

[8] The reasoning behind using the term "gate" is that these functions serve to control electronic signals in much the same way that farmyard gates can be used to control animals.

Using Transistors to Build Primitive Logic Functions

There are several different families of transistors available to designers and, although the actual implementations vary, each can be used to construct primitive logic gates. This book concentrates on the *metal-oxide semiconductor field-effect transistors (MOSFETs)* introduced in Chapter 4, because their symbols, construction, and operation are easier to understand than are *bipolar junction transistors (BJTs)*.

Logic gates can be created using only NMOS or only PMOS transistors; however, a popular implementation called *complementary metal-oxide semiconductor (CMOS)*[1] makes use of both NMOS and PMOS transistors connected in a complementary manner.

CMOS gates operate from two voltage levels, which are usually given the labels V_{DD} and V_{SS}. To some extent the actual values of V_{DD} and V_{SS} are irrelevant as long as V_{DD} is sufficiently more positive than V_{SS}. There are also two conventions known as *positive logic* and *negative logic*.[2] Under the positive logic convention used throughout this book, the more positive V_{DD} is assigned the value of logic 1, and the more negative V_{SS} is assigned the value of logic 0. In Chapter 5 it was noted that truth table assignments can be specified using the abstract values FALSE and TRUE. However, for reasons that are more fully examined in Appendix B, electronic designers usually represent FALSE and TRUE as 0 and 1, respectively.

NOT and BUF Gates

The simplest logic function to implement in CMOS is a NOT gate (Figure 6-1). The small circle, or bobble, on the control input of transistor Tr_1 indicates a PMOS transistor. The bobble is used to indicate that this transistor

---

[1] In conversation, the term CMOS is pronounced "C-MOS"; that is, spelling the letter "C" followed by "MOS" to rhyme with "boss".

[2] The positive- and negative-logic conventions are discussed in detail in Appendix B.

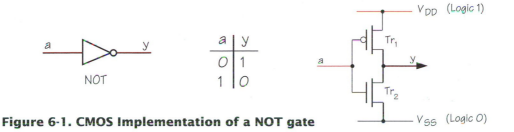

Figure 6-1. CMOS Implementation of a NOT gate

has an *active-low*[3] control, which means that a logic 0 applied to the control input turns the transistor ON and a logic 1 turns it OFF. The lack of a bobble on the control input of transistor Tr_2 indicates an NMOS transistor. The lack of a bobble says that this transistor has an *active-high*[4] control, which means that a logic 1 applied to the control input turns the transistor ON and a logic 0 turns it OFF.

Thus, when a logic 0 is applied to input a, transistor Tr_1 is turned ON, transistor Tr_2 is turned OFF, and output y is connected to logic 1 via Tr_1. Similarly, when a logic 1 is applied to input a, transistor Tr_1 is turned OFF, transistor Tr_2 is turned ON, and output y is connected to logic 0 via Tr_2.

Don't worry if all this seems a bit confusing at first. The main points to remember are that a logic 0 applied to its control input turns the PMOS transistor ON and the NMOS transistor OFF, while a logic 1 turns the PMOS transistor OFF and the NMOS transistor ON. It may help to visualize the NOT gate's operation in terms of switches rather than transistors (Figure 6-2).

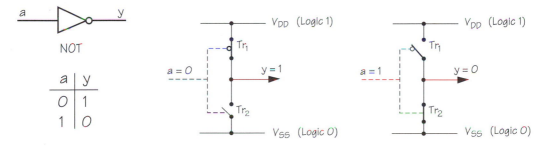

Figure 6-2. NOT gate's operation represented in terms of switches

---

[3] The "low" comes from the fact that, under the commonly used positive-logic system, logic 0 is more negative (conceptually "lower") than logic 1.

[4] The "high" comes from the fact that, under the commonly used positive-logic system, logic 1 is more positive (conceptually "higher") than logic 0.

Surprisingly, a non-inverting BUF gate is more complex than an inverting NOT gate. This is due to the fact that a BUF gate is constructed from two NOT gates connected in *series* (one after the other), which means that it requires four transistors (Figure 6-3).

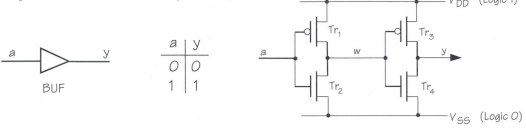

Figure 6-3. CMOS implementation of a BUF gate

The first NOT gate is formed from transistors Tr_1 and Tr_2, while the second is formed from transistors Tr_3 and Tr_4. A logic 0 applied to input a is inverted to a logic 1 on w, and then inverted back again to a logic 0 on output y. Similarly, a logic 1 on a is inverted to a logic 0 on w, and then inverted back again to a logic 1 on y.

Around this stage it is not unreasonable to question the need for BUF gates in the first place—after all, their logical function could be achieved using a simple piece of wire. But there's method to our madness, because BUF gates may actually be used for a number of reasons: for example, to isolate signals, to provide increased drive capability, or to add an element of delay.

NAND and AND Gates

The implementations of the NOT and BUF gates shown above illustrate an important point, which is that it is generally easier to implement an inverting function than its non-inverting equivalent. In the same way that a NOT is easier to implement than a BUF, a NAND is easier to implement than an AND, and a NOR is easier to implement than an OR. More significantly, inverting functions typically require fewer transistors and operate faster than their non-inverting counterparts. This can obviously be an important design consideration. Consider a 2-input NAND gate, which requires four transistors (Figure 6-4).[5]

---

[5] A 3-input version could be constructed by adding an additional PMOS transistor in parallel with Tr_1 and Tr_2, and an additional NMOS transistor in series with Tr_3 and Tr_4.

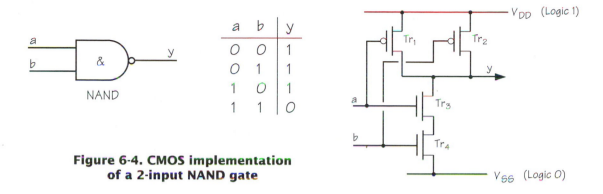

Figure 6-4. CMOS implementation of a 2-input NAND gate

When both *a* and *b* are presented with logic 1s, transistors Tr_1 and Tr_2 are turned OFF, transistors Tr_3 and Tr_4 are turned ON, and output *y* is connected to logic 0 via Tr_3 and Tr_4. Any other combination of inputs results in one or both of Tr_3 and Tr_4 being turned OFF, one or both of Tr_1 and Tr_2 being turned ON, and output *y* being connected to logic 1 via Tr_1 and/or Tr_2. Once again, it may help to visualize the gate's operation in terms of switches rather than transistors (Figure 6-5).

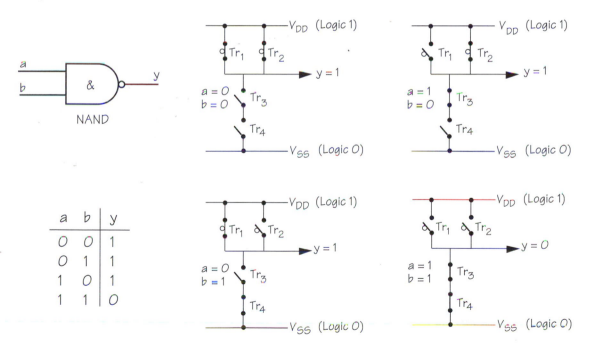

Figure 6-5. NAND gate's operation represented in terms of switches

Now consider an AND gate. This is formed by inverting the output of a NAND with a NOT, which means that a 2-input AND requires six transistors (Figure 6-6).[6]

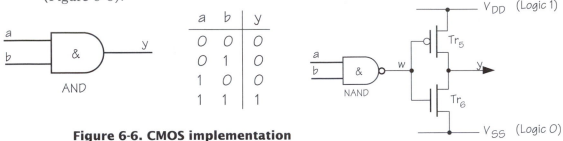

Figure 6-6. CMOS implementation of a 2-input AND gate

a	b	y
0	0	0
0	1	0
1	0	0
1	1	1

NOR and OR Gates

A similar story occurs in the case of NOR gates and OR gates. First, consider a 2-input NOR, which requires four transistors (Figure 6-7).[7]

When both a and b are set to logic 0, transistors Tr_3 and Tr_4 are turned OFF, transistors Tr_1 and Tr_2 are turned ON, and output y is connected to logic 1 via Tr_1 and Tr_2. Any other combination of inputs results in one or both of Tr_1 and Tr_2 being turned OFF, one or both of Tr_3 and Tr_4 being turned ON, and output y being connected to logic 0 via Tr_3 and/or Tr_4.

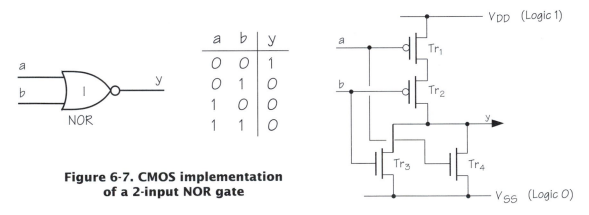

Figure 6-7. CMOS implementation of a 2-input NOR gate

a	b	y
0	0	1
0	1	0
1	0	0
1	1	0

---

[6] Remember that electronic designers are cunning little devils with lots of tricks up their sleeves. In fact, it's possible to create an AND gate using only one transistor and a resistor (see the discussions on *Pass-transistor Logic* later in this chapter).

[7] A 3-input version could be constructed by adding an additional PMOS transistor in series with Tr_1 and Tr_2, and an additional NMOS transistor in parallel with Tr_3 and Tr_4.

Once again, an OR gate is formed by inverting the output of a NOR with a NOT, which means that a 2-input OR requires six transistors (Figure 6-8).

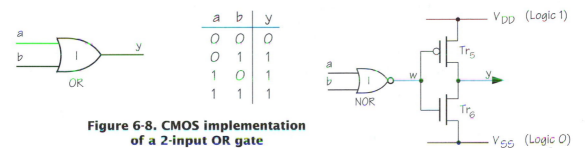

a	b	y
0	0	0
0	1	1
1	0	1
1	1	1

**Figure 6-8. CMOS implementation
of a 2-input OR gate**

XNOR and XOR Gates

The concepts of NAND, AND, NOR, and OR are relatively easy to understand because they map onto the way we think in everyday life. For example, a textual equivalent of a NOR could be: *"If it's windy or if it's raining then I'm not going out."*

By comparison, the concepts of XOR and XNOR can be a little harder to grasp because we don't usually consider things in these terms. A textual equivalent of an XOR could be: *"If it is windy and it's not raining, or if it's not windy and it is raining, then I will go out."* Although this does make sense (in a strange sort of way), we don't often find ourselves making decisions in this manner.

For this reason, it is natural to assume that XNOR and XOR gates would be a little more difficult to construct. However, these gates are full of surprises, both in the way in which they work and the purposes for which they can be used. For example, a 2-input XNOR can be implemented using only four transistors (Figure 6-9).[8]

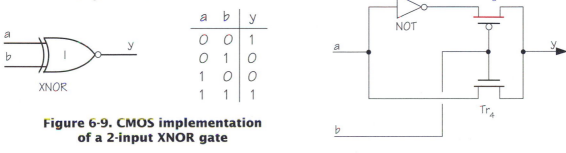

a	b	y
0	0	1
0	1	0
1	0	0
1	1	1

**Figure 6-9. CMOS implementation
of a 2-input XNOR gate**

---

[8] Unlike AND, NAND, OR, and NOR gates, there are no such beasts as XNOR or XOR primitives with more than two inputs. However, equivalent functions with more than two inputs can be formed by connecting a number of 2-input primitives together.

The NOT gate would be constructed in the standard way using two transistors as described above, but the XNOR differs from the previous gates in the way that transistors Tr_3 and Tr_4 are utilized. First, consider the case where input b is presented with a logic 0: transistor Tr_4 is turned OFF, transistor Tr_3 is turned ON, and output y is connected to the output of the NOT gate via Tr_3. Thus, when input b is logic 0, output y has the *inverse* of the value on input a. Now consider the case where input b is presented with a logic 1: transistor Tr_3 is turned OFF, transistor Tr_4 is turned ON, and output y is connected to input a via Tr_4. Thus, when input b is logic 1, output y has the *same* value as input a. The end result of all these machinations is that wiring the transistors together in this way does result in a function that satisfies the requirements of the XNOR truth table.

Unlike the other complementary gates, it is not necessary to invert the output of the XNOR to form an XOR (although we could if we wanted to, of course). A little judicious rearranging of the components results in a 2-input XOR that also requires only four transistors (Figure 4-10).

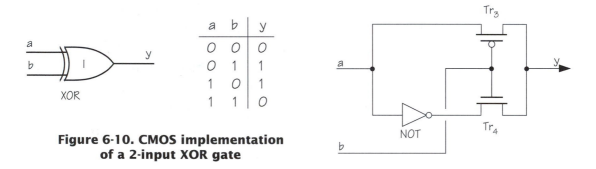

a	b	y
0	0	0
0	1	1
1	0	1
1	1	0

**Figure 6-10. CMOS implementation
of a 2-input XOR gate**

First, consider the case where input b is presented with a logic 0: transistor Tr_4 is turned OFF, transistor Tr_3 is turned ON, and output y is connected to input a via Tr_3. Thus, when input b is logic 0, output y has the *same* value as input a. Now consider the case where input b is presented with a logic 1: transistor Tr_3 is turned OFF, transistor Tr_4 is turned ON, and output y is connected to the output of the NOT gate via Tr_4. Thus, when input b is logic 1, output y has the *inverse* of the value on input a. Once again, this results in a junction that satisfies the requirements of the XOR truth table.

Pass-transistor Logic

In the BUF, NOT, AND, NAND, OR, and NOR gates described earlier, the input signals and internal data signals are only used to drive control terminals on the transistors. By comparison, transistors Tr_3 and Tr_4 in the XOR and XNOR gates shown above are connected so that input and internal data signals pass between their data terminals. This technique is known as *pass-transistor logic*. It can be attractive in that it minimizes the number of transistors required to implement a function, but it's not necessarily the best approach. Strange and unexpected effects can ensue if you're not careful and you don't know what you're doing.

An alternative solution for an XOR is to invert the output of the XNOR shown above with a NOT. Similarly, an XNOR can be constructed by inverting the output of the XOR shown above with a NOT. Although these new implementations each now require six transistors rather than four, they are more robust because the NOT gates buffer the outputs and provide a higher drive capability. In many cases, XORs and XNORs are constructed from combinations of the other primitive gates. This increases the transistor count still further, but once again results in more robust solutions.

Having said all this, pass-transistor logic can be applicable in certain situations for designers who do know what they're doing. In the discussions above, it was noted that it is possible to create an AND using a single transistor and a resistor. Similarly, it's possible to create an OR using a single transistor and a resistor, and to create an XOR or an XNOR using only two transistors and a resistor. If you're feeling brave, try to work out how to achieve these minimal implementations for yourself (solutions are given in Appendix G).

Alternative Numbering Systems

Decimal (Base-10)

The commonly used decimal numbering system is based on ten digits: 0, 1, 2, 3, 4, 5, 6, 7, 8 and 9. The name decimal comes from the Latin *decem*, meaning "ten." The symbols used to represent these digits arrived in Europe around the thirteenth century from the Arabs, who in turn borrowed them from the Hindus (and never gave them back). As the decimal system is based on ten digits, it is said to be *base-10* or *radix-10*, where the term *radix* comes from the Latin word meaning "root."

With the exception of specialist requirements such as computing, base-10 numbering systems have been adopted almost universally—this is almost certainly due to the fact that humans happen to have ten fingers.[1] If mother nature had dictated six fingers on each hand, the outcome would most probably have been the common usage of base-12 numbering systems.

The decimal system is a *place-value* system, which means that the value of a particular digit depends both on the digit itself and its position within the number (Figure 7-1).

Every column in a place-value number has a "weight" associated with it, and each digit is combined with its column's weight to determine the final value of the number (Figure 7-2).

Counting in decimal commences at 0 and progresses

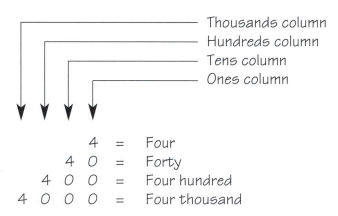

Figure 7-1. Place-value number systems

---

[1] Including thumbs.

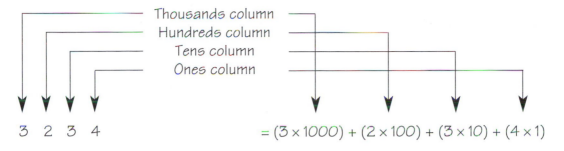

$$3 \quad 2 \quad 3 \quad 4 \qquad\qquad = (3 \times 1000) + (2 \times 100) + (3 \times 10) + (4 \times 1)$$

Figure 7-2. Combining digits with column weights in decimal

up to 9, at which point all of the available digits have been used. Thus, the next count causes the first column to be reset to 0 and the second column to be incremented, resulting in 10. Similarly, when the count reaches 99, the next count causes the first column to be reset to zero and the second column to be incremented. But, as the second column already contains a 9, this causes *it* to be reset to 0 and the third column to be incremented resulting in 100 (Figure 7-3).

Although base-10 systems are anatomically convenient, they have few other advantages to recommend them. In fact, depending on your point of view, almost any other base (with the possible exception of nine) would be as good as, or better than, base-10, which is only wholly divisible by 2 and 5. For many arithmetic operations, the use of a base that is wholly divisible by many numbers, especially the smaller values, conveys certain advantages. An educated layman may well prefer a base-12 system on the basis that 12 is wholly divisible by 2, 3, 4 and 6. For their own esoteric purposes, some mathematicians would ideally prefer a system with a prime number as a base; for example, seven or eleven.

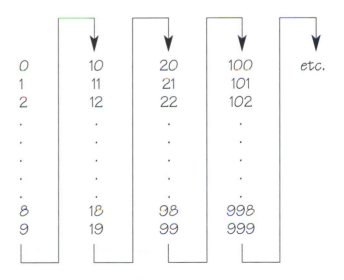

Figure 7-3. Counting in decimal

Duo-Decimal (Base-12)

Number systems with bases other than ten have sprouted up like weeds throughout history. Some cultures made use of duo-decimal (base-12) systems; instead of counting fingers they counted finger-joints. The thumb is used to point to the relevant joint and, as each finger has three joints, the remaining fingers can be used to represent values from 1 through 12 (Figure 7-4).

Thumb is used to point to relevant joint

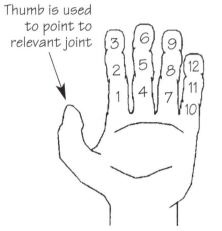

Figure 7-4. Using finger joints to count in duo-decimal

This system is particularly useful if one wishes to count up to twelve while still maintaining a free hand to throw a spear at someone whom, we may assume, is not a close friend.

This form of counting may explain why the ancient Sumerians, Assyrians, and Babylonians divided their days into twelve periods, six for day and six for night. The lengths of the periods were adjusted to the seasons (since the length of daylight compared to night time varies throughout the year), but were approximately equal to two of our hours. In fact, the Chinese use a form of this system to the present day (Figure 7-5).

11:00pm	⇨	1:00am	≡	Hour of the Rat
1:00am	⇨	3:00am	≡	Hour of the Ox
3:00am	⇨	5:00am	≡	Hour of the Tiger
5:00am	⇨	7:00am	≡	Hour of the Hare
7:00am	⇨	9:00am	≡	Hour of the Dragon
9:00am	⇨	11:00am	≡	Hour of the Snake
11:00am	⇨	1:00pm	≡	Hour of the Horse
1:00pm	⇨	3:00pm	≡	Hour of the Ram
3:00pm	⇨	5:00pm	≡	Hour of the Monkey
5:00pm	⇨	7:00pm	≡	Hour of the Cock
7:00pm	⇨	9:00pm	≡	Hour of the Dog
9:00pm	⇨	11:00pm	≡	Hour of the Boar

Figure 7-5. The Chinese twelve-hour day

If a similar finger-joint counting strategy is applied to both hands, the counter can represent values from 1 through 24 (Figure 7-6).

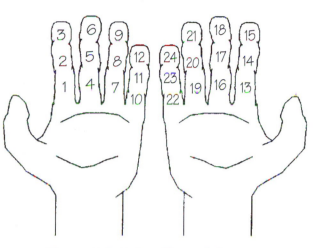

This may explain why the ancient Egyptians divided their days into twenty-four periods, twelve for day and twelve for night. Once again, the lengths of the periods were adjusted to the seasons, but were approximately equal to one of our hours.

Figure 7-6. Using finger joints on both hands to count to twenty-four

Sexagesimal (Base-60)

The ancient Babylonians used a sexagesimal (base-60) numbering system. This system, which appeared between 1900 BC and 1800 BC, is also credited as being the first known place-value number system. While there is no definite proof as to the origins of the sexagesimal base, it's possible that this was an extension of the finger-joint counting schemes discussed above (Figure 7-7).

The finger joints of the left hand are still used to represent the values one through twelve; however, instead of continuing directly with the finger joints of the right hand, the thumb and fingers on the right hand are used to keep track of each count of twelve. When all of the right hand digits are extended the count is sixty ($5 \times 12 = 60$).

Although sixty may appear to be a large value to have as a base, it does

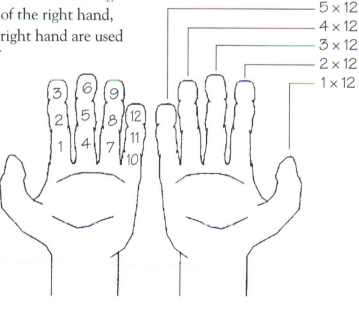

---

[2] After reading this book, one of the author's friends found this technique useful to keep track of her child's finishing position in a cross-country run.

Figure 7-7. Using fingers to count to sixty[2]

convey certain advantages. Sixty is the smallest number that can be wholly divided by 2 through 6; in fact, sixty can be wholly divided by 2, 3, 4, 5, 6, 10, 12, 15, 20 and 30. Just to increase their fun, in addition to using base sixty the Babylonians also made use of six and ten as sub-bases.

The Concepts of Zero and Negative Numbers

The concept of numbers like 1, 2, and 3 developed a long time before the concept of zero. In the original Babylonian system a zero was simply represented by a space; the decimal equivalent would be as shown in Figure 7-8.

"126904"	would be written as	"1269 4"
"102056"	would be written as	"1 2 56"
"160014"	would be written as	"16 14"

Figure 7-8. Representing zeros using spaces

It is easy to see how this can lead to a certain amount of confusion, especially when attempting to portray multiple zeros next to each other. The problems can only be exacerbated if one is using a sexagesimal system and writing on tablets of damp clay in a thunderstorm.

After more than 1500 years of potentially inaccurate calculations, the Babylonians finally began to use a special sign for zero, which first appeared around 300 BC. Some say that this was one of the most significant inventions in the history of mathematics. However, the Babylonians only used the zero symbol as a place holder to separate digits—they didn't have the concept of zero as an actual value. If we were to use the 'Δ' character to represent the Babylonian place holder, the decimal equivalent to the clay tablet accounting records of the time would read something like that shown in Figure 7-9.

The water clock, or Clepsydra, is an ancient instrument for measuring time which is believed to have originated in Egypt. These clocks consisted of a container with a small hole through which the water escaped. Units of time were marked on the side of the container or on the side of a bowl used to collect the water. The length of the units corresponding to day and night could be adjusted by varying the shape of the container; for example, by having the top wider than the bottom.

Original Amount	−	Fish Distributed	=	Fish Remaining
1Δ52 fish	−	45 fish to Steph	=	1ΔΔ7 fish
	−	2Δ fish to Max	=	987 fish
	−	4Δ7 fish to Lucie	=	58Δ fish
	−	176 fish to Drew	=	4Δ4 fish
	−	4Δ4 fish to Abby	=	"No fish left"

Figure 7-9. Representing zeros using a place-holder

The last entry in the "Fish Remaining" column is particularly revealing. Due to the fact that the Babylonian zero was only a place holder and not a value, the accounting records had to say *"No fish left"* rather than *"Δ fish."* In fact, it was not until around 600 AD that the use of zero as an actual value first appeared in India.

Vigesimal (Base-20)

The Mayans, Aztecs, and Celts developed vigesimal (base-20) systems by counting using both fingers and toes. The Eskimos of Greenland, the Tamanas of Venezuela, and the Ainu of northern Japan are three of the many other groups of people who also make use of vigesimal systems. For example, to say fifty-three, the Greenland Eskimos would use the expression *"Inup pinga-jugsane arkanek-pingasut,"* which translates as *"Of the third man, three on the first foot."* [3] This means that the first two men contribute twenty each (ten fingers and ten toes), and the third man contributes thirteen (ten fingers and three toes).

To this day we bear the legacies of almost every number system our ancestors experimented with. From the duo-decimal systems we have twenty-four hours in a day (2×12), twelve inches in a foot, and special words such as *dozen* (12) and *gross* (144). Similarly, the Chinese have twelve hours in a day and twenty-four seasons in a year. From the sexagesimal systems we have sixty seconds in a minute, sixty minutes in an hour, and 360 degrees in a circle. [4] From the base twenty systems we have special words like *score* (20), as in *"Four score and seven*

---

[3] George Ifrah: *From One to Zero (A Universal History of Numbers)*.

[4] The 360 degrees is derived from the product of the Babylonian's main base (sixty) and their sub-base (six); $6 \times 60 = 360$.

years ago, our fathers brought forth upon this continent a new nation . . ." This all serves to illustrate that number systems with bases other than ten are not only possible, but positively abound throughout history.

Quinary (Base Five)

One system that is relatively easy to understand is quinary (base-5), which uses the digits 0, 1, 2, 3 and 4. This system is particularly interesting in that a quinary finger-counting scheme is still in use today by merchants in the Indian state of Maharashtra near Bombay.

As with any place-value system, each column in a quinary number has a weight associated with it, where the weights are derived from the base. Each digit is combined with its column's weight to determine the final value of the number (Figure 7-10).

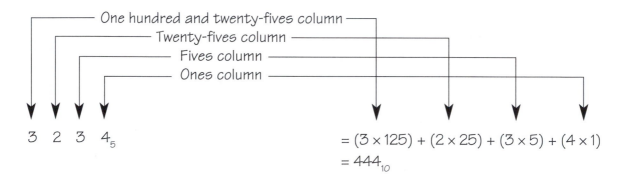

Figure 7-10. Combining digits with column weights in quinary

When using systems with bases other than ten, subscripts are used to indicate the relevant base; for example, $3234_5 = 444_{10}$ ($3234_{\text{QUINARY}} = 444_{\text{DECIMAL}}$). By convention, any value without a subscript is assumed to be in decimal.

Counting in quinary commences at 0 and progresses up to 4_5, at which point all the available digits have been used. Thus, the next count causes the first column to be reset to 0 and the second column to be incremented resulting in 10_5. Similarly, when the count reaches 44_5, the next count causes the first column to be reset to zero and the second column to be incremented. But, as the second column already contains a 4, this causes it to be reset to 0 and the third column to be incremented resulting in 100_5 (Figure 7-11).

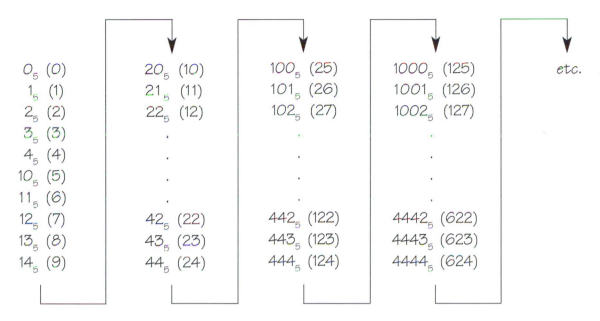

Figure 7-11. Counting in quinary

Binary (Base-2)

Digital systems are constructed out of logic gates that can only represent two states; thus, computers are obliged to make use of a number system comprising only two digits. Base-2 number systems are called *binary* and use the digits 0 and 1. As usual, each column in a binary number has a weight derived from the base, and each digit is combined with its column's weight to determine the final value of the number (Figure 7-12).

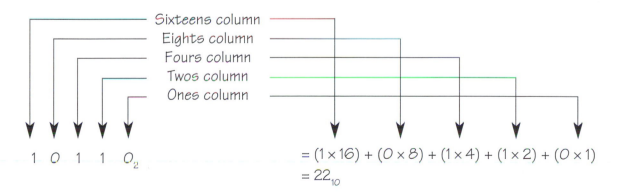

Figure 7-12. Combining digits with column weights in binary

Once again, subscripts are used to indicate the relevant base; for example, $10110_2 = 22_{10}$ ($10110_{\text{BINARY}} = 22_{\text{DECIMAL}}$). Sometime in the late 1940s, the American chemist-turned-topologist-turned-statistician John Wilder Tukey realized that computers and the binary number system were destined to become important. In addition to coining the word "software," Tukey decided that saying "binary digit" was a bit of a mouthful, so he started to look for an alternative. He considered a variety of options like *binit* and *bigit*, but he eventually settled on *bit*, which is elegant in its simplicity and is used to this day. Thus, the binary value 10110_2 would be said to be 5 bits wide. Additionally, a group of 4 bits is known as a *nybble* (sometimes called a nibble), and a group of 8 bits is known as a *byte*. The idea that "two nybbles make a byte" in is the way of being an engineer's idea of a joke, which shows that they do have a sense of humor (it's just not a particularly sophisticated one).[5]

Counting in binary commences at 0 and rather quickly progresses up to 1_2, at which point all the available digits have been used. Thus, the next count causes the first column to be reset to 0 and the second column to be incremented resulting in 10_2. Similarly, when the count reaches 11_2, the next count causes the first column to be reset to zero and the second column to be incremented. But, as the second column already contains a 1, this causes it to be reset to 0 and the third column to be incremented resulting in 100_2 (Figure 7-13).

Although binary mathematics is fairly simple, humans tend to find it difficult at first because the numbers are inclined to be long and laborious to manipulate. For example, the binary value 11010011_2 is relatively difficult to conceptualize, while its decimal equivalent of 211 is comparatively easy. On the other hand, working in binary has its advantages. For example, if you can remember . . .

$$0 \times 0 = 0$$
$$0 \times 1 = 0$$
$$1 \times 0 = 0$$
$$1 \times 1 = 1$$

. . . then you've just memorized the entire binary multiplication table!

---

[5] Continuing the theme, there have been sporadic attempts to construct terms for bit-groups of other sizes; for example, *tayste* or *crumb* for a 2-bit group, *playte* or *chawmp* for a 16-bit group, *dynner* or *gawble* for a 32-bit group, and *tayble* for a 64-bit group. But you can only take a joke so far, and using anything other than the standard terms *nybble* and *byte* is extremely rare. Having said this, the term *word* is commonly used as discussed in Chapter 15.

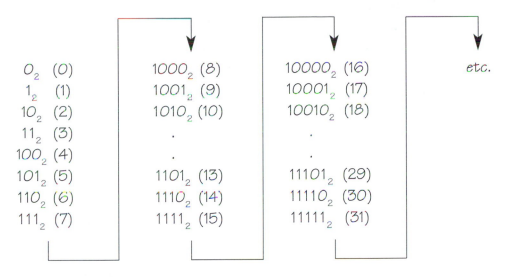

Figure 7-13. Counting in binary

Octal (Base-8) and Hexadecimal (Base-16)

Any number system having a base that is a power of two (2, 4, 8, 16, 32, etc.) can be easily mapped into its binary equivalent and vice versa. For this reason, electronics engineers typically make use of either the octal (base-8) or hexadecimal (base-16) systems.

As a base-8 system, octal requires eight individual symbols to represent all of its digits. This isn't a problem because we can simply use the symbols 0 through 7 that we know and love so well. However, as a base-16 system, hexadecimal requires sixteen individual symbols to represent all of its digits. This does pose something of a problem because there are only ten Hindu-Arabic symbols available (0 through 9). One solution would be to create some new symbols, but some doubting Thomases (and Thomasinas) regard this as less than optimal because it would necessitate the modification of existing typewriters and computer keyboards. As an alternative, the first six letters of the alphabet are brought into play (Figure 7-14).

Hexadecimal	0	1	2	3	4	5	6	7	8	9	A	B	C	D	E	F
Decimal	0	1	2	3	4	5	6	7	8	9	10	11	12	13	14	15

Figure 7-14. The sixteen hexadecimal digits

The rules for counting in octal and hexadecimal are the same as for any other place-value system—when all the digits in a column are exhausted, the next count sets that column to zero and increments the column to the left (Figure 7-15).

Although not strictly necessary, binary, octal and hexadecimal numbers are often prefixed by leading zeros to pad them to whatever width is required. This padding is usually performed to give some indication as to the physical number of bits used to represent that value within a computer.

Each octal digit can be directly mapped onto three binary digits, and each hexadecimal digit can be directly mapped onto four binary digits (Figure 7-16).

Decimal	Binary	Octal	Hexadecimal
0	00000000	000	00
1	00000001	001	01
2	00000010	002	02
3	00000011	003	03
4	00000100	004	04
5	00000101	005	05
6	00000110	006	06
7	00000111	007	07
8	00001000	010	08
9	00001001	011	09
10	00001010	012	0A
11	00001011	013	0B
12	00001100	014	0C
13	00001101	015	0D
14	00001110	016	0E
15	00001111	017	0F
16	00010000	020	10
17	00010001	021	11
18	00010010	022	12
:	:	:	:
etc.	etc.	etc.	etc.

Figure 7-15. Counting in octal and hexadecimal

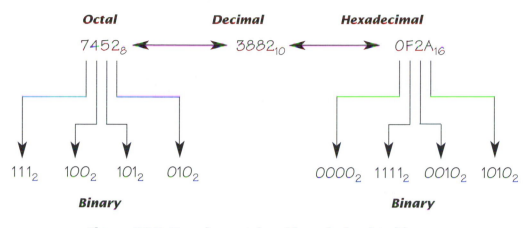

Figure 7-16. Mapping octal and hexadecimal to binary

In the original digital computers, data paths were often 9 bits, 12 bits, 18 bits, or 24 bits wide, which provides one reason for the original popularity of the octal system. Due to the fact that each octal digit maps directly to three binary bits, these data-path values were easily represented in octal. More recently, digital computers have standardized on data-path widths that are integer multiples of 8 bits; for example, 8 bits, 16 bits, 32 bits, 64 bits, and so forth. Because each hexadecimal digit maps directly to four binary bits, these data-path values are more easily represented in hexadecimal. This may explain the decline in popularity of the octal system and the corresponding rise in popularity of the hexadecimal system.

Representing Numbers Using Powers

An alternative way of representing numbers is by means of powers; for example, 10^3, where 10 is the base value and the superscripted 3 is known as the *power* or *exponent*. We read 10^3 as *"Ten to the power of three."* The power specifies how many times the base value must be multiplied by itself; thus, 10^3 represents $10 \times 10 \times 10$. Any value can be used as a base (Figure 7-17).

Any base to the power of one is equal to itself; for example, $8^1 = 8$. Strictly speaking, a power of zero is not part of the series, but, by convention, any base to the power of zero equals one; for example, $8^0 = 1$. Powers provide a convenient way to represent column-weights in place-value systems (Figure 7-18).

Decimal (Base-10)

10^0	$= 1$	$=$	1_{10}
10^1	$= 10$	$=$	10_{10}
10^2	$= 10 \times 10$	$=$	100_{10}
10^3	$= 10 \times 10 \times 10$	$=$	1000_{10}

Binary (Base-2)

2^0	$= 1$	$=$	1_2	$=$	1_{10}
2^1	$= 2$	$=$	10_2	$=$	2_{10}
2^2	$= 2 \times 2$	$=$	100_2	$=$	4_{10}
2^3	$= 2 \times 2 \times 2$	$=$	1000_2	$=$	8_{10}

Octal (Base-8)

8^0	$= 1$	$=$	1_8	$=$	1_{10}
8^1	$= 8$	$=$	10_8	$=$	8_{10}
8^2	$= 8 \times 8$	$=$	100_8	$=$	64_{10}
8^3	$= 8 \times 8 \times 8$	$=$	1000_8	$=$	512_{10}

Hexadecimal (Base-16)

16^0	$= 1$	$=$	1_{16}	$=$	1_{10}
16^1	$= 16$	$=$	10_{16}	$=$	16_{10}
16^2	$= 16 \times 16$	$=$	100_{16}	$=$	256_{10}
16^3	$= 16 \times 16 \times 16$	$=$	1000_{16}	$=$	4096_{10}

Figure 7-17. Representing numbers using powers

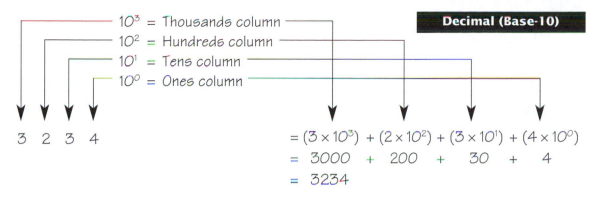

Figure 7-18a. Using powers to represent column weights in decimal

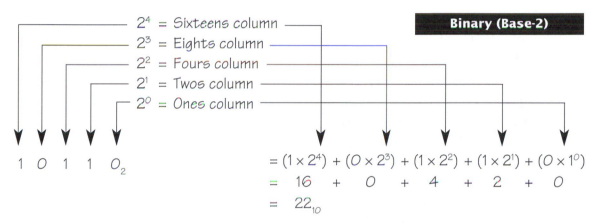

Figure 7-18b. Using powers to represent column weights in binary

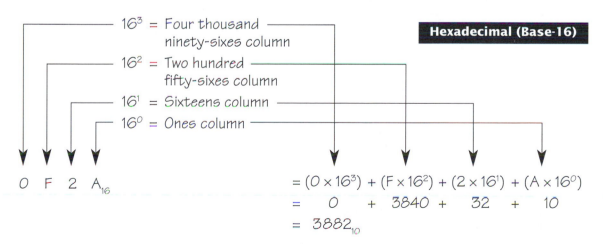

Figure 7-18c. Using powers to represent column weights in hexadecimal

Tertiary Logic

And finally, for reasons that become a little involved, communications theory tells us that optimal data transfer rates can be achieved if each data element represents three states. Now engineers wouldn't be the guys and gals they are if they weren't prepared to accept a challenge, and some experiments have been performed with *tertiary logic*. This refers to logic gates that are based on three distinct voltage levels. In this case, the three voltages are used to represent the tertiary (base-3) values of 0, 1, and 2, and their logical equivalents FALSE, TRUE, and MAYBE.[6]

However, while it's relatively easy to use transistors to generate two distinct voltage levels, it's harder to generate a consistent intermediate voltage to represent a third value. Similarly, while it's relatively easy to use transistors to interpret (detect) two voltage levels, it's harder to interpret a third, intermediate value.[7] Additionally, creating and using an equivalent to Boolean algebra[8] that works with the three logic states FALSE, TRUE, and MAYBE is enough to make even the strongest among us quail. Thankfully, tertiary logic is currently of academic interest only (otherwise this book might have been substantially longer). Still, there's an old saying *"What goes around comes around,"* and it's not beyond the realm of possibility that tertiary logic or an even bigger relative will rear its ugly head sometime in the future.

> **Apropos of nothing at all, man has been interested in the properties of odd numbers since antiquity, often ascribing mystical and magical properties to them, for example, *"lucky seven"* and *"unlucky thirteen."* As Pliny the Elder (AD 23–79) is reported to have said, *"Why is it that we entertain the belief that for every purpose odd numbers are the most effectual?"***

---

[6] A tertiary digit is known as a "trit".

[7] Actually, if the truth were told, it's really not too difficult to generate and detect three voltage levels. The problem is that you can't achieve this without using so many transistors that any advantages of using a tertiary system are lost.

[8] Boolean Algebra is introduced in Chapter 9.

Chapter

8

Binary Arithmetic

Due to the fact that digital computers are constructed from logic gates that can represent only two states, they are obliged to make use of the binary number system with its two digits: 0 and 1. Unlike calculations on paper where both decimal and binary numbers can be of any size—limited only by the size of your paper, the endurance of your pencil, and your stamina—the numbers manipulated within a computer have to be mapped onto a physical system of logic gates and wires. Thus, the maximum value of a number inside a computer is dictated by the width of its data path; that is, the number of bits used to represent that number.[1]

Unsigned Binary Numbers

Unsigned binary numbers can only be used to represent positive values. Consider the range of numbers that can be represented using 8 bits (Figure 8-1).

Each 'x' character represents a single bit; the right-hand bit is known as the *least significant bit (LSB)* because it represents the smallest value. Similarly, the left hand bit is known as the *most significant bit (MSB)* because it represents the largest value.

In computing it is usual to commence indexing things from zero, so the least significant bit is referred to as *bit 0*, and the most significant bit (of an 8-bit value) is referred to as *bit 7*. Every bit can be individually assigned a value of 0 or 1, so a group of 8 bits can be assigned $2^8 = 256$ unique combinations of 0s and 1s. This means that an 8-bit unsigned binary number can be used to represent values in the range 0_{10} through $+255_{10}$.

---

[1] Actually, this isn't strictly true, because there are tricks we can use to represent large numbers by splitting them into smaller "chunks" and re-using the same bits over and over again, but that's beyond the scope of what we're looking at here.

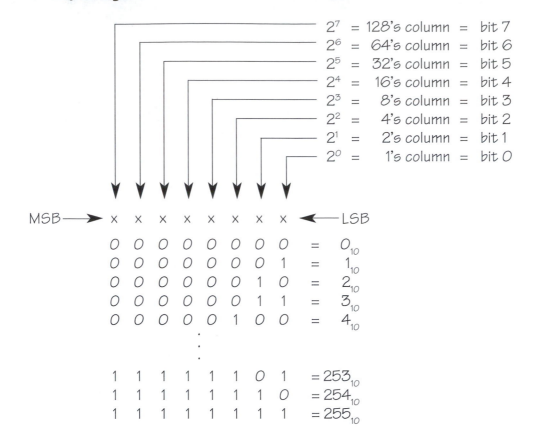

$2^7 = 128$'s column = bit 7
$2^6 = 64$'s column = bit 6
$2^5 = 32$'s column = bit 5
$2^4 = 16$'s column = bit 4
$2^3 = 8$'s column = bit 3
$2^2 = 4$'s column = bit 2
$2^1 = 2$'s column = bit 1
$2^0 = 1$'s column = bit 0

MSB ⟶ x x x x x x x x ⟵ LSB

0	0	0	0	0	0	0	0	$= 0_{10}$
0	0	0	0	0	0	0	1	$= 1_{10}$
0	0	0	0	0	0	1	0	$= 2_{10}$
0	0	0	0	0	0	1	1	$= 3_{10}$
0	0	0	0	0	1	0	0	$= 4_{10}$

⋮

1	1	1	1	1	1	0	1	$= 253_{10}$
1	1	1	1	1	1	1	0	$= 254_{10}$
1	1	1	1	1	1	1	1	$= 255_{10}$

Figure 8-1. Unsigned binary numbers

Binary Addition

Two binary numbers may be added together using an identical process to that used for decimal addition. First, the two least significant bits are added together to give a sum and, possibly, a carry-out to the next stage. This process is repeated for the remaining bits progressing towards the most significant. For each of the remaining bits, there may be a carry-in from the previous stage and a carry-out to the next stage. To fully illustrate this process, consider the step-by-step addition of two 8-bit binary numbers (Figure 8-2).

We commence with the least significant bits in step (a), $1 + 0 = 1$. In step (b) we see $0 + 1 = 1$, followed by $0 + 0 = 0$ in step (c). The first carry-out occurs in step (d), where $1 + 1 = 2$ or, in binary, $1_2 + 1_2 = 10_2$; thus, in this instance, the sum is 0 and there is a carry-out of 1 to the next stage. The second

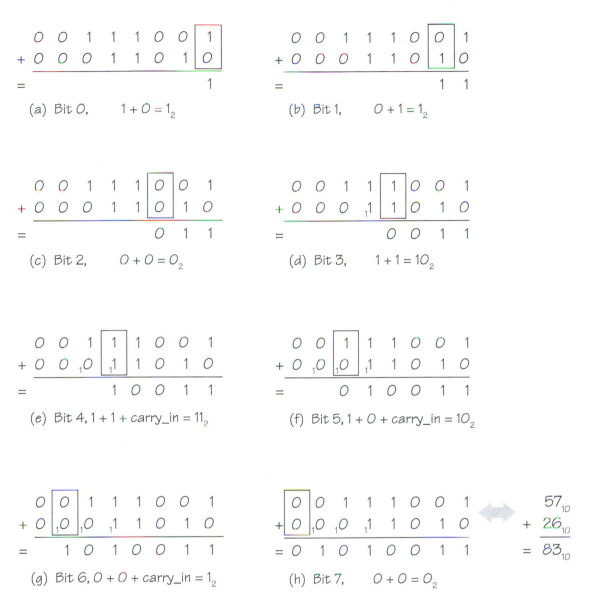

Figure 8-2. Binary addition

carry-out occurs in step (e), where the carry-in from the previous stage results in $1 + 1 + 1 = 3$ or, in binary, $1_2 + 1_2 + 1_2 = 11_2$; thus, in this instance, the sum is 1 and there is a carry-out of 1 to the next stage. The third and final carry-out occurs in step (f), and it's plain sailing from there on in.

Binary Subtraction

Unsigned binary numbers may be subtracted from each other using an identical process to that used for decimal subtraction. However, for reasons of efficiency, computers rarely perform subtractions in this manner; instead, these operations are typically performed by means of complement techniques.

There are two forms of complement associated with every number system, the *radix complement* and the *diminished radix complement*, where radix refers to the base of the number system. Under the decimal (base-10) system, the radix complement is also known as the *tens complement* and the diminished radix complement is known as the *nines complement*. First, consider a decimal subtraction performed using the nines complement technique—a process known in ancient times as *"Casting out the nines"* (Figure 8-3).

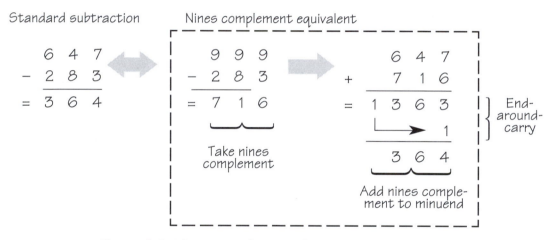

Figure 8-3. Nines complement decimal subtraction

The standard way of performing the operation would be to subtract the subtrahend (283) from the minuend (647), which, as in this example, may require the use of one or more borrow operations. To perform the equivalent operation in nines complement, each of the digits of the subtrahend is first subtracted from a 9. The resulting nines complement value is added to the minuend, and then an end-around-carry operation is performed. The advantage of the nines complement technique is that it is never necessary to perform a borrow operation.[2]

---

[2] This made this technique extremely popular in the days of yore, when the math skills of the general populace weren't particularly high.

Now consider the same subtraction performed using the tens complement technique (Figure 8-4).

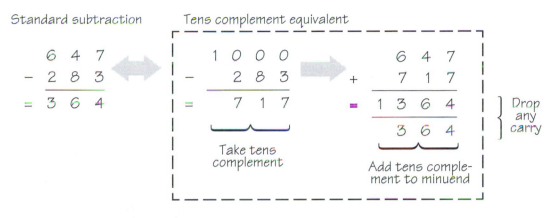

Figure 8-4. Tens complement decimal subtraction

The advantage of the tens complement is that it is not necessary to perform an end-around-carry; any carry-out resulting from the addition of the most significant digits is simply dropped from the final result. The disadvantage is that, during the process of creating the tens complement, it is necessary to perform a borrow operation for *every* digit in the subtrahend. This problem could be solved by first taking the nines complement of the subtrahend, adding one to the result, and then performing the remaining operations as for the tens complement.

Similar techniques may be employed with binary (base-2) numbers, where the radix complement is known as the *twos complement* and the diminished radix complement is known as the *ones complement*. First, consider a binary subtraction performed using the ones complement technique (Figure 8-5).

Once again, the standard way of performing the operation would be to subtract the subtrahend (00011110_2) from the minuend (00111001_2), which may require the use of one or more borrow operations. To perform the equivalent operation in ones complement, each of the digits of the subtrahend is first subtracted from a 1. The resulting ones complement value is added to the minuend, and then an end-around-carry operation is performed. The advantage of the ones complement technique is that it is never necessary to perform a borrow operation. In fact, it isn't even necessary to perform a

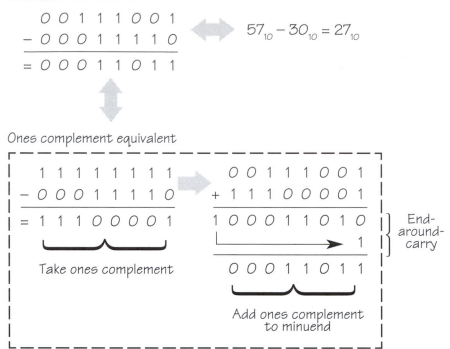

Figure 8-5. Ones complement binary subtraction

subtraction, because the ones complement of a binary number can be simply generated by inverting all of its bits; that is, by exchanging all the 0s with 1s and vice versa.

Now consider the same binary subtraction performed using the twos complement technique (Figure 8-6).

As before, the advantage of the twos complement is that it is not necessary to perform an end-around-carry; any carry-out resulting from the addition of the two most significant bits is simply dropped from the final result. The disadvantage is that, during the process of creating the twos complement, it is necessary to perform a borrow operation for almost every digit in the subtrahend. This problem can be solved by first taking the ones complement of the subtrahend, adding one to the result, and then performing the remaining operations as for the twos complement.

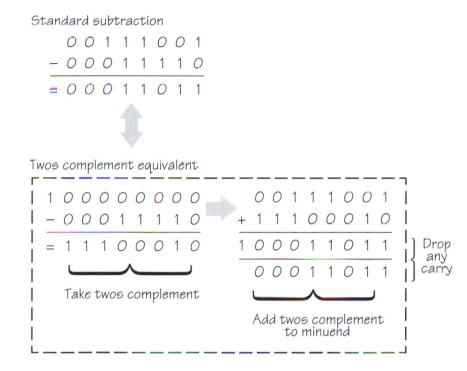

Figure 8-6. Twos complement binary subtraction

As fate would have it, there is a short-cut approach available to generate the twos complement of a binary number. Commencing with the least significant bit of the value to be complemented, each bit up to and including the first 1 is copied directly, and the remaining bits are inverted (Figure 8-7).

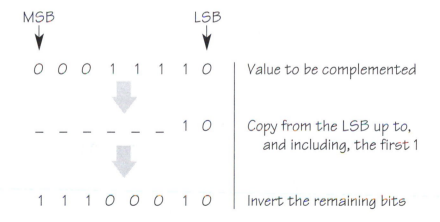

Figure 8-7. Shortcut for generating a twos complement

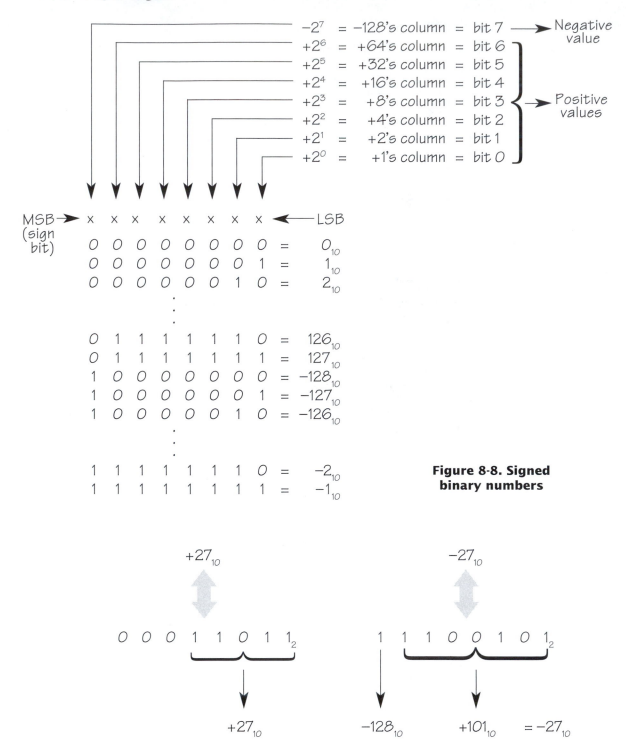

Figure 8-8. Signed binary numbers

Figure 8-9. Comparison of positive and negative signed binary numbers

Unfortunately, all of the previous examples will return incorrect results if a larger value is subtracted from a smaller value; that is, for these techniques to work, the final result must be greater than or equal to zero. In the case of unsigned binary numbers, the reason is clear because, by definition, an unsigned binary number can only be used to represent a positive value, but subtracting a larger value from a smaller value results in a negative value. It would obviously be somewhat inconvenient if computers could only be used to generate positive values, so we need some way to represent negative numbers.

Signed Binary Numbers

Signed binary numbers can be used to represent both positive and negative values, and they do this in a rather cunning way. In standard decimal arithmetic, negative numbers are typically represented in a form known as *sign-magnitude*,[3] which means prefixing values with plus or minus signs. For example, values of plus and minus twenty-seven would be shown as +27 and –27, respectively. However, for reasons of efficiency, computers rarely employ the sign-magnitude form, and instead use the *signed binary* format, in which the most significant bit is also called the *sign bit* (Figure 8-8).

The least significant bits continue to represent the same positive quantities as for unsigned binary numbers, but the sign bit is used to represent a *negative quantity*. In the case of a signed 8-bit number, a 1 in the sign bit represents -2^7 = -128, and the remaining bits are used to represent positive values in the range 0_{10} through $+127_{10}$. Thus, when the value represented by the sign bit is combined with the values represented by the remaining bits, an 8-bit signed binary number can be used to represent values in the range -128_{10} through $+127_{10}$.

To illustrate the differences between the sign-magnitude and signed binary formats, first consider a positive sign-magnitude decimal number and its negative equivalent: for example, +27 and –27. As we see, the digits are identical for both cases and only the sign changes. Now consider the same values represented as signed binary numbers (Figure 8-9).

In this case, the bit patterns of the two binary numbers are very different. This is because the sign bit represents an actual quantity (-128_{10}) rather than a simple plus or minus; thus, the signed equivalent of -27_{10} is formed by combining -128_{10} with $+101_{10}$. Now pay attention because this is the clever part:

---

[3] Sometimes written as *sign+magnitude*.

closer investigation reveals that each bit pattern is in fact the twos complement of the other! To put this another way, taking the twos complement of a positive signed binary value returns its negative equivalent and vice versa.

The end result is that using signed binary numbers greatly reduces the complexity of operations within a computer. To illustrate why this is so, first consider one of the simplest operations, that of addition. Compare the additions of positive and negative decimal values in sign-magnitude form with their signed binary counterparts (Figure 8-10).

Examine the standard decimal calculations on the left. The one at the top is easy to understand because it's a straightforward addition of two positive values. However, even though you are familiar with decimal addition, you probably found the other three a little harder because you had to decide exactly what to do with the negative values. By comparison, the signed binary

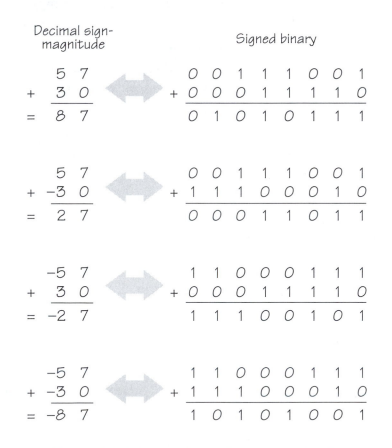

Figure 8-10. Comparison of sign-magnitude versus signed binary additions

calculations on the right are all simple additions, regardless of whether the individual values are positive or negative.

If computers were forced to use a binary version of the sign-magnitude form to perform additions, they would have to perform a sequence of operations (Figure 8-11).

First compare the signs of the two values.	
IF THE SIGNS ARE THE SAME: Add the values. Note that the result will always have the same sign as the original values.	IF THE SIGNS ARE DIFFERENT: Subtract the smaller value from the larger value, then attach the correct sign to the result.

Figure 8-11. Steps required for sign-magnitude additions

As well as being time consuming, performing all these operations would require a substantial number of logic gates. Thus, the advantages of using the signed binary format for addition operations are apparent: signed binary numbers can always be directly added together to provide the correct result in a single operation, regardless of whether they represent positive or negative values. That is, the operations $a + b$, $a + (-b)$, $(-a) + b$, and $(-a) + (-b)$ are all performed in exactly the same way, by simply adding the two values together. This results in fast computers that can be constructed using a minimum number of logic gates.

Now consider the case of subtraction. We all know that $10 - 3 = 7$ in decimal arithmetic, and that the same result can be obtained by negating the right-hand value and inverting the operation: that is, $10 + (-3) = 7$. This technique is also true for signed binary arithmetic, although the negation of the right hand value is performed by taking its twos complement rather than by changing its sign. For example, consider a generic signed binary subtraction[4]

---

[4] For the sake of simplicity, only the case of $a - b$ is discussed here. However, the operations $a - b$, $a - (-b)$, $(-a) - b$, and $(-a) - (-b)$ are all performed in exactly the same way, by simply taking the twos complement of b and adding the result to a, regardless of whether a and b represent positive or negative values.

represented by $a - b$. Generating the twos complement of b results in $-b$, allowing the operation to be performed as an addition: $a + (-b)$. This means that computers do not require different blocks of logic to add and subtract; instead, they only require an adder and a twos complementor. As well as being faster than a subtractor, a twos complementor requires significantly fewer logic gates.

Binary Multiplication

One technique for performing multiplication in any number base is by means of repeated addition; for example, in decimal, $6 \times 4 = 6 + 6 + 6 + 6 = 24$. However, even though computers can perform millions of operations every second, the repeated addition technique is time-consuming when the values to be multiplied are large. As an alternative, binary numbers may by multiplied together by means of a shift-and-add technique known as Booth's Algorithm (Figure 8-12).

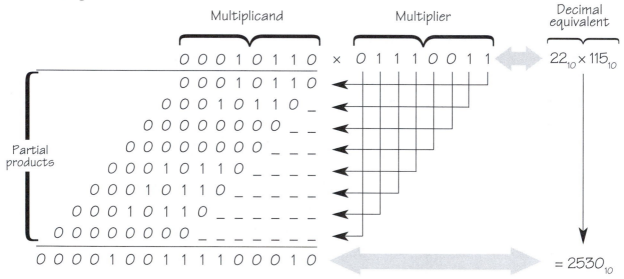

Figure 8-12. Binary multiplication using Booth's Algorithm

Using Booth's algorithm, a *partial product* is generated for every bit in the multiplier. If the value of the multiplier bit is 0, its corresponding partial product consists only of 0s; if the value of the bit is 1, its corresponding partial product is a copy of the multiplicand. Additionally, each partial product is left-shifted as a function of the multiplier bit with which it is associated; for example, the partial product associated with bit 0 in the multiplier is left-shifted zero bits, the partial product associated with bit 1 is left-shifted one bit, etc. All of the partial products are then added together to generate the result, whose width is equal to the sum of the widths of the two values being multiplied together.

There are several ways to implement a multiplier based on Booth's technique. In one implementation, all of the partial products are generated simultaneously and then added together. This requires a lot of logic gates, but the resulting multiplication is extremely fast. Unfortunately, Booth's Algorithm only works with unsigned binary values. However, this problem can be overcome by taking the twos complement of any negative values before feeding them into the multiplier. If the signs of the two values are the same, both positive or both negative, then no further action need be taken.[5] However, if the signs are different, then the result returned from the multiplier must be negated by transforming it into its twos complement.

Last but not least, long division is just about as much fun in binary as it is in decimal, which is to say "Not a lot!" For this reason, binary division is best left to computers because they are in no position to argue about it.

---

[5] A negative multiplied by a negative equals a positive; for example, $(-3) \times (-4) = +12$.

Boolean Algebra

One of the most significant mathematical tools available to electronics designers was actually invented for quite a different purpose. Around the 1850s, a British mathematician, George Boole (1815–1864), developed a new form of mathematics that is now known as *Boolean Algebra*. Boole's intention was to use mathematical techniques to represent and rigorously test logical and philosophical arguments. His work was based on the following: a *statement* is a sentence that asserts or denies an attribute about an object or group of objects:

Statement: *Your face resembles a cabbage.*

Depending on how carefully you choose your friends, they may either agree or disagree with the sentiment expressed; therefore, this statement cannot be proved to be either true or false.

By comparison, a *proposition* is a statement that is either true or false with no ambiguity:

Proposition: *I just tipped a bucket of burning oil into your lap.*

This proposition may be true or it may be false, but it is definitely one or the other and there is no ambiguity about it.

Propositions can be combined together in several ways; a proposition combined with an AND operator is known as a *conjunction*:

> **Conjunction:** *You have a parrot on your head AND you have a fish in your ear.*
>
> The result of a conjunction is true if all of the propositions comprising that conjunction are true.

A proposition combined with an OR operator is known as a *disjunction*:

> **Disjunction:** *You have a parrot on your head OR you have a fish in your ear.*
>
> The result of a disjunction is true if at least one of the propositions comprising that disjunction is true.

From these humble beginnings, Boole established a new mathematical field known as *symbolic logic*, in which the logical relationship between propositions can be represented symbolically by such means as equations or truth tables. Sadly, this work found little application outside the school of symbolic logic for almost one hundred years.

In fact, the significance of Boole's work was not fully appreciated until the late 1930s, when a graduate student at MIT, Claude Shannon, submitted a master's thesis that revolutionized electronics. In this thesis, Shannon showed that Boolean Algebra offered an ideal technique for representing the logical operation of digital systems. Shannon had realized that the Boolean concepts of FALSE and TRUE could be mapped onto the binary digits 0 and 1, and that both could be easily implemented by means of electronic circuits.

Logical functions can be represented using graphical symbols, equations, or truth tables, and these views can be used interchangeably (Figure 9-1).

There are a variety of ways to represent Boolean equations. In this book, the symbols &, |, and ^ are used to represent AND, OR, and XOR respectively; a negation, or NOT, is represented by a horizontal line, or bar, over the portion of the equation to be negated.

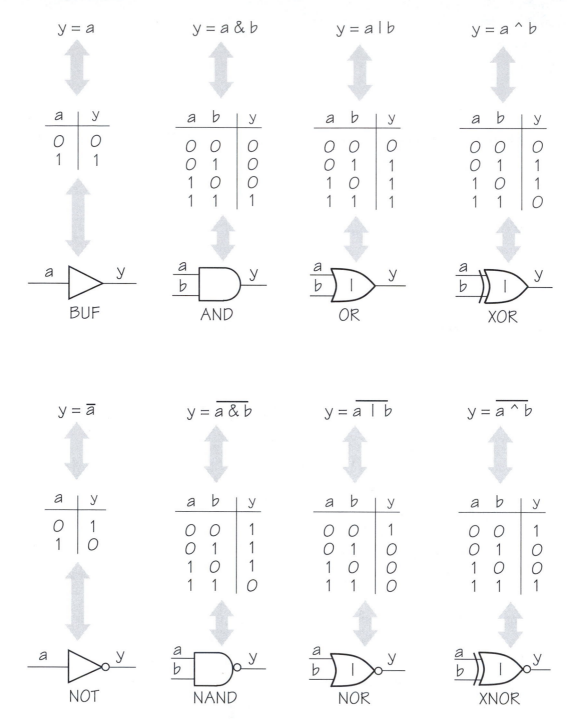

Figure 9-1. Summary of primitive logic functions

Combining a Single Variable with Logic 0 or Logic 1

A set of simple but highly useful rules can be derived from the combination of a single variable with a logic 0 or logic 1 (Figure 9-2).[1,2]

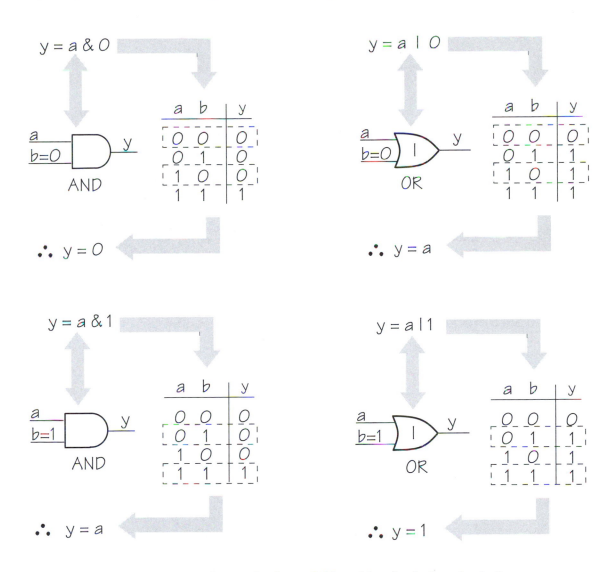

Figure 9-2. Combining a single variable with a logic 0 or logic 1

---

[1] Note that, throughout these discussions, the results from the NAND and NOR functions would be the inverse of those from the AND and OR functions, respectively.

[2] Note that the symbol $\therefore$ shown in the equations in Figure 9-2 means "*therefore.*"

The Idempotent Rules

The rules derived from the combination of a single variable with itself are known as the *idempotent rules* (Figure 9-3).

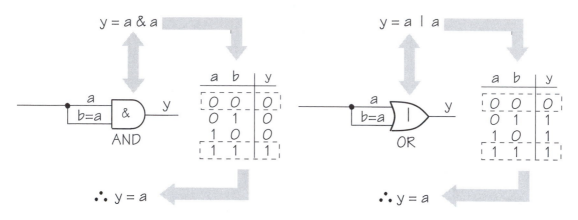

Figure 9-3. The idempotent rules

The Complementary Rules

The rules derived from the combination of a single variable with the inverse of itself are known as the *complementary rules* (Figure 9-4).

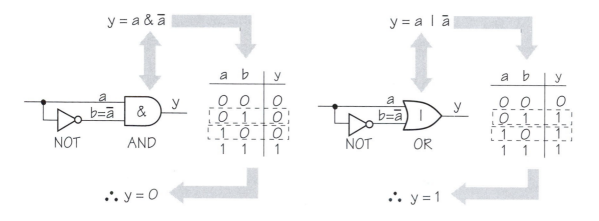

Figure 9-4. The complementary rules

The Involution Rule

The involution rule states that an even number of inversions cancel each other out; for example, two NOT functions connected in series generate an identical result to that of a BUF function (Figure 9-5).

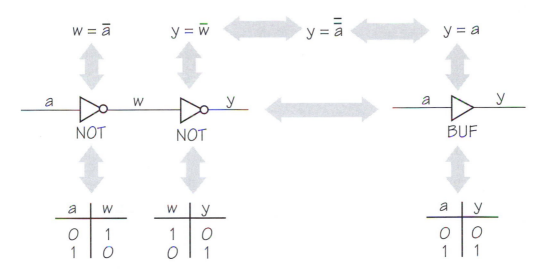

Figure 9-5. The involution rules

The Commutative Rules

The commutative rules state that the order in which variables are specified will not affect the result of an AND or OR operation (Figure 9-6).

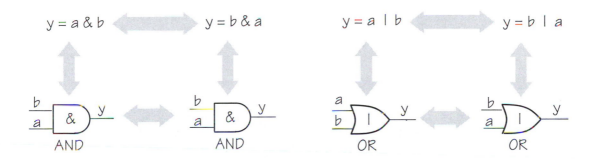

Figure 9-6. The commutative rules

The Associative Rules

The associative rules state that the order in which pairs of variables are associated together will not affect the result of multiple AND or OR operations (Figure 9-7).

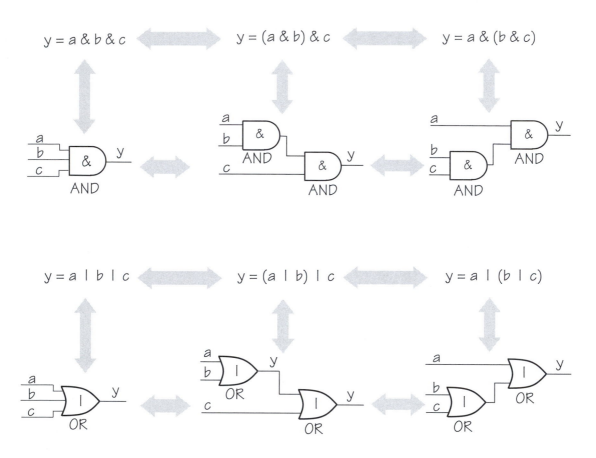

Figure 9-7. The associative rules

In addition to recognizing the application of Boolean algebra to electronic design, Shannon is also credited with the invention of the rocket-powered frisbee, and is famous for riding down the corridors at Bell Laboratories on a unicycle while simultaneously juggling four balls.

Precedence of Operators

In standard arithmetic, the multiplication operator is said to have a higher *precedence* than the addition operator. This means that, if an equation contains both multiplication and addition operators without parenthesis, then the multiplication is performed before the addition; for example:[3]

$$6 + 2 \times 4 \equiv 6 + (2 \times 4)$$

Similarly, in Boolean Algebra, the & (AND) operator has a higher precedence than the | (OR) operator:

$$a \mid b \& c \equiv a \mid (b \& c)$$

Due to the similarities between these arithmetic and logical operators, the & (AND) operator is known as a *logical multiplication* or *product*, while the | (OR) operator is known as a *logical addition* or *sum*. To avoid any confusion as to the order in which logical operations will be performed, this book will always make use of parentheses.

The first true electronic computer, ENIAC (Electronic Numerical Integrator and Calculator), was constructed at the University of Pennsylvania between 1943 and 1946. In many ways ENIAC was a monster; it occupied 30 feet by 50 feet of floor space, weighed approximately 30 tons, and used more than 18,000 vacuum tubes which required 150 kilowatts of power—enough to light a small town. One of the big problems with computers built from vacuum tubes was reliability; 90% of ENIAC's down-time was attributed to locating and replacing burnt-out tubes. Records from 1952 show that approximately 19,000 vacuum tubes had to be replaced in that year alone; that averages out to about 50 tubes a day!

---

[3] Note that the symbol $\equiv$ shown in these equations indicates *"is equivalent to"* or *"is the same as."*

The First Distributive Rule

In standard arithmetic, the multiplication operator will distribute over the addition operator because it has a higher precedence; for example:

$$6 \times (5 + 2) \equiv (6 \times 5) + (6 \times 2)$$

Similarly, in Boolean Algebra, the & (AND) operator will distribute over an | (OR) operator because it has a higher precedence; this is known as the first distributive rule (Figure 9-8).

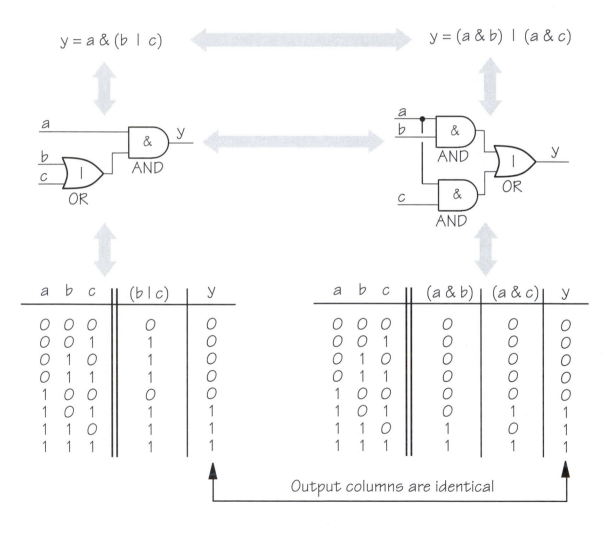

Figure 9-8. The first distributive rule

The Second Distributive Rule

In standard arithmetic, the addition operator will not distribute over the multiplication operator because it has a lower precedence:[4]

$$6 + (5 \times 2) \neq (6 + 5) \times (6 + 2)$$

However, Boolean Algebra is special in this case. Even though the | (OR) operator has lower precedence than the & (AND) operator, it will still distribute over the & operator; this is known as the second distributive rule (Figure 9-9).

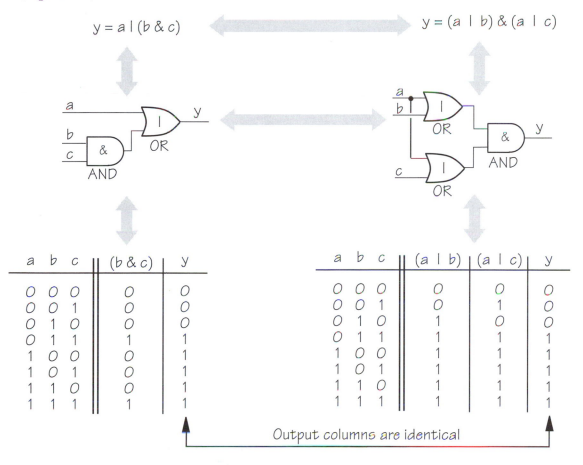

Output columns are identical

Figure 9-9. The second distributive rule

---

[4] Note that the symbol ≠ shown in the equation indicates *"is not equal to."*

The Simplification Rules

There are a number of simplification rules that can be used to reduce the complexity of Boolean expressions. As the end result is to reduce the number of logic gates required to implement the expression, the process of simplification is also known as *minimization* (Figure 9-10).

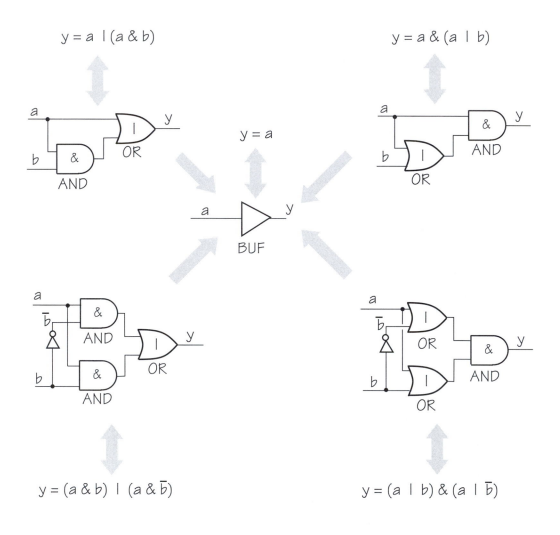

Figure 9-10. The simplification rules

– *continued on next page*

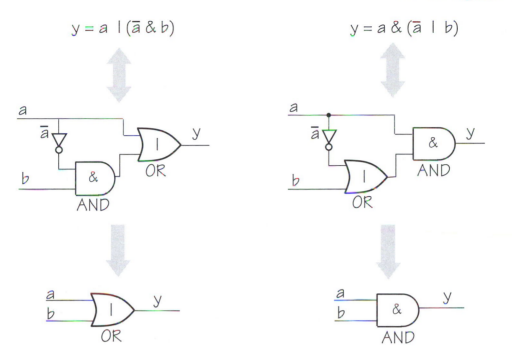

$$y = a \mid (\bar{a} \,\&\, b)$$

$$y = a \,\&\, (\bar{a} \mid b)$$

Figure 9-10 *(continued).* The simplification rules

Few now know that, before elbowing his way up the ladder of corporate success, the author of this epic volume once aspired to be a world-famous explorer. Picture him, if you will, as he strode across the veldt (hair by Vidal Sassoon®, eyelashes by Max Factor®, and loin cloths by Fruit of the Loom™). The name *Tracker Max,* as he then styled himself, was oft bandied by the denizens of the dim and dingy taverns in the deepest, direst domains of the dark continent.

Today his Arnold Schwarzenegger look-alike body is compressed into an executive-style business suit (in a fashion reminiscent of years gone by) and he passes his twilight years quaffing the odd fermented coconut and penning trail-blazing books such as this one. Now that time has healed the deeper wounds, some in unusual and interesting places that he is only too happy to display (for a nominal fee), he feels more able to tell the tale of those far off times ... unfortunately we don't have enough space here!

DeMorgan Transformations

A contemporary of Boole's, Augustus DeMorgan (1806–1871), also made significant contributions to the field of symbolic logic, most notably a set of rules which facilitate the conversion of Boolean expressions into alternate and often more convenient forms. A *DeMorgan Transformation* comprises four steps:

1. Exchange all of the & operators for | operators and vice versa.
2. Invert all the variables; also exchange 0s for 1s and vice versa.
3. Invert the entire function.
4. Reduce any multiple inversions.

Consider the DeMorgan Transformation of a 2-input AND function (Figure 9-11). Note that the NOT gate on the output of the new function can be combined with the OR to form a NOR.

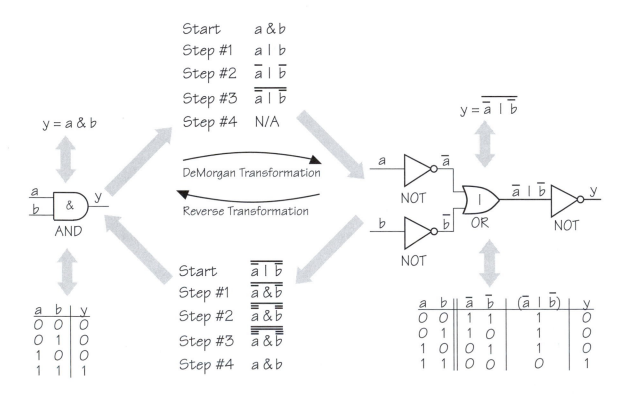

Figure 9-11. DeMorgan Transformation of an AND function

Similar transformations can be performed on the other primitive functions (Figure 9-12).

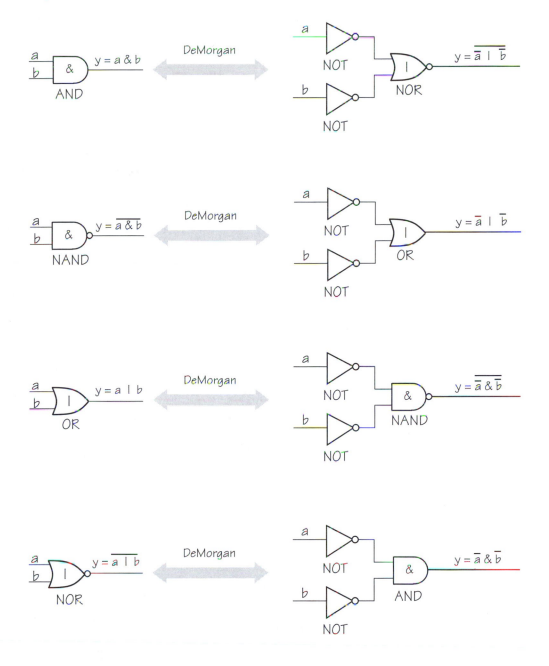

Figure 9-12. DeMorgan Transformations of AND, NAND, OR, and NOR functions

Minterms and Maxterms

For each combination of inputs to a logical function, there is an associated *minterm* and an associated *maxterm*. Consider a truth table with three inputs: a, b, and c (Figure 9-13).

a	b	c	minterms	maxterms
0	0	0	$(\bar{a} \& \bar{b} \& \bar{c})$	$(a \mid b \mid c)$
0	0	1	$(\bar{a} \& \bar{b} \& c)$	$(a \mid b \mid \bar{c})$
0	1	0	$(\bar{a} \& b \& \bar{c})$	$(a \mid \bar{b} \mid c)$
0	1	1	$(\bar{a} \& b \& c)$	$(a \mid \bar{b} \mid \bar{c})$
1	0	0	$(a \& \bar{b} \& \bar{c})$	$(\bar{a} \mid b \mid c)$
1	0	1	$(a \& \bar{b} \& c)$	$(\bar{a} \mid b \mid \bar{c})$
1	1	0	$(a \& b \& \bar{c})$	$(\bar{a} \mid \bar{b} \mid c)$
1	1	1	$(a \& b \& c)$	$(\bar{a} \mid \bar{b} \mid \bar{c})$

Figure 9-13. Minterms and maxterms

The minterm associated with each input combination is the & (AND), or product, of the input variables, while the maxterm is the | (OR), or sum, of the inverted input variables. Minterms and maxterms are useful for deriving Boolean equations from truth tables as discussed below.

Sum-of-Products and Product-of-Sums

A designer will often specify portions of a design using truth tables, and determine how to implement these functions as logic gates later. The designer may start by representing a function as a "black box"[5] with an associated truth table (Figure 9-14). Note that the values assigned to the output y in the truth table shown in Figure 9-14 were selected randomly, and have no significance beyond the purposes of this example.

There are two commonly used techniques for deriving Boolean equations from a truth table. In the first technique, the minterms corresponding to each line in the truth table for which the output is a logic 1 are extracted and combined using | (OR) operators; this method results in an equation said to be

---

[5] A "black box" is so-called because initially we don't know exactly what's going to be in it.

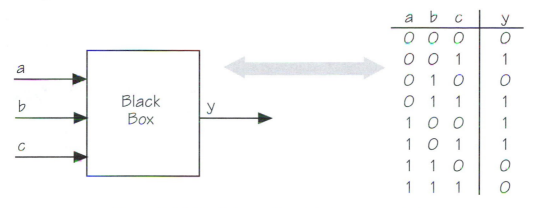

Figure 9-14. Black box with associated truth table

in *sum-of-products* form. In the second technique, the maxterms corresponding to each line in the truth table for which the output is a logic 0 are combined using & (AND) operators; this method results in an equation said to be in *product-of-sums* form (Figure 9-15).

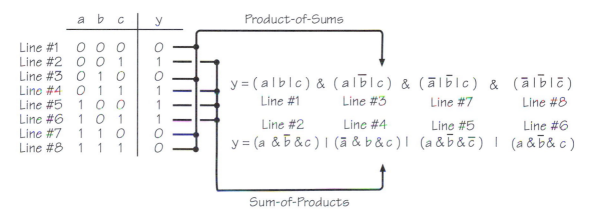

Figure 9-15. Sum-of-products versus product-of-sums equations

For a function whose output is logic 1 fewer times than it is logic 0, it is generally easier to extract a sum-of-products equation. Similarly, if the output is logic 0 fewer times than it is logic 1, it is generally easier to extract a product-of-sums equation. The sum-of-products and product-of-sums forms complement each other and return identical results. An equation in either form can be transformed into its alternative form by means of the appropriate DeMorgan Transformation.

Once an equation has been obtained in the required form, the designer would typically make use of the appropriate simplification rules to minimize the number of logic gates required to implement the function. However, neglecting any potential minimization, the equations above could be translated directly into their logic gate equivalents (Figure 9-16).

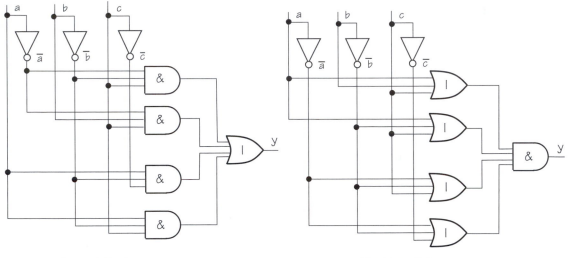

Sum-of-Products   Product-of-Sums

Figure 9-16. Sum-of-products versus product-of-sums implementations

Canonical Forms

In a mathematical context, the term *canonical form* is taken to mean a generic or basic representation. Canonical forms provide the means to compare two expressions without falling into the trap of trying to compare "apples" with "oranges." The sum-of-products and product-of-sums representations are different canonical forms. Thus, to compare two Boolean equations, both must first be coerced into the same canonical form; either sum-of-products or product-of-sums.

Karnaugh Maps

In 1953, Maurice Karnaugh (pronounced "car-no") invented a form of logic diagram[1] called a *Karnaugh Map*, which provides an alternative technique for representing Boolean functions; for example, consider the Karnaugh Map for a 2-input AND function (Figure 10-1).

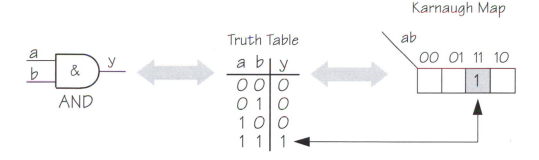

Figure 10-1. Karnaugh Map for a 2-input AND function

The Karnaugh Map comprises a box for every line in the truth table. The binary values above the boxes are those associated with the a and b inputs. Unlike a truth table, in which the input values typically follow a binary sequence, the Karnaugh Map's input values must be ordered such that the values for adjacent columns vary by only a single bit: for example, 00_2, 01_2, 11_2, and 10_2. This ordering is known as a *Gray code*,[2] and it is a key factor in the way in which Karnaugh Maps work.

The y column in the truth table shows all the 0 and 1 values associated with the gate's output. Similarly, all of the output values could be entered into the Karnaugh Map's boxes. However, for reasons of clarity, it is common for only a single set of values to be used (typically the 1s).

---

[1] The topic of logic diagrams in general is discussed in more detail in Chapter 22.

[2] Gray codes are discussed in more detail in Appendix D.

Similar maps can be constructed for 3-input and 4-input functions. In the case of a 4-input map, the values associated with the *c* and *d* inputs must also be ordered as a Gray code: that is, they must be ordered in such a way that the values for adjacent rows vary by only a single bit (Figure 10-2).

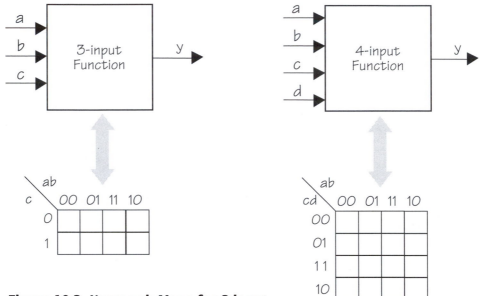

**Figure 10-2. Karnaugh Maps for 3-input
and 4-input functions**

Minimization Using Karnaugh Maps

Karnaugh Maps often prove useful in the simplification and minimization of Boolean functions. Consider an example 3-input function represented as a black box with an associated truth table (Figure 10-3).[3]

The equation extracted from the truth table in sum-of-products form contains four minterms,[4] one for each of the 1s assigned to the output. Algebraic simplification techniques could be employed to minimize this equation, but this would necessitate every minterm being compared to each of the others, which can be somewhat time-consuming.

---

[3] The values assigned to output y in the truth table were selected randomly and have no significance beyond the purposes of this example.

[4] The concepts of *minterms* and *maxterms* were introduced in Chapter 9.

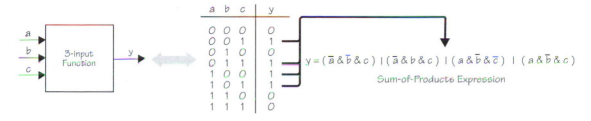

Figure 10-3. Example 3-input function

This is where Karnaugh Maps enter the game. The 1s assigned to the map's boxes represent the same minterms as the 1s in the truth table's output column; however, as the input values associated with each row and column in the map differ by only one bit, any pair of horizontally or vertically adjacent boxes corresponds to minterms that differ by only a single variable. Such pairs of minterms can be grouped together and the variable that differs can be discarded (Figure 10-4).

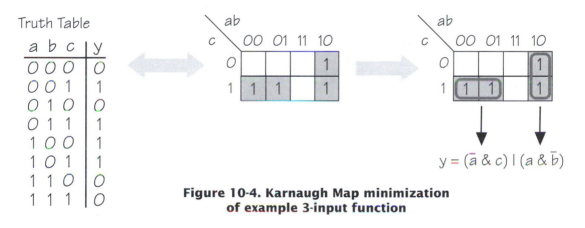

Figure 10-4. Karnaugh Map minimization of example 3-input function

In the case of the horizontal group, input *a* is 0 for both boxes, input *c* is 1 for both boxes, and input *b* is 0 for one box and 1 for the other. Thus, for this group, changing the value on *b* does not affect the value of the output. This means that *b* is redundant and can be discarded from the equation representing this group. Similarly, in the case of the vertical group, input *a* is 1 for both boxes, input *b* is 0 for both boxes, and input *c* is 0 for one box and 1 for the other. Thus, input *c* is redundant for *this* group and can be discarded.

Grouping Minterms

In the case of a 3-input Karnaugh Map, any two horizontally or vertically adjacent minterms, each composed of three variables, can be combined to form a new product term composed of only two variables. Similarly, in the case of a 4-input map, any two adjacent minterms, each composed of four variables, can be combined to form a new product term composed of only three variables. Additionally, the 1s associated with the minterms can be used to form multiple groups. For example, consider the 3-input function shown in Figure 10-5, in which the minterm corresponding to $a = 1$, $b = 1$, and $c = 0$ is common to three groups.

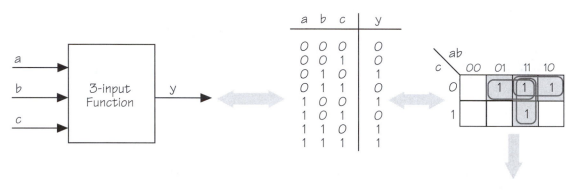

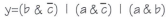

$$y=(b \& \bar{c}) \mid (a \& \bar{c}) \mid (a \& b)$$

Figure 10-5. Karnaugh Map minterms used to form multiple groups

Groupings can also be formed from four adjacent minterms, in which case two redundant variables can be discarded; consider some 4-input Karnaugh Map examples (Figure 10-6).

In fact, any group of 2^n adjacent minterms can be gathered together where n is a positive integer. For example, 2^1 = two minterms, $2^2 = 2 \times 2$ = four minterms, $2^3 = 2 \times 2 \times 2$ = eight minterms, etc.

As was noted earlier, Karnaugh Map input values are ordered so that the values associated with adjacent rows and columns differ by only a single bit. One result of this ordering is that the top and bottom rows are also separated by only a single bit (it may help to visualize the map rolled into a horizontal cylinder such that the top and bottom edges are touching). Similarly, the left and right columns are separated by only a single bit (in this case it may help to

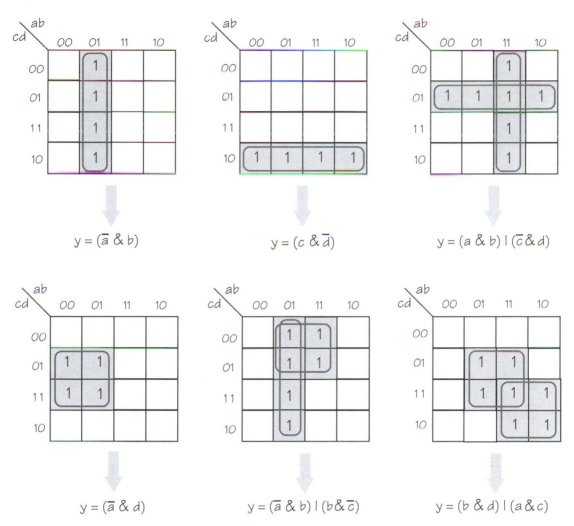

Figure 10-6. Karnaugh Map groupings of four adjacent minterms

visualize the map rolled into a vertical cylinder such that the left and right edges are touching). This leads to some additional groupings, a few of which are shown in Figure 10-7.

Note especially the last example. Diagonally adjacent minterms generally cannot be used to form a group: however, remembering that the left-right columns and the top-bottom rows are logically adjacent, this means that the four corner minterms are also logically adjacent, which in turn means that they can be used to form a single group.

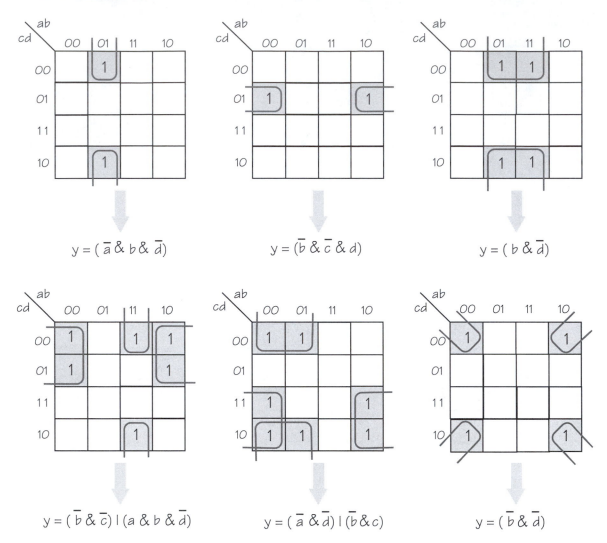

Figure 10-7. Additional Karnaugh Map grouping possibilities

Incompletely Specified Functions

In certain cases a function may be incompletely specified: that is, the output may be undefined for some of the input combinations. For example, if the designer knows that certain input combinations will never occur, then the value assigned to the output for these combinations is irrelevant. Alternatively, for some input combinations the designer may simply not care about the value on the output. In both cases, the designer can represent the output values

associated with the relevant input combinations as
question marks in the map (Figure 10-8).

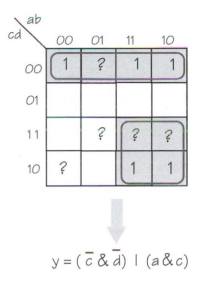

Figure 10-8. Karnaugh Map for an incompletely specified function

$$y = (\bar{c}\,\&\,\bar{d}) \mid (a\,\&\,c)$$

**Figure 10-8. Karnaugh Map
for an incompletely
specified function**

The *?* characters indicate *don't care* states, which
can be considered to represent either 0 or 1 values at
the designer's discretion. In the example shown in
Figure 10-8, we have no interest in the *?* character at
$a = 0$, $b = 0$, $c = 1$, $d = 0$ or the *?* character at $a = 0$,
$b = 1$, $c = 1$, $d = 1$, because neither of these can be
used to form a larger group. However, if we decide
that the other three *?* characters are going to represent 1 values, then they can be used to form larger
groups, which allows us to minimize the function to
a greater degree than would otherwise be possible.

It should be noted that many electronics references use X characters to represent don't care states.
Unfortunately, this may lead to confusion as design
tools such as logic simulators use X characters to
represent *don't know* states. Unless otherwise indicated, this book will use *?* and
X to represent *don't care* and *don't know* states, respectively.

Populating Maps Using 0s Versus 1s

When we were extracting Boolean equations from truth tables in the
previous chapter, we noted that in the case of a function whose output is logic 1
fewer times than it is logic 0, it is generally easier to extract a *sum-of-products*
equation. Similarly, if the output is logic 0 fewer times than it is logic 1, it is
generally easier to extract a *product-of-sums* equation.

The same thing applies to a Karnaugh Map. If the output is logic 1 fewer
times than it is logic 0, then it's probably going to be a lot easier to populate
the map using logic 1's. Alternatively, if the output is logic 0 fewer times than
it is logic 1, then populating the map using logic 0s may not be a bad idea.

When a Karnaugh Map is populated using the 1s assigned to the truth
table's output, the resulting Boolean expression is extracted from the map in
sum-of-products form. By comparison, if the Karnaugh Map is populated using
the 0s assigned to the truth table's output, then the groupings of 0s are used to
generate expressions in product-of-sums form (Figure 10-9).

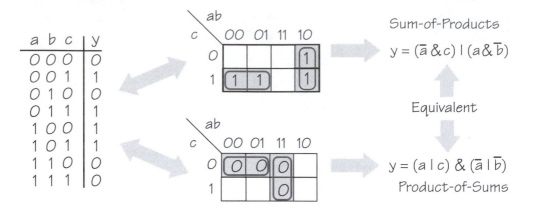

Figure 10-9. Populating Karnaugh Maps with 0s versus 1s

Although the sum-of-products and product-of-sums expressions appear to be somewhat different, they do produce identical results. The expressions can be shown to be equivalent using algebraic means, or by constructing truth tables for each expression and comparing the outputs.

Karnaugh Maps are most often used to represent 3-input and 4-input functions. It is possible to create similar maps for 5-input and 6-input functions, but these maps can quickly become unwieldy and difficult to use. Thus, the Karnaugh technique is generally not considered to have any application for functions with more than six inputs.

Using Primitive Logic Functions to Build More Complex Functions

The primitive functions NOT, AND, OR, NAND, NOR, XOR, and XNOR can be connected together to build more complex functions which may, in turn, be used as building blocks in yet more sophisticated systems. The examples introduced in this chapter were selected because they occur commonly in designs, are relatively simple to understand, and will prove useful in later discussions.

Scalar versus Vector Notation

A single signal carrying one bit of binary data is known as a *scalar* entity. A set of signals carrying similar data can be gathered together into a group known as a *vector* (see also the glossary definition of "vector").

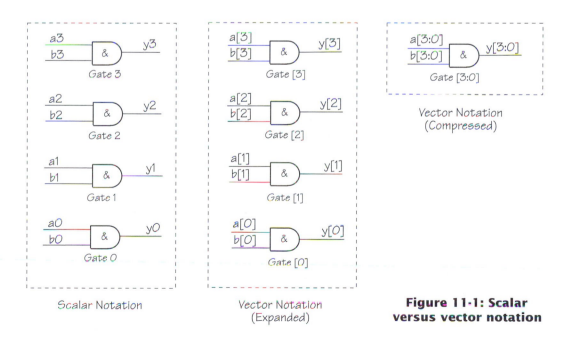

Scalar Notation

Vector Notation
(Expanded)

Vector Notation
(Compressed)

Figure 11-1: Scalar versus vector notation

Consider the circuit fragments shown in Figure 11-1. Each of these fragments represents four 2-input AND gates. In the case of the scalar notation, each signal is assigned a unique name: for example, a3, a2, a1, and a0. By comparison, when using vector notation, a single name is applied to a group of signals, and individual signals within the group are referenced by means of an index: for example, a[3], a[2], a[1], and a[0]. This means that if we were to see a schematic (circuit) diagram containing two signals called a3 and a[3], we would understand this to represent two completely different signals (the former being a scalar named "a3" and the latter being the third element of a vector named "a").

A key advantage of vector notation is that it allows all of the signals comprising the vector to be easily referenced in a single statement: for example, a[3:0], b[3:0], and y[3:0]. Thus, vector notation can be used to reduce the size and complexity of a circuit diagram while at the same time increasing its clarity.

Equality Comparators

In some designs it may be necessary to compare two sets of binary values to see if they contain the same data. Consider a function used to compare two 4-bit vectors: a[3:0] and b[3:0]. A scalar output called *equal* is to be set to logic 1 if each bit in a[3:0] is equal to its corresponding bit in b[3:0]: that is, the vectors are equal if a[3] = b[3], a[2] = b[2], a[1] = b[1], and a[0] = b[0] (Figure 11-2).

The values on a[3] and b[3] are compared using a 2-input XNOR gate. As we know from Chapters 5 and 6, if the values on its inputs are the same (both 0s or both 1s), the output of an XNOR will be 1, but if the values on its inputs are different, the output will be 0. Similar comparisons are performed between the other inputs: a[2] with b[2], a[1] with b[1], and a[0] with b[0]. The final AND gate is used to gather the results of the individual comparisons. If all the inputs to the AND gate are 1, the two vectors are the same, and the output of the AND gate will be 1. Correspondingly, if any of the inputs to the AND gate are 0, the two vectors are different, and the output of the AND gate will be 0.

Note that a similar result could have been obtained by replacing the XNORs with XORs and the AND with a NOR, and that either of these implementations could be easily extended to accommodate input vectors of greater width.

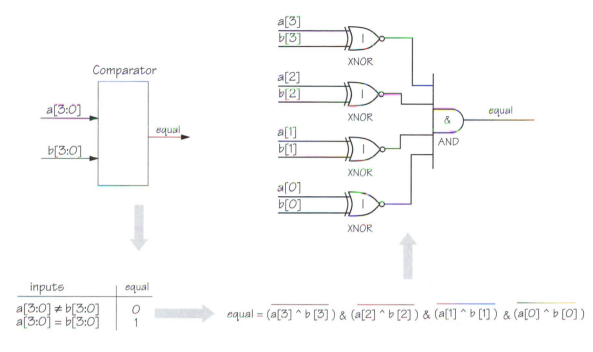

$$equal = \overline{(a[3] \wedge b[3])} \ \& \ \overline{(a[2] \wedge b[2])} \ \& \ \overline{(a[1] \wedge b[1])} \ \& \ \overline{(a[0] \wedge b[0])}$$

Figure 11-2. Equality comparator

Multiplexers

A multiplexer uses a binary value, or *address*, to select between a number of inputs and to convey the data from the selected input to the output. For example, consider a 2:1 ("two-to-one") multiplexer (Figure 11-3).

The *0* and *1* annotations on the multiplexer symbol represent the possible values of the *select* input and are used to indicate which data input will be selected.

The *?* characters in the truth table indicate *don't care* states. When the *select* input is presented with a *0*, the output from the function depends only on the value of the *d0* data input, and we don't care about the value on *d1*. Similarly, when *select* is presented with a 1, the output from the function depends only on the value of the *d1* data input, and we don't care about the value on *d0*. The use of don't care states reduces the size of the truth table, better represents the operation of this particular function, and simplifies the extraction of the sum-of-products expression because the don't cares are ignored.

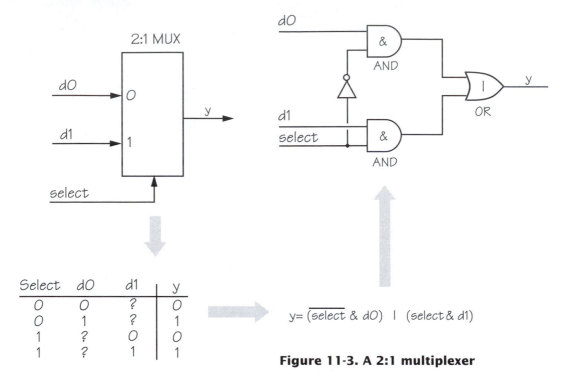

Select	d0	d1	y
0	0	?	0
0	1	?	1
1	?	0	0
1	?	1	1

$$y = (\overline{\text{select}} \, \& \, d0) \; | \; (\text{select} \, \& \, d1)$$

Figure 11-3. A 2:1 multiplexer

An identical result could have been achieved using a full truth table combined with a Karnaugh Map minimization (Figure 11-4).[1]

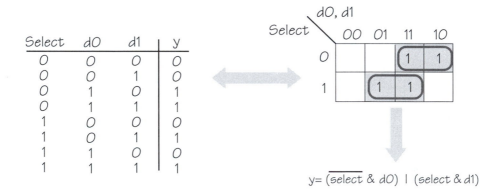

Select	d0	d1	y
0	0	0	0
0	0	1	0
0	1	0	1
0	1	1	1
1	0	0	0
1	0	1	1
1	1	0	0
1	1	1	1

$$y = (\overline{\text{select}} \, \& \, d0) \; | \; (\text{select} \, \& \, d1)$$

Figure 11-4. Deriving the 2:1 multiplexer equation by means of a Karnaugh Map

---

[1] Karnaugh Map minimization techniques were introduced in Chapter 10.

Larger multiplexers are also common in designs: for example, 4:1 multiplexers with four data inputs feeding one output and 8:1 multiplexers with eight data inputs feeding one output. In the case of a 4:1 multiplexer, we will require two select inputs to choose between the four data inputs (using binary patterns of 00, 01, 10, and 11). Similarly, in the case of an 8:1 multiplexer, we will require three select inputs to choose between the eight data inputs (using binary patterns of 000, 001, 010, 011, 100, 101, 110, and 111).

Decoders

A decoder uses a binary value, or *address*, to select between a number of outputs and to assert the selected output by placing it in its active state. For example, consider a 2:4 ("two-to-four") decoder (Figure 11-5).

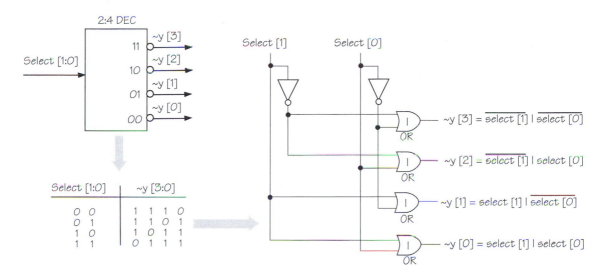

Figure 11-5. A 2:4 decoder with active-low outputs

The 00, 01, 10 and 11 annotations on the decoder symbol represent the possible values that can be applied to the select[1:0] inputs and are used to indicate which output will be asserted.

The truth table shows that when a particular output is selected, it is asserted to a 0, and when that output is not selected, it returns to a 1. Because the outputs are asserted to 0s, this device is said to have *active-low* outputs.

An *active-low* signal is one whose active state is considered to be logic 0.[2] The active-low nature of this particular function is also indicated by the bobbles (small circles) associated with the symbol's outputs and by the tilde ("~")[3] characters in the names of the output signals. Additionally, from our discussions in Chapters 9 and 10, we know that as each output is 0 for only one input combination, it is simpler to extract the equations in product-of-sums form.

Larger decoders are also commonly used in designs: for example, 3:8 decoders with three select inputs and eight outputs, 4:16 decoders with four select inputs and sixteen outputs, etc.

Tri-State Functions

There is a special category of gates called tri-state functions whose outputs can adopt three states: 0, 1, and Z. Lets first consider a simple tri-state buffer (Figure 11-6).

The tri-state buffer's symbol is based on a standard buffer with an additional control input known as the *enable*. The active-low nature of this particular

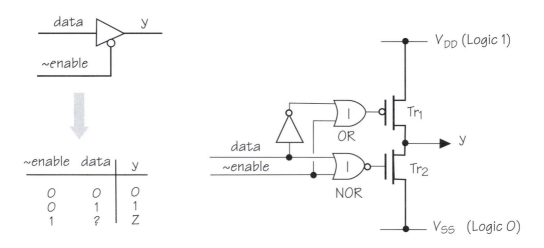

Figure 11-6. Tri-state buffer with active-low enable

---

[2] Similar functions can be created with *active-high* outputs, which means that when an output is selected it is asserted to a logic 1.

[3] The tilde '~' characters prefixing the output names ~y[3], ~y[2], ~y[1], and ~y[0] are used to indicate that these signals are active-low. The use of tilde characters is discussed in more detail in Appendix A.

function's enable is indicated by the bobble associated with this input on the symbol and by the tilde character in its name, ~enable. (Similar functions with active-high enables are also commonly used in designs.)

The Z character in the truth table represents a state known as *high-impedance*, in which the gate is not driving either of the standard 0 or 1 values. In fact, in the high-impedance state the gate is effectively disconnected from its output.

Although Boolean algebra is not well equipped to represent the Z state, the implementation of the tri-state buffer is relatively easy to understand. When the ~enable input is presented with a 1 (its inactive state), the output of the OR gate is forced to 1 and the output of the NOR gate is forced to 0, thereby turning both the Tr_1 and Tr_2 transistors OFF, respectively. With both transistors turned OFF, the output y is disconnected from V_{DD} and V_{SS}, and is therefore in the high-impedance state.

When the ~enable input is presented with a 0 (its active state), the outputs of the OR and NOR gates are determined by the value on the *data* input. The circuit is arranged so that only one of the Tr_1 and Tr_2 transistors can be ON at any particular time. If the *data* input is presented with a 1, transistor Tr_1 is turned ON, thereby connecting output y to V_{DD} (which equates to logic 1).

By comparison, if the *data* input is presented with a 0, transistor Tr_2 is turned ON, thereby connecting output y to V_{SS} (which equates to logic 0).

Tri-state buffers can be used in conjunction with additional control logic to allow the outputs of multiple devices to drive a common signal. For example, consider the simple circuit shown in Figure 11-7.

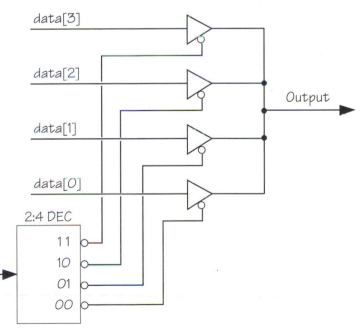

Figure 11-7. Multiple devices driving a common signal

The use of a 2:4 decoder with active-low outputs ensures that only one of the tri-state buffers is enabled at any time. The enabled buffer will propagate the data on its input to the common output, while the remaining buffers will be forced to their tri-state condition.

With hindsight it now becomes obvious that the standard primitive gates (AND, OR, NAND, NOR, etc.) depend on internal Z states to function (when any transistor is turned OFF, its output effectively goes to a Z state). However, the standard primitive gates are constructed in such a way that at least one of the transistors connected to the output is turned ON, which means that the output of a standard gate is always driving either 0 or 1.

Combinational versus Sequential Functions

Logic functions are categorized as being either *combinational* (sometimes referred to as *combinatorial*) or *sequential*. In the case of a *combinational* function, the logic values on that function's outputs are directly related to the current *combination* of values on its inputs. All of the previous example functions have been of this type.

In the case of a sequential function, the logic values on that function's outputs depend not only on its current input values, but also on previous input values. That is, the output values depend on a *sequence* of input values. Because sequential functions remember previous input values, they are also referred to as *memory elements*.

RS Latches

One of the simpler sequential functions is that of an RS latch, which can be implemented using two NOR gates connected in a back-to-back configuration (Figure 11-8). In this NOR implementation, both *reset* and *set* inputs are active-high as indicated by the lack of bobbles associated with these inputs on the symbol. The names of these inputs reflect the effect they have on the q output; when *reset* is active q is reset to 0, and when *set* is active q is set to 1.

The q and ~q outputs are known as the *true* and *complementary* outputs, respectively.[4] In the latch's normal mode of operation, the value on ~q is the

---

[4] In this case, the tilde '~' character prefixing the output name ~q is used to indicate that this signal is a complementary output. Once again, the use of tilde characters is discussed in detail in Appendix A.

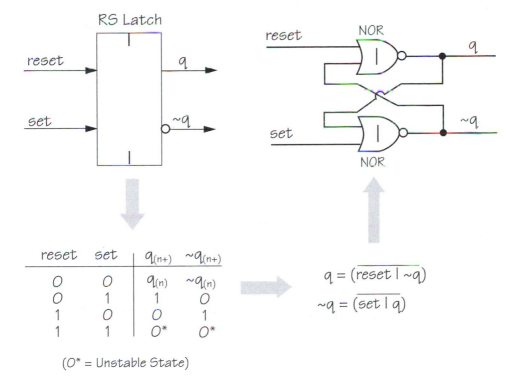

Figure 11-8. NOR implementation of an RS latch

inverse, or complement, of the value on q. This is also indicated by the bobble associated with the ~q output on the symbol. The only time ~q is not the inverse of q occurs when both *reset* and *set* are active at the same time (this unstable state is discussed in more detail below).

The truth table column labels $q_{(n+)}$ and $\sim q_{(n+)}$ indicate that these columns refer to the future values on the outputs. The n+ subscripts represent some future time, or "now-plus." By comparison, the labels $q_{(n)}$ and $\sim q_{(n)}$ used in the body of the truth table indicate the current values on the outputs. In this case the n subscripts represent the current time, or "now." Thus, the first row in the truth table indicates that when both *reset* and *set* are in their inactive states (logic Os), the future values on the outputs will be the same as their current values.

The secret of the RS latch's ability to remember previous input values is based on a technique known as *feedback*. This refers to the feeding back of the outputs as additional inputs into the function. In order to see how this works, let's assume that both the *reset* and *set* inputs are initially in their inactive

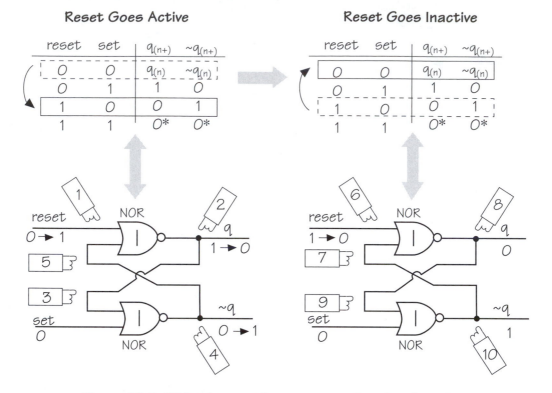

Figure 11-9. RS latch: reset input goes active then inactive

states, but that some previous input sequence placed the latch in its *set condition*; that is, q is 1 and $\sim q$ is 0. Now consider what occurs when the *reset* input is placed in its active state and then returns to its inactive state (Figure 11-9).

As a reminder, if *any* input to a NOR is 1, its output will be forced to 0, and it's only if *both* inputs to a NOR are 0 that the output will be 1. Thus, when *reset* is placed in its active (logic 1) state ⌐ 1 ⌐, the q output from the first gate is forced to 0 ⌐ 2 ⌐. This 0 on q is fed back into the second gate ⌐ 3 ⌐ and, as *both* inputs to this gate are now 0, the $\sim q$ output is forced to 1 ⌐ 4 ⌐. The key point to note is that the 1 on $\sim q$ is now fed back into the first gate ⌐ 5 ⌐.

When the *reset* input returns to its inactive (logic 0) state ⌐ 6 ⌐, the 1 from the $\sim q$ output continues feeding back into the first gate ⌐ 7 ⌐, which means that the q output continues to be forced to 0 ⌐ 8 ⌐. Similarly, the 0 on q continues feeding back into the second gate ⌐ 9 ⌐, and as both of this gate's inputs are now at 0, the $\sim q$ output continues to be forced to 1 ⌐ 10 ⌐. The end result is that the 1 from ⬚ 7 ⬚ causes the 0 at ⬚ 8 ⬚ which is fed back to ⬚ 9 ⬚, and the 0 on the *set* input combined with the 0 from ⬚ 9 ⬚ causes the 1 at ⬚ 10 ⬚ which is fed back to ⬚ 7 ⬚.

Thus, the latch[5] has now been placed in its *reset condition,* and a self-sustaining loop has been established. Even though both the *reset* and *set* inputs are now inactive, the *q* output remains at 0, indicating that *reset* was the last input to be in its active state. Once the function has been placed in its *reset condition,* any subsequent activity on the *reset* input will have no effect on the outputs, which means that the only way to affect the function is by means of its *set* input.

Now consider what occurs when the *set* input is placed in its active state and then returns to its inactive state (Figure 11-10).

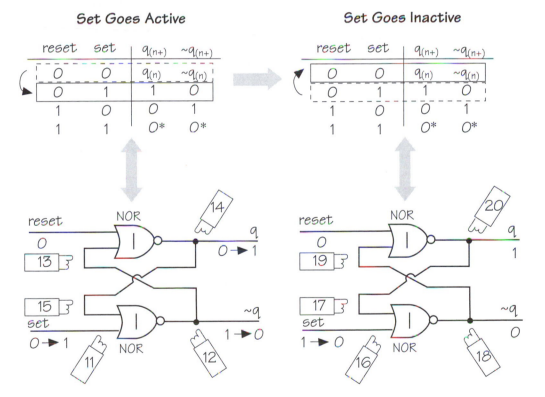

Figure 11-10. RS latch: set input goes active then inactive

---

[5] The term *latch*—which is commonly associated with a fastening for a door or gate—comes from the Old English *lacchen,* meaning "to seize."

When *set* is placed in its active (logic 1) state ⟦ 11 ⟧, the ~q output from the second gate is forced to 0 ⟦ 12 ⟧. This 0 on ~q is fed back into the first gate ⟦ 13 ⟧ and, as both inputs to this gate are now 0, the q output is forced to 1 ⟦ 14 ⟧. The key point to note is that the 1 on q is now fed back into the second gate ⟦ 15 ⟧.

When the *set* input returns to its inactive (logic 0) state ⟦ 16 ⟧, the 1 from the q output continues feeding back to the second gate ⟦ 17 ⟧ and the ~q output continues to be forced to 0 ⟦ 18 ⟧. Similarly, the 0 on the ~q output continues feeding back into the first gate ⟦ 19 ⟧, and the q output continues to be forced to 1 ⟦ 20 ⟧. The end result is that the 1 at ⟦ 17 ⟧ causes the 0 at ⟦ 18 ⟧ which is fed back to ⟦ 19 ⟧, and the 0 on the *reset* input combined with the 0 at ⟦ 19 ⟧ causes the 1 at ⟦ 20 ⟧ which is fed back to ⟦ 17 ⟧.

Thus, the latch has been returned to its *set condition* and, once again, a self-sustaining loop has been established. Even though both the *reset* and *set* inputs are now inactive, the q output remains at 1, indicating that *set* was the last input to be in its active state. Once the function has been placed in its *set condition*, any subsequent activity on the *set* input will have no effect on the outputs, which means that the only way to affect the function is by means of its *reset* input.

The unstable condition indicated by the fourth row of the RS latch's truth table occurs when both the *reset* and *set* inputs are active at the same time. Problems occur when both *reset* and *set* return to their inactive states simultaneously or too closely together (Figure 11-11).

When both *reset* and *set* are active at the same time, the 1 on *reset* ⟦ 21 ⟧ forces the q output to 0 ⟦ 22 ⟧ and the 1 on *set* ⟦ 23 ⟧ forces the ~q output to 0 ⟦ 24 ⟧. The 0 on q is fed back to the second gate ⟦ 25 ⟧, and the 0 on ~q is fed back to the first gate ⟦ 26 ⟧.

Now consider what occurs when *reset* and *set* go inactive simultaneously (⟦ 27 ⟧ and ⟦ 28 ⟧, respectively). When the new 0 values on *reset* and *set* are combined with the 0 values fed back from q ⟦ 29 ⟧ and ~q ⟦ 30 ⟧, each gate initially sees both of its inputs at 0 and therefore both gates attempt to drive their outputs to 1. After any delays associated with the gates have been satisfied, both of the outputs will indeed go to 1.

When the output of the first gate goes to 1, this value is fed back to the input of the second gate. While this is happening, the output of the second gate

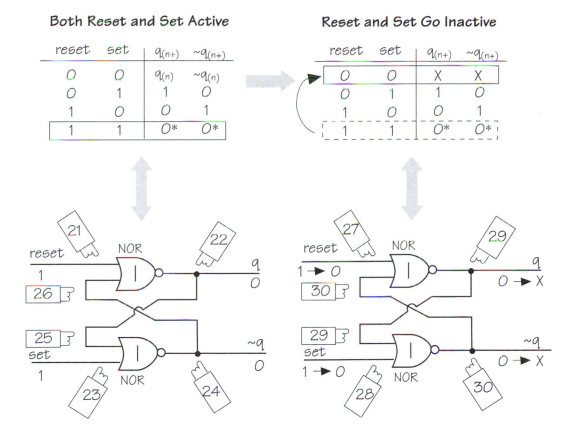

Figure 11-11. RS latch: the reset and set inputs go inactive simultaneously

goes to 1, and this value is fed back to the input of the first gate. Each gate now has its fed-back input at 1, and both gates therefore attempt to drive their outputs to 0. As we see, the circuit has entered a *metastable* condition in which the outputs oscillate between 0 and 1 values.

If both halves of the function were exactly the same, these metastable oscillations would continue indefinitely. But there will always be some differences (no matter how small) between the gates and their delays, and the function will eventually collapse into either its *reset condition* or its *set condition*. As there is no way to predict the final values on the q and $\sim q$ outputs, they are indicated as being in X, or *don't know*, states ([29] and [30]). These X states will persist until a valid input sequence occurs on either the *reset* or *set* inputs.

An alternative implementation for an RS latch can be realized using two NAND gates connected in a back-to-back configuration (Figure 11-12).

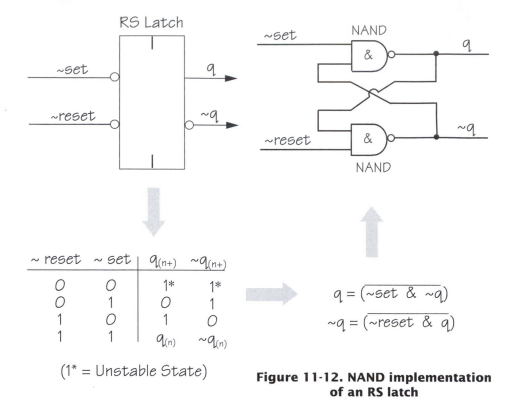

$$q = \overline{(\sim set\ \&\ \sim q)}$$
$$\sim q = \overline{(\sim reset\ \&\ q)}$$

Figure 11-12. NAND implementation of an RS latch

In a NAND implementation, both the ~reset and ~set inputs are active low, as is indicated by the bobbles associated with these inputs on the symbol and by the tilde characters in their names. As a reminder, if *any* input to a NAND is 0, the output is 1, and it's only if *both* inputs to a NAND are 1 that the output will be 0. Working out how this version of the latch works is left as an exercise to the reader.[6]

D-Type Latches

A more sophisticated function called a D-type ("data-type") latch can be constructed by attaching two ANDs and a NOT to the front of an RS latch (Figure 11-13).

---

[6] This is where we see if you've been paying attention <grin>.

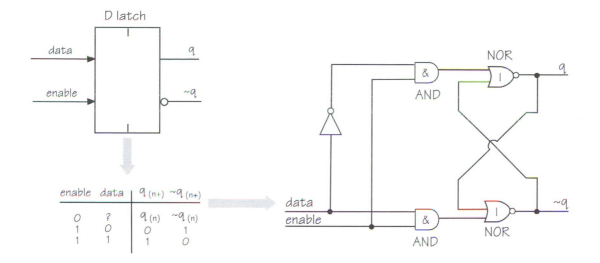

Figure 11-13. D-type latch with active-high enable

The *enable* input is active high for this configuration, as is indicated by the lack of a bobble on the symbol. When *enable* is placed in its active (logic 1) state, the true and inverted versions of the *data* input are allowed to propagate through the AND gates and are presented to the back-to-back NOR gates. If the *data* input changes while *enable* is still active, the outputs will respond to reflect the new value.

When enable returns to its inactive (logic 0) state, it forces the outputs of both ANDs to 0, and any further changes on the *data* input have no effect. Thus, the back-to-back NOR gates remember the last value they saw from the *data* input prior to the *enable* input going inactive.

Consider an example waveform (Figure 11-14). While the *enable* input is in its active state, whatever value is presented to the *data* input appears on the *q* output and an inverted version appears on the ~q output. As usual, there will always be some element of delay between changes on the inputs and corresponding responses on the outputs. When *enable* goes inactive, the outputs remember their previous values and no longer respond to any changes on the *data* input. As the operation of the device depends on the logic value, or level, on *enable*, this input is said to be *level-sensitive*.

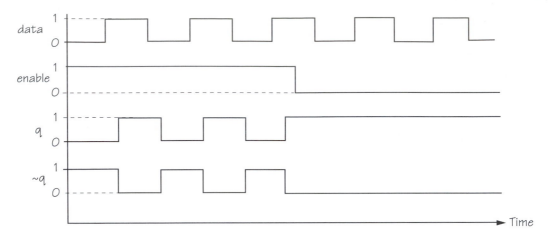

Figure 11-14. Waveform for a D-type latch with active-high enable

D-Type Flip-flops

In the case of a D-type flip-flop (which may also be referred to as a *register*), the data *appears* to be loaded when a transition, or edge, occurs on the *clock* input, which is therefore said to be *edge-sensitive* (the reason we say *"appears to be loaded when an edge occurs"* is discussed in the sidebar on the next page). A transition from *0* to 1 is known as a *rising-edge* or a *positive-edge*, while a transition from 1 to *0* is known as a *falling-edge* or a *negative-edge*. A D-type flip-flop's *clock* input may be positive-edge or negative-edge triggered (Figure 11-15).

The chevrons (arrows ">") associated with the clock inputs on the symbols indicate

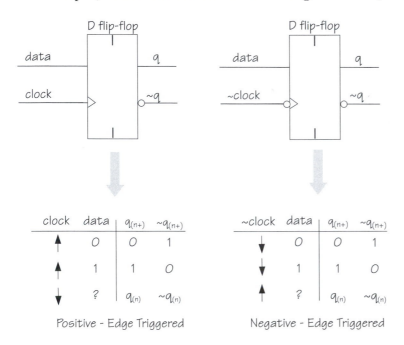

Figure 11-15. Positive-edge and negative-edge D-type flip-flops

that these are edge-sensitive inputs. A chevron without an associated bobble indicates a positive-edge clock, and a chevron with a bobble indicates a negative-edge clock. The last rows in the truth tables show that an inactive edge on the clock leaves the contents of the flip-flops unchanged (these cases are often omitted from the truth tables).

Consider an example waveform for a positive-edge triggered D-type flip-flop (Figure 11-16). As the observer initially has no knowledge as to the contents of the flop-flop, the *q* and ~*q* outputs are initially shown as having X, or *don't know*, values.

There are a number of ways to implement a D-Type flip-flop. The most understandable from our point of view would be to use two D-type latches in series (one after the other). The first latch could have an active-low enable and the second could have an active-high enable. Both of these enables would be connected together, and would be known as the *clock* input to the outside world.

This is known as a *master-slave* relationship, where the first latch is the "master" and the second is the "slave."

When the *clock* input is 0, the master latch is enabled and passes whatever value is presented to its data input through to its outputs (only its *q* output is actually used in this example). Meanwhile, the slave latch is disabled and continues to store (and output) its existing contents.

When the *clock* input is subsequently driven to a 1, the master latch is disabled and continues to store (and output) its existing contents. Meanwhile the slave latch is now enabled and passes whatever value is presented to *its* data input (the value from the output of the master latch) through to *its* outputs.

Thus, everything is really controlled by voltage *levels*, but from the outside world it *appears* that the flip-flop was loaded by a *rising-edge* on the *clock* input.

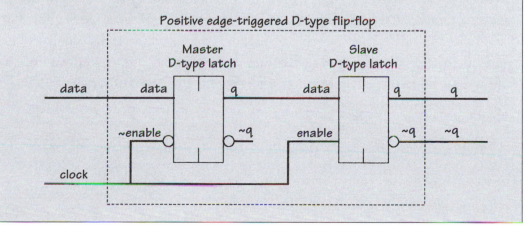

Positive edge-triggered D-type flip-flop

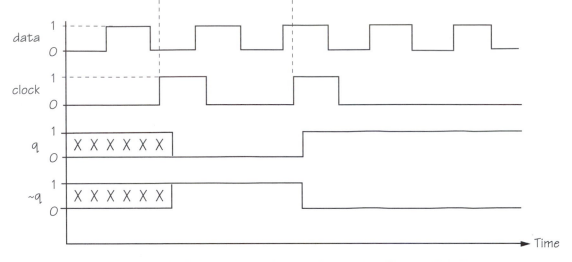

Figure 11-16. Waveform for positive-edge D-type flip-flop

The first rising edge of the clock loads the *0* on the *data* input into the flip-flop, which (after a small delay) causes *q* to change to 0 and *~q* to change to 1. The second rising edge of the clock loads the 1 on the *data* input into the flip-flop; *q* goes to 1 and *~q* goes to *0*.

Some flip-flops have an additional input called *~clear* or *~reset* which forces *q* to *0* and *~q* to 1, irrespective of the value on the *data* input (Figure 11-17). Similarly, some flip-flops have a *~preset* or *~set* input, which forces *q* to 1 and *~q* to *0*, and some have both *~clear* and *~preset* inputs.

The examples shown in Figure 11-17 reflect active-low *~clear* inputs, but active-high equivalents are also available. Furthermore, as is illustrated in Figure 11-17, these inputs may be either *asynchronous* or *synchronous*. In the more common asynchronous case, the effect of *~clear* going active is immediate and overrides both the *clock* and *data* inputs (the "asynchronous" qualifier reflects the fact that the effect of this input is *not* synchronized to the *clock*). By comparison, in the synchronous case the effect of *~clear* *is* synchronized to the active edge of the *clock*.[7]

---

[7] The component symbols used in this book are relatively traditional and simple. One disadvantage of this is that, as seen in Figure 11-17, there's no way to tell if a *clear* or *preset* input is synchronous or asynchronous without also looking at its truth table. There are more modern and sophisticated symbol standards that do cover all eventualities, but their complexity is beyond the scope of this book to explain.

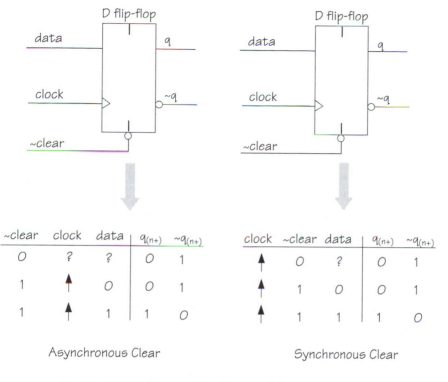

~clear	clock	data	$q_{(n+)}$	$\sim q_{(n+)}$
0	?	?	0	1
1	↑	0	0	1
1	↑	1	1	0

Asynchronous Clear

clock	~clear	data	$q_{(n+)}$	$\sim q_{(n+)}$
↑	0	?	0	1
↑	1	0	0	1
↑	1	1	1	0

Synchronous Clear

Figure 11-17. D-type flip-flops with asynchronous and synchronous clear inputs

JK and T Flip-flops

The majority of examples in this book are based on D-type flip-flops. However, for the sake of completeness, it should be noted that there are several other flavors of flip-flops available. Two common types are the JK and T (for Toggle) flip-flops (Figure 11-18).

The first row of the JK flip-flop's truth table shows that when both the j and k (data) inputs are 0, an active edge on the *clock* input leaves the contents of the flip-flop unchanged. The two middle rows of the truth table show that if the j and k inputs have opposite values, an active edge on the *clock* input will effectively load the flip-flop (the q output) with the value on j (the ~q output will take the complementary value). The last line of the truth table shows that when both the j and k inputs are 1, an active edge on the *clock* causes the outputs to toggle to the inverse of their previous values.[8] By comparison, the

---

[8] This may be the origin of the term "flip-flop," because the outputs *flip* and *flop* back and forth.

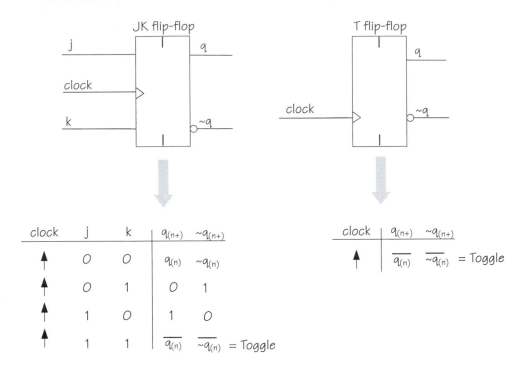

Figure 11-18. JK and T flip-flops

T flip-flop doesn't have any data inputs; the outputs simply toggle to the inverse of their previous values on each active edge of the *clock* input.

Shift Registers

As was previously noted, another term for a flip-flop is "register." Functions known as *shift registers*—which facilitate the shifting of binary data one bit at a time—are commonly used in digital systems. Consider a simple 4-bit shift register constructed using D-type flip-flops (Figure 11-19).

This particular example is based on positive-edge triggered D-type flip-flops with active-low ~clear inputs (in this case we're only using each register's q output). Also, this example is classed as a *serial-in-parallel-out* (SIPO) shift register, because data is loaded in serially (one after the other) and read out in parallel (side by side).

When the ~clear input is set to 1 (its inactive state), a positive-edge on the clock input loads the value on the serial_in input into the first flip-flop, dff[0]. At the same time, the value that used to be in dff[0] is loaded into dff[1], the

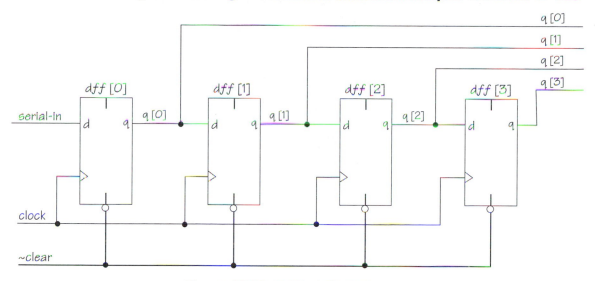

Figure 11-19. SIPO shift register

value that used to be in dff[1] is loaded into dff[2], and the value that used to be in dff[2] is loaded into dff[3].

This may seem a bit weird and wonderful the first time you see it, but the way in which this works is actually quite simple (and of course capriciously cunning). Each flip-flop exhibits a delay between seeing an active edge on its clock input and the ensuing response on its q output. These delays provide sufficient time for the next flip-flop in the chain to load the value from the previous stage before that value changes. Consider an example waveform where a single logic 1 value is migrated through the shift register (Figure 11-20).

Initially all of the flip-flops contain don't know X values. When the ~clear input goes to its active state (logic 0), all of the flip-flops are cleared to 0. When the first active edge occurs on the clock input, the serial_in input is 1, so this is the value that's loaded into the first flip-flop. At the same time, the original 0 value from the first flip-flop is loaded into the second, the original 0 value from the second flip-flop is loaded into the third, and the original 0 value from the third flip-flop is loaded into the fourth.

When the next active edge occurs on the clock input, the serial_in input is 0, so this is the value that's loaded into the first flip-flop. At the same time, the original 1 value from the first flip-flop is loaded into the second, the 0 value from the second flip-flop is loaded into the third, and the 0 value from the third flip-flop is loaded into the fourth.

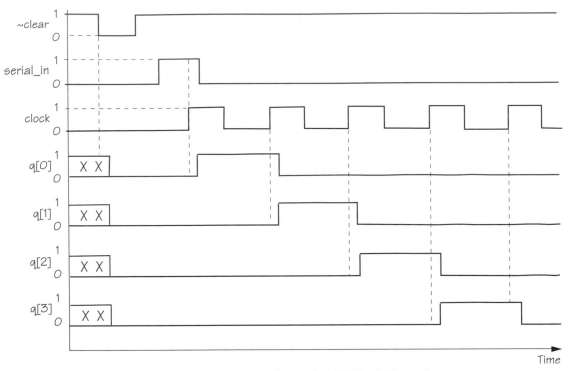

Figure 11-20. Waveform for SIPO shift register

Similarly, when the next active edge occurs on the clock input, the serial_in input is still 0, so this is the value that's loaded into the first flip-flop. At the same time, the 0 value from the first flip-flop is loaded into the second, the 1 value from the second flip-flop is loaded into the third, and the 0 value from the third flip-flop is loaded into the fourth. And so it goes . . .

Other common shift register variants are the *parallel-in-serial-out (PISO)*, and the *serial-in-serial-out (SISO)*; for example, consider a 4-bit SISO shift register (Figure 11-21).

Counters

Counter functions are also commonly used in digital systems. The number of states that the counter will sequence through before returning to its original value is called the *modulus* of the counter. For example, a function that counts from 0000_2 to 1111_2 in binary (or 0 to 15 in decimal) has a modulus of sixteen and would be called a modulo-16, or mod-16, counter. Consider a modulo-16 counter implemented using D-type flip-flops (Figure 11-22).

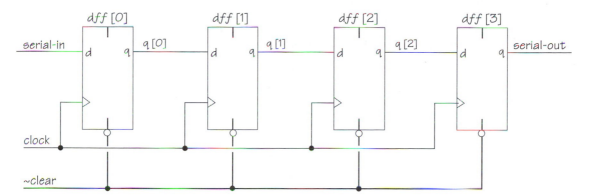

Figure 11-21. SISO shift register

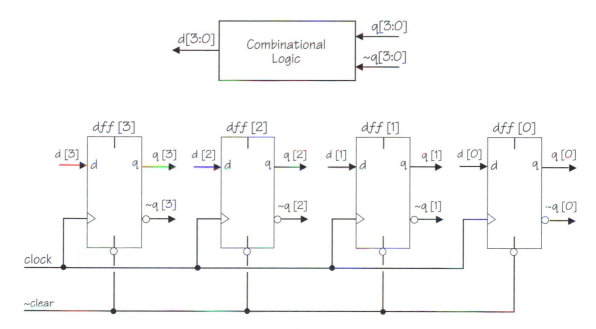

Figure 11-22. Modulo-16 binary counter

This particular example is based on positive-edge triggered D-type flip-flops with active-low ~clear inputs. The four flip-flops are used to store the current count value which is displayed on the q[3:0] outputs. When the ~clear input is set to 1 (its inactive state), a positive-edge on the clock input causes the counter to load the next value in the count sequence.

A block of combinational logic is used to generate the next value, *d[3:0]*, which is based on the current value *q[3:0]* (Figure 11-23). Note that there is no need to create the inverted versions of *q[3:0]*, because these signals are already available from the flip-flops as ~*q[3:0]*.

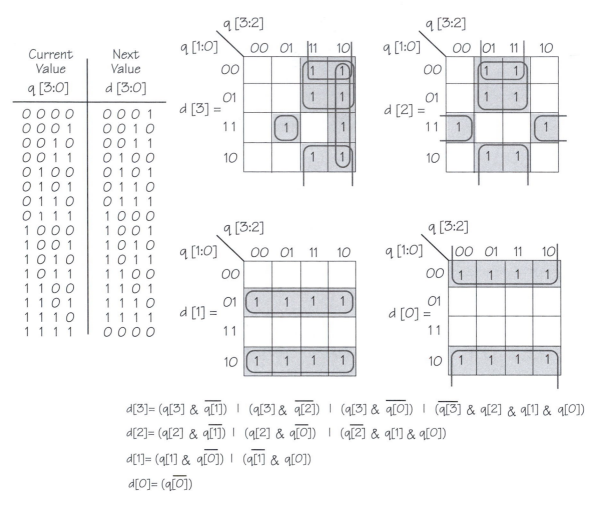

Current Value q [3:0]	Next Value d [3:0]
0 0 0 0	0 0 0 1
0 0 0 1	0 0 1 0
0 0 1 0	0 0 1 1
0 0 1 1	0 1 0 0
0 1 0 0	0 1 0 1
0 1 0 1	0 1 1 0
0 1 1 0	0 1 1 1
0 1 1 1	1 0 0 0
1 0 0 0	1 0 0 1
1 0 0 1	1 0 1 0
1 0 1 0	1 0 1 1
1 0 1 1	1 1 0 0
1 1 0 0	1 1 0 1
1 1 0 1	1 1 1 0
1 1 1 0	1 1 1 1
1 1 1 1	0 0 0 0

$d[3] = (q[3] \,\&\, \overline{q[1]}) \;|\; (q[3] \,\&\, \overline{q[2]}) \;|\; (q[3] \,\&\, \overline{q[0]}) \;|\; (\overline{q[3]} \,\&\, q[2] \,\&\, q[1] \,\&\, q[0])$

$d[2] = (q[2] \,\&\, \overline{q[1]}) \;|\; (q[2] \,\&\, \overline{q[0]}) \;|\; (\overline{q[2]} \,\&\, q[1] \,\&\, q[0])$

$d[1] = (q[1] \,\&\, \overline{q[0]}) \;|\; (\overline{q[1]} \,\&\, q[0])$

$d[0] = (\overline{q[0]})$

Figure 11-23. Generating the next count value

Setup and Hold Times

One point we've glossed over thus far is the fact that there are certain timing requirements associated with flip-flops. In particular, there are two parameters called the setup and hold times which describe the relationship between the flip-flop's data and clock inputs (Figure 11-24).

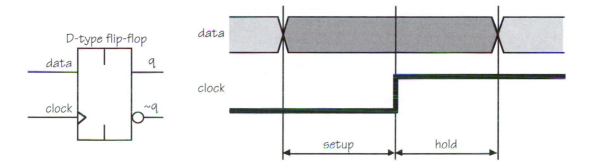

Figure 11-24. Setup and hold times

The waveform shown here is a little different to those we've seen before. What we're trying to indicate is that when we start (on the left-hand side), the value presented to the *data* input may be a *0* or a *1*, and it can change back and forth as often as it pleases. However, it must settle one way or the other before the *setup* time; otherwise when the active edge occurs on the *clock* we can't guarantee what will happen. Similarly, the value presented to the *data* input must remain stable for the *hold* time following the *clock*, or once again we can't guarantee what will happen. In our illustration, the period for which the value on the *data* input must remain stable is shown as being the darker gray.

The setup and hold times shown above are reasonably understandable. However things can sometimes become a little confusing, especially in the case of today's deep submicron (DSM) integrated circuit technologies.[9] The problem is that we may sometimes see so-called *negative* setup and hold times (Figure 11-25).

Once again, the periods for which the value on the *data* input must remain stable are shown as being the darker gray. These effects, which may seem a little strange at first, are caused by internal delay paths inside the flip-flop.

Last but not least, we should note that there will also be setup and hold times between the *clear* (or *reset*) and *preset* (or *set*) inputs and the *clock* input. Also, there will be corresponding setup and hold times between the *data*, *clear* (or *reset*) and *preset* (or *set*) inputs and the *enable* input on D-type latches (phew!).

---

[9] Integrated circuits (and DSM technologies) are introduced in Chapter 14.

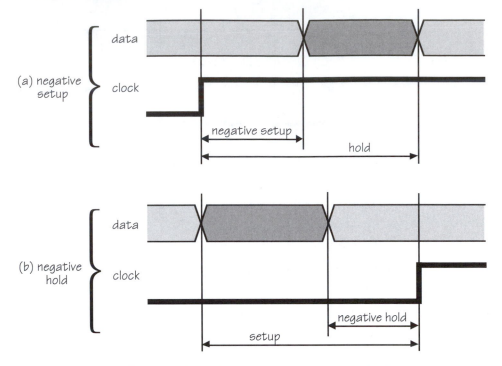

Figure 11-25. Negative setup and hold times

Brick by Brick

Let us pause here for a brief philosophical moment. Consider, if you will, a brick formed from clay. Now, there's not a lot you can do with a single brick, but when you combine thousands and thousands of bricks together you can create the most tremendous structures. At the end of the day, the Great Wall of China is no more than a pile of bricks molded by man's imagination.[10]

In the world of the electronics engineer, transistors are the clay, primitive logic gates are the bricks, and the functions described above are simply building blocks.[11] Any digital system, even one as complex as a supercomputer, is constructed from building blocks like comparators, multiplexers, shift registers, and counters. Once you understand the building blocks, there are no ends to the things you can achieve!

---

[10] But at approximately 2,400 km in length, it's a very impressive pile of bricks (the author has walked—and climbed—a small portion of the beast and it fair took his breath away).

[11] We might also note that clay and transistors share something else in common—they both consist predominantly of silicon!

State Diagrams, State Tables, and State Machines

Consider a coin-operated machine that accepts nickels and dimes[1] and, for the princely sum of fifteen cents, dispenses some useful article called a "gizmo" that the well-dressed man-about-town could not possibly be without. We may consider such a machine to comprise three main blocks: a *receiver* that accepts money, a *dispenser* that dispenses the "gismo" and any change, and a *controller* that oversees everything and makes sure things function as planned (Figure 12-1).

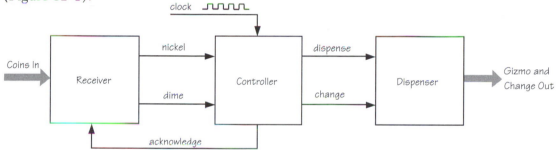

Figure 12-1. Block diagram of a coin-operated machine

The connections marked nickel, dime, dispense, change, and acknowledge represent digital signals carrying logic 0 and 1 values. The user can deposit nickels and dimes into the receiver in any order, but may only deposit one coin at a time. When a coin is deposited, the receiver determines its type and sets the corresponding signal (nickel or dime) to a logic 1.

The operation of the controller is synchronized by the clock signal. On a rising edge of the clock, the controller examines the nickel and dime inputs to see if any coins have been deposited. The controller keeps track of the amount of money deposited and determines if any actions are to be performed.

---

[1] For the benefit of those readers who do not reside in the United States, nickels and dimes are American coins worth five and ten cents, respectively.

Every time the controller inspects the *nickel* and *dime* signals, it sends an *acknowledge* signal back to the receiver. The *acknowledge* signal informs the receiver that the coin has been accounted for, and the receiver responds by resetting the *nickel* and *dime* signals to 0 and awaiting the next coin. The *acknowledge* signal can be generated in a variety of ways which are not particularly relevant here.

When the controller decides that sufficient funds have been deposited, it instructs the dispenser to dispense a "gizmo" and any change (if necessary) by setting the *dispense* and *change* signals to 1, respectively.

State Diagrams

A useful level of abstraction for a function such as the controller is to consider it as consisting of a set of *states* through which it sequences. The *current state* depends on the *previous state* combined with the *previous* values on the *nickel* and *dime* inputs. Similarly, the *next state* depends on the *current state* combined with the *current* values on the *nickel* and *dime* inputs. The operation of the controller may be represented by means of a *state diagram*, which offers a way to view the problem and to describe a solution (Figure 12-2).

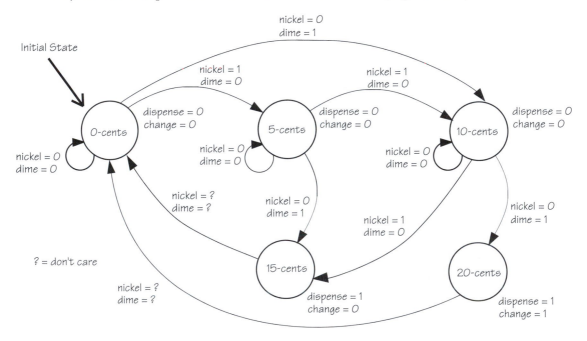

Figure 12-2. State diagram for the controller

The states are represented by the circles labeled *0-cents, 5-cents, 10-cents, 15-cents,* and *20-cents,* and the values on the *dispense* and *change* outputs are associated with these states. The arcs connecting the states are called *state transitions* and the values of the *nickel* and *dime* inputs associated with the state transitions are called *guard conditions.* The controller will only sequence between two states if the values on the *nickel* and *dime* inputs match the guard conditions.

Let's assume that the controller is in its initial state of *0-cents.* The values of the *nickel* and *dime* inputs are tested on every rising edge on the clock.[2] As long as no coins are deposited, the *nickel* and *dime* inputs remain at *0* and the controller remains in the *0-cents* state. Once a coin is deposited, the next rising edge on the clock will cause the controller to sequence to the *5-cents* or the *10-cents* states depending on the coin's type. It is at this point that the controller sends an *acknowledge* signal back to the receiver instructing it to reset the *nickel* and *dime* signals back to *0* and to await the next coin.

Note that the *0-cents, 5-cents,* and *10-cents* states have state transitions that loop back into them (the ones with associated *nickel = 0* and *dime = 0* guard conditions). These indicate that the controller will stay in whichever state it is currently in until a new coin is deposited.

So at this stage of our discussions, the controller is either in the *5-cents* or the *10-cents* state depending on whether the first coin was a nickel or dime, respectively. What happens when the next coin is deposited? Well this depends on the state we're in and the type of the new coin. If the controller is in the *5-cents* state, then a nickel or dime will move it to the *10-cents* or *15-cents* states, respectively. Alternatively, if the controller is in the *10-cents* state, then a nickel or dime will move it to the *15-cents* or *20-cents* states, respectively.

When the controller reaches either the *15-cents* or *20-cents* states, the next clock will cause it to dispense a "gizmo" and return to its initial *0-cents* state (in the case of the *20-cents* state, the controller will also dispense a nickel in change).

---

[2] The controller is known to sequence between states only on the rising edge of the clock, so displaying this signal on every state transition would be redundant.

State Tables

Another form of representation is that of a *state table*. This is similar to a truth table (inputs on the left and corresponding outputs on the right), but it also includes the *current state* as an input and the *next state* as an output (Figure 12-3).

Current state	clock	nickel	dime	dispense	change	Next state
0-cents	↑	0	0	0	0	0-cents
0-cents	↑	1	0	0	0	5-cents
0-cents	↑	0	1	0	0	10-cents
5-cents	↑	0	0	0	0	5-cents
5-cents	↑	1	0	0	0	10-cents
5-cents	↑	0	1	0	0	15-cents
10-cents	↑	0	0	0	0	10-cents
10-cents	↑	1	0	0	0	15-cents
10-cents	↑	0	1	0	0	20-cents
15-cents	↑	?	?	1	0	0-cents
20-cents	↑	?	?	1	1	0-cents

Figure 12-3. State table for the controller

In this instance the *clock* signal has been included for purposes of clarity (it's only when there's a rising edge on the *clock* that the outputs are set to the values shown in that row of the table). However, as for the state diagram, displaying this signal is somewhat redundant and it is often omitted.

State Machines

The actual implementation of a function such as the controller is called a *state machine*. In fact, when the number of states is constrained and finite, this is more usually called a *finite state machine (FSM)*. The heart of a state machine consists of a set of registers[3] known as the *state variables*. Each state, 0-cents, 5-cents, 10-cents, . . . is assigned a unique binary pattern of 0s and 1s, and the pattern representing the *current state* is stored in the state variables.

---

[3] For the purposes of these discussions we'll assume that these registers are D-type flip-flops as were introduced in Chapter 11.

The two most common forms of synchronous, or clocked, state machines are known as *Moore* and *Mealy* machines after the men who formalized them. A Moore machine is distinguished by the fact that the outputs are derived only from the values in the state variables (Figure 12-4). The controller function featured in this discussion is a classic example of a Moore machine.

By comparison, the outputs from a Mealy machine may be derived from a combination of the values in the state variables and one or more of the inputs (Figure 12-5).

In both of the Moore and Mealy forms, the input logic consists of primitive gates such as AND, NAND, OR, and NOR. These combine the values on the inputs with the *current state* (which is fed back from the state variables) to generate the pattern of 0s and 1s representing the *next state*. This new pattern of 0s and 1s is presented to the inputs of the state variables and will be loaded into them on the next rising edge of the *clock*.

The output logic also consists of standard primitive logic gates that generate the appropriate values on the outputs from the *current state* stored in the state variables.

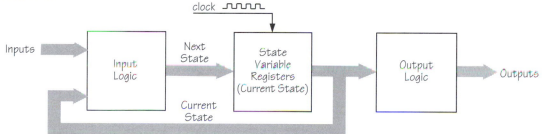

Figure 12-4. Block diagram of a Moore machine

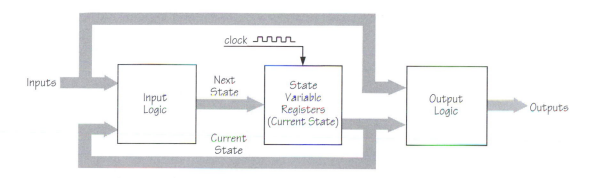

Figure 12-5. Block diagram of a Mealy machine

State Assignment

A key consideration in the design of a state machine is that of *state assignment*, which refers to the process by which the states are assigned to the binary patterns of 0s and 1s that are to be stored in the state variables.

A common form of state assignment requiring the minimum number of registers is known as *binary encoding*. Each register can only contain a single binary digit, so it can only be assigned a value of 0 or 1. Two registers can be assigned four binary values (00, 01, 10, and 11), three registers can be assigned eight binary values (000, 001, 010, 011, 100, 101, 110, and 111), and so forth. The controller used in our coin-operated machine consists of five unique states, and therefore requires a minimum of three state variable registers.

The actual process of binary encoded state assignment is a nontrivial problem. In the case of our controller function, there are 6,720 possible combinations[4] by which five states can be assigned to the eight binary values provided by three registers. Each of these solutions may require a different arrangement of primitive gates to construct the input and output logic, which in turn affects the maximum frequency that can be used to drive the system clock. Additionally, the type of registers used to implement the state variables also affects the supporting logic; the following discussions are based on the use of D-type flip-flops.

Assuming that full use is made of *don't care* states, an analysis of the various binary encoded solutions for our controller yields the following . . .

138 solutions requiring 7 product terms
852 solutions requiring 8 product terms
1,876 solutions requiring 9 product terms
3,094 solutions requiring 10 product terms
570 solutions requiring 11 product terms
190 solutions requiring 12 product terms

. . . where a *product term* is a group of *literals* linked by & (AND) operators—for example, (a & b & c)—and a *literal* is any true or inverted variable. Thus, the product term (a & b & c) contains three literals (a, b, and c).

---

[4] This number would be somewhat reduced if all the mirror-image combinations were taken into account, but that would not significantly lessen the complexity problem of determining the optimal combination.

But wait, there's more! A further analysis of the 138 solutions requiring seven product terms yields the following:

66 solutions requiring 17 literals
24 solutions requiring 18 literals
48 solutions requiring 19 literals

Thus, the chances of a random assignment resulting in an optimal solution is relatively slight. Fortunately, there are computer programs available to aid designers in this task.[5] One solution resulting in the minimum number of product terms and literals is shown in Figure 12-6.

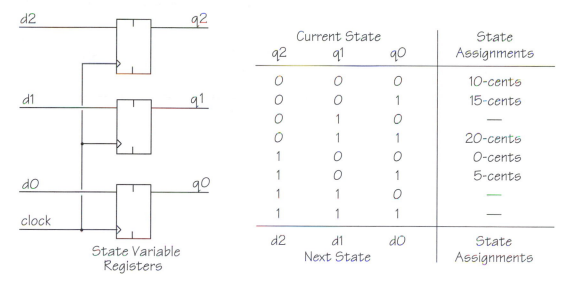

	Current State			State
q2	q1	q0		Assignments
0	0	0		10-cents
0	0	1		15-cents
0	1	0		—
0	1	1		20-cents
1	0	0		0-cents
1	0	1		5-cents
1	1	0		—
1	1	1		—
d2	d1	d0		State
	Next State			Assignments

State Variable
Registers

Figure 12-6. Example binary encoded state assignment

A truth table for the controller function can now be derived from the state table shown in Figure 12-3 by replacing the assignments in the *current state* column with the corresponding binary patterns for the state variable outputs ($q2$, $q1$, and $q0$), and replacing the assignments in the *next state* column with the corresponding binary patterns for the state variable inputs ($d2$, $d1$, and $d0$). The resulting equations can then be derived from the truth table by means of standard algebraic or Karnaugh map techniques. As an alternative, a computer

---

[5] The author used the program BOOL, which was created by his friend Alon Kfir (a man with a size-16 brain if ever there was one).

program can be used to obtain the same results in less time with far fewer opportunities for error.[6] Whichever technique is employed, the state assignments above lead to the following minimized Boolean equations:

$$d0 \ = \ (\overline{q0} \ \& \ \overline{q2} \ \& \ dime) \ | \ (q0 \ \& \ q2 \ \& \ \overline{nickel}) \ | \ (\overline{q0} \ \& \ nickel)$$

$$d1 \ = \ (\overline{q0} \ \& \ \overline{q2} \ \& \ dime)$$

$$d2 \ = \ (q0 \ \& \ \overline{q2}) \ | \ (q2 \ \& \ \overline{nickel} \ \& \ \overline{dime}) \ | \ (\overline{q0} \ \& \ q2 \ \& \ dime)$$

$$dispense \ = \ (q0 \ \& \ \overline{q2})$$

$$change \ = \ (q1)$$

The product terms shown in bold appear in multiple equations. However, regardless of the number of times a product term appears, it is only counted once because it only has to be physically implemented once. Similarly, the literals used to form product terms that appear in multiple equations are only counted once.

Another common form of state assignment is known as *one-hot encoding*, in which each state is represented by an individual register. In this case, our controller with its five states would require five register bits. The one-hot technique typically requires a greater number of logic gates than does binary encoding. However, as the logic gates are used to implement simpler equations, the one-hot method results in faster state machines that can operate at higher clock frequencies.

Don't Care States, Unused States, and Latch-Up Conditions

It was previously noted that the analysis of the binary encoded state assignment made full use of *don't care* states.[7] This allows us to generate a solution that uses the least number of logic gates, but there are additional considerations that must now be discussed in more detail.

The original definition of our coin-operated machine stated that it is only possible for a single coin to be deposited at a time. Assuming this to be true,

---

[6] Once again, the author used BOOL (*"What's the point of barking if you have a dog?"* as they say in England).

[7] The concept of *don't care* states was introduced in Chapter 10.

then the nickel and dime signals will never be assigned 1 values simultaneously. The designer (or a computer program) can use this information to assign *don't care* states to the outputs for any combination of inputs that includes a 1 on both nickel and dime signals.

Additionally, the three binary encoded state variable registers provide eight possible binary patterns, of which only five were used. The analysis above was based on the assumption that *don't care* states can be assigned to the outputs for any combination of inputs that includes one of the unused patterns on the state variables. This assumption also requires further justification.

When the coin-operated machine is first powered-up, each state variable register can potentially initialize with a random logic 0 or 1 value. The controller could therefore power-up with its state variables containing any of the eight possible patterns of 0s and 1s. For some state machines this would not be an important consideration, but this is not true in the case of our coin-operated machine. For example, the controller could power-up in the 20-cents state, in which case it would immediately dispense a "gizmo" and five cents change. The owner of such a machine may well be of the opinion that this was a less than ideal feature.

Alternatively, the controller could power-up with its state variables in one of the unused combinations. Subsequently, the controller could sequence directly—or via one or more of the other unused combinations—to any of the defined states. In a worst-case scenario, the controller could remain in the unused combination indefinitely or sequence endlessly between unused combinations; these worst-case scenarios are known as *latch-up conditions*.

One method of avoiding latch-up conditions is to assign additional, dummy states to each of the unused combinations and to define state transitions from each of these dummy states to the controller's initialization state of 0-cents. Unfortunately, in the case of our coin-operated machine, this technique would not affect the fact that the controller could wake up in a valid state other than 0-cents. An alternative is to provide some additional circuitry to generate a *power-on reset* signal—for example, a single pulse that occurs only when the power is first applied to the machine. The power-on reset can be used to force the state variable registers into the pattern associated with the 0-cents state. The analysis above assumed the use of such a power-on-reset.

Analog-to-Digital and Digital-to-Analog

As we began our discussions in Chapter 1 by separating the analog and digital views of the world, it seems appropriate to close this section of the book by reuniting them. While some systems operate solely on digital data, others have to interact with the analog world. It may be necessary to convert an analog input into a form that can be manipulated by the digital system, or to transform an output from a digital system into the analog realm. These tasks are performed by analog-to-digital (A/D) and digital-to-analog (D/A) converters, respectively (Figure 13-1).

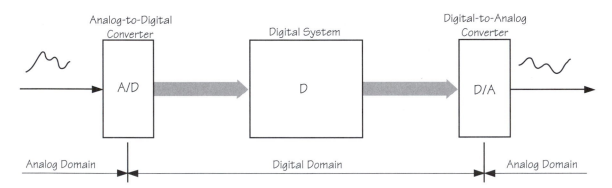

Figure 13-1. Analog-to-digital (A/D) and digital-to-analog (D/A) converters

Analog-to-Digital

A *transducer* is a device that converts input energy of one form into output energy of another. Analog effects can manifest themselves in a variety of different ways such as heat and pressure. In order to be processed by a digital system, the analog quantity must be detected and converted into a suitable form by means of an appropriate transducer called a *sensor*. For example, a microphone is a sensor that detects sound and converts it into a corresponding voltage. The analog-to-digital conversion process can be represented as shown in Figure 13-2.

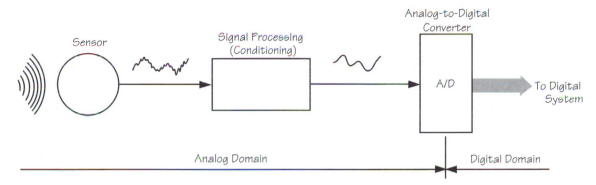

Figure 13-2. Analog-to-digital conversion process

The output from the sensor typically undergoes some form of signal processing such as filtering and amplification before being passed to the A/D converter. This signal processing is generically referred to as *conditioning*. The A/D converter accepts the conditioned analog voltage and converts it into a series of equivalent digital values by *sampling* and *quantization* (Figure 13-3).

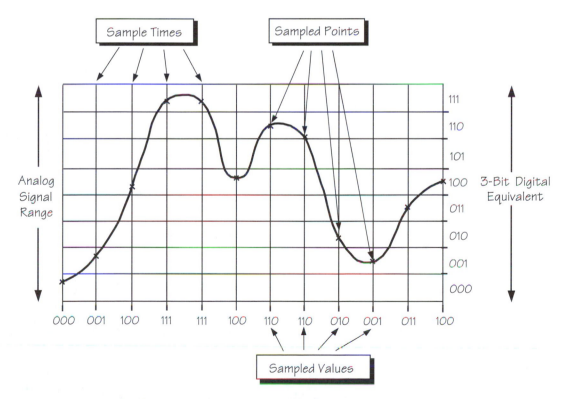

Figure 13-3. The sampling and quantization of a digital signal

The sampling usually occurs at regular time intervals and is triggered by the digital part of the system. The complete range of values that the analog signal can assume is divided into a set of discrete bands or quanta. At each sample time, the A/D converter determines which band the analog signal falls into (this is the "quantization" part of the process) and outputs the equivalent binary code for that band.

The main factor governing the accuracy of the conversion is the number of bands used. For example, a 3-bit code can represent only eight bands, each encompassing 12.5% of the analog signal's range, while a 12-bit code can represent 4,096 bands, each encompassing 0.025% of the signal's range.

Digital-to-Analog

A D/A converter accepts a digital code and transforms it into a corresponding analog current or voltage suitable for use by means of an appropriate transducer called an *actuator*. For example, a loudspeaker is an actuator that converts an electrical signal into sound. The digital-to-analog conversion process can be represented as shown in Figure 13-4.

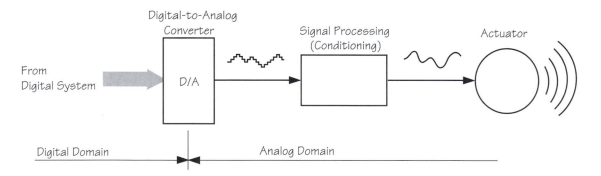

Figure 13-4. Digital-to-analog conversion process

The conversions usually occur at regular time intervals and are triggered by a clock signal from the digital part of the system. The output from the D/A converter typically undergoes some form of conditioning before being passed to the actuator. For example, in the case of an audio system, the "staircase-like" signal coming out of the D/A converter will be "smoothed" before being passed to an amplifier (not shown in Figure 13-4) and, ultimately, to the loudspeaker.

Integrated Circuits (ICs)

In the 1950s, transistors and other electronic components were available only in individual packages. These discrete components were laid out on a circuit board and hand connected using separate wires. At that time, an electronic gate capable of storing a single binary bit of data cost more than $2. By comparison, in the early 1990s, enough gates to store 5,000 bits of data cost less than a cent. This vast reduction in price was primarily due to the invention of the *integrated circuit (IC)*.[1]

A functional electronic circuit requires transistors, resistors, diodes, etc. and the connections between them. A monolithic integrated circuit (the "monolithic" qualifier is usually omitted) has all of these components formed on the surface layer of a sliver, or chip, of a single piece of semiconductor; hence the term *monolithic*, meaning "seamless." Although a variety of semiconductor materials are available, the most commonly used is silicon, and integrated circuits are popularly known as "silicon chips." Unless otherwise noted, the remainder of these discussions will assume integrated circuits based on silicon as the semiconductor.

An Overview of the Fabrication Process

The construction of integrated circuits requires one of most exacting production processes ever developed. The environment must be at least a thousand times cleaner than that of an operating theater, and impurities in materials have to be so low as to be measured in parts per billion.[2] The process begins with the growing of a single crystal of pure silicon in the form of a cylinder with a diameter that can be anywhere up to 300 mm.[3] The cylinder is cut into paper-thin slices called *wafers*, which are approximately 0.2 mm thick (Figure 14-1).

---

[1] In conversation, IC is pronounced by spelling it out as "I-C".

[2] If you took a bag of flour and added a grain of salt, this would be impure by comparison.

[3] This 300 mm value was true as of 2002. However, in 1995 (when the first edition of this tome hit the streets), the maximum diameter was 200 mm, so who knows what it will be in the future?

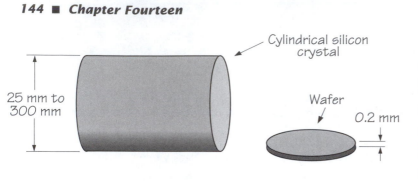

Cylindrical silicon crystal

25 mm to 300 mm

Wafer

0.2 mm

Figure 14-1. Creating silicon wafers

The thickness of the wafer is determined by the requirement for sufficient mechanical strength to allow it to be handled without damage. The actual thickness necessary for the creation of the electronic components is less than 10 µm (ten-millionths of a meter). After the wafers have been sliced from the cylinder, they are polished to a smoothness rivaling the finest mirrors.

The most commonly used fabrication process is *optical lithography,* in which ultraviolet light (UV) is passed through a stencil-like[4] object called a *photo-mask,* or just *mask* for short. This square or rectangular mask carries patterns formed by areas that are either transparent or opaque to ultraviolet frequencies (similar in concept to a black and white photographic negative) and the resulting image is projected onto the surface of the wafer. By means of some technical wizardry that we'll consider in the next section, we can use the patterns of ultraviolet light to grow corresponding structures in the silicon. The simple patterns shown in the following diagrams were selected for reasons of clarity; in practice, a mask can contain millions of fine lines and geometric shapes (Figure 14-2).

Each wafer can contain hundreds or thousands of identical integrated circuits. The pattern projected onto the wafer's surface corresponds to a

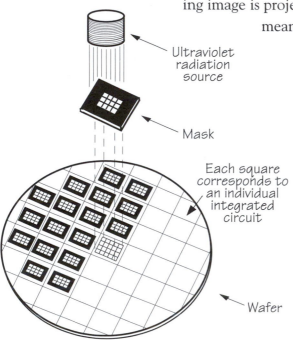

Ultraviolet radiation source

Mask

Each square corresponds to an individual integrated circuit

Wafer

Figure 14-2. The opto-lithographic step-and-repeat process

---

[4] The term "stencil" comes from the Middle English word *stencelled,* meaning "adorned brightly."

single integrated circuit, which is typically in the region of 1mm × 1mm to 10mm × 10mm, but may be even larger. After the area corresponding to one integrated circuit has been exposed, the wafer is moved and the process is repeated until the pattern has been replicated across the whole of the wafer's surface. This technique for duplicating the pattern is called a *step-and-repeat* process.

As we shall see, multiple layers are required to construct the transistors (and other components), where each layer requires its own unique mask. Once all of the transistors have been created, similar techniques are used to lay down the tracking (wiring) layers that connect the transistors together.

A More Detailed Look at the Fabrication Process

To illustrate the manufacturing process in more detail, we will consider the construction of a single NMOS transistor occupying an area far smaller than a speck of dust. For reasons of electronic stability, the majority of processes begin by lightly doping the entire wafer to form either N-type or, more commonly,

P-type silicon. However, for the purposes of this discussion, we will assume a process based on a pure silicon wafer (Figure 14-3).

Assume that the small area of silicon shown here is sufficient to accommodate a single transistor in the middle of one of the integrated circuits residing some-where on the wafer. During the fabrication process the wafer is often referred to as the *substrate*, meaning "base layer." A common first stage is to either grow or deposit a thin layer of silicon dioxide (glass) across the entire surface of the wafer by exposing it to oxygen in a high-temperature oven (Figure 14-4).

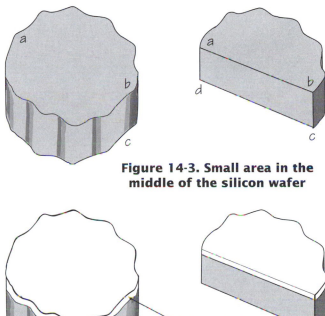

Figure 14-3. Small area in the middle of the silicon wafer

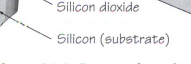

Silicon dioxide

Silicon (substrate)

Figure 14-4. Grow or deposit a layer of silicon dioxide

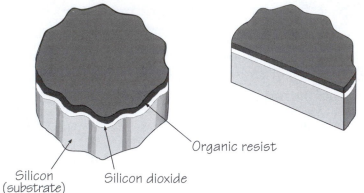

Organic resist

Silicon
(substrate)   Silicon dioxide

Figure 14-5. Apply a layer of organic resist

Ultraviolet
radiation source

Mask

Organic resist

Silicon
(substrate)   Silicon dioxide

**Figure 14-6. The exposed resist is degraded
by the ultraviolet light**

After the wafer has cooled, it is coated with a thin layer of organic resist,[5] which is first dried and then baked to form an impervious layer (Figure 14-5).

A mask is created and ultraviolet light is applied. The ionizing ultraviolet radiation passes through the transparent areas of the mask into the resist, silicon dioxide, and silicon. The ultraviolet breaks down the molecular structure of the resist, but does not have any effect on the silicon dioxide or the pure silicon (Figure 14-6).

As was previously noted, the small area of the mask shown here is associated with a single transistor. The full mask for an integrated circuit can consist of millions of similar patterns. After the area under the mask has been exposed, the wafer is moved, and the process is repeated until the pattern has been replicated across the wafer's entire surface, once for each integrated circuit. The wafer is then bathed in an organic solvent to dissolve the degraded resist. Thus, the

---

[5] The term "organic" is used because this type of resist is a carbon-based compound, and carbon is the key element for life as we know it.

pattern on the mask has been transferred to a series of corresponding patterns in the resist (Figure 14-7).

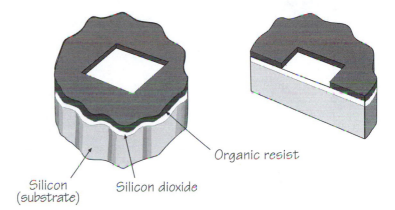

Figure 14-7. The degraded resist is dissolved with an organic solvent

A process in which ultraviolet light passing through the transparent areas of the mask causes the resist to be degraded is known as a *positive-resist* process; *negative-resist* processes are also available. In a negative-resist process, the ultraviolet radiation passing through the trans-parent areas of the mask is used to cure the resist, and the remaining uncured areas are then removed using an appropriate solvent.

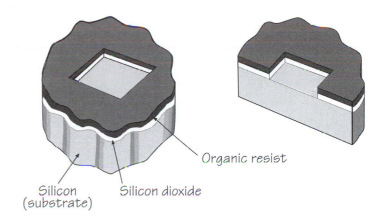

Figure 14-8. Etch the exposed silicon dioxide

After the unwanted resist has been removed, the wafer undergoes a process known as *etching*, in which an appropriate solvent is used to dissolve any exposed silicon dioxide without having any effect on the organic resist or the pure silicon (Figure 14-8).

The remaining resist is then removed using an appropriate solvent, and the wafer is placed in a high temperature oven where it is exposed to a gas contain-ing the selected dopant (a P-type dopant in this case). The atoms in the gas diffuse into the substrate resulting in a region of doped silicon (Figure 14-9).[6]

---

[6] In some processes, diffusion is augmented with *ion implantation* techniques, in which beams of ions are directed at the wafer to alter the type and conductivity of the silicon in selected regions.

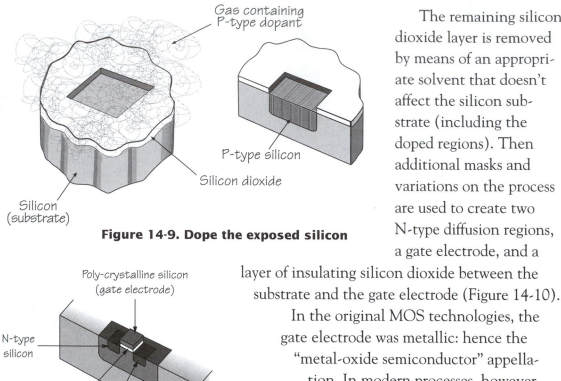

Figure 14-9. Dope the exposed silicon

Figure 14-10. Add n-type diffusion regions and the gate electrode

The remaining silicon dioxide layer is removed by means of an appropriate solvent that doesn't affect the silicon substrate (including the doped regions). Then additional masks and variations on the process are used to create two N-type diffusion regions, a gate electrode, and a layer of insulating silicon dioxide between the substrate and the gate electrode (Figure 14-10). In the original MOS technologies, the gate electrode was metallic: hence the "metal-oxide semiconductor" appellation. In modern processes, however, the gate electrode is formed from *poly-crystalline silicon* (often abbreviated to *polysilicon* or even just *poly*), which is also a good conductor.

The N-type diffusions form the transistor's *source* and *drain* regions (you might wish to refer back to Figure 4-9a in Chapter 4 to refresh your memory at this point). The gap between the source and drain is called the *channel*. To provide a sense of scale, the length of the channel may be in the order of 0.1 μm (one-tenth of one-millionth of a meter), and the thickness of the silicon dioxide layer between the gate electrode and the substrate may be in the order of 0.05 μm (see also the discussions on *device geometries* later in this chapter).

Another layer of insulating silicon dioxide is now grown or deposited across the surface of the wafer. Using similar lithographic techniques to those described above, holes are etched through the silicon dioxide in areas in which it is desired to make connections, and a metalization layer of aluminum interconnections (think of them as wires) called *tracks* is deposited (Figure 14-11).

The end result is an NMOS transistor; a logic 1 on the track connected to the *gate* will turn the transistor ON, thereby enabling current to flow between its *source* and *drain* terminals. An equivalent PMOS transistor could have been formed by exchanging the P-type and N-type diffusion regions. By varying the structures

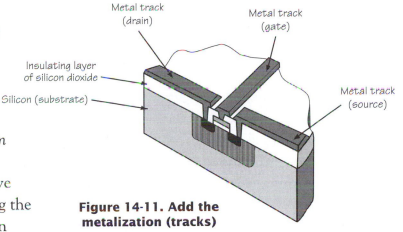

Figure 14-11. Add the metalization (tracks)

created by the masks, components such as resistors and diodes can be fabricated at the same time as the transistors. The tracks are used to connect groups of transistors to form primitive logic gates and to connect groups of these gates to form more complex functions.

An integrated circuit contains three distinct levels of conducting material: the *diffusion* layer at the bottom, the *polysilicon* layers in the middle, and the *metalization* layers at the top. In addition to forming components, the diffusion layer may also be used to create embedded wires. Similarly, in addition to forming gate electrodes, the polysilicon may also be used to interconnect components. There may be several layers of polysilicon and several layers of metalization, with each pair of adjacent layers separated by an insulating layer of silicon dioxide. The layers of silicon dioxide are selectively etched with holes known as *vias*, which allow connections to be made between the various tracking layers.

Early integrated circuits typically supported only two layers of metalization. The tracks on the first layer predominantly ran in a "North-South" direction, while the tracks on the second predominantly ran "East-West."[7] As the number

---

[7] In 2001, a group of companies announced a new chip interconnect concept called *X Architecture* (www.xinitiative.org) in which logic functions on chips are wired together using diagonal tracks (as opposed to traditional North-South and East-West tracking layers). Initial evaluations apparently show that this diagonal interconnect strategy can increase chip performance by 10% and reduce power consumption by 20%. However, it may take awhile before design tools and processes catch up . . . watch this space!

of transistors increased, engineers required more and more tracking layers. The problem is that when a layer of insulating silicon dioxide is deposited over a tracking layer, you end up with slight "bumps" where the tracks are (like snow falling over a snoozing polar bear—you end up with a bump).

After a few tracking layers, the bumps are pronounced enough that you can't continue. The answer is to re-planarize the wafer (smooth the bumps out) after each tracking and silicon dioxide layer combo has been created. This is achieved by means of a process called *chemical mechanical polishing* (CMP), which returns the wafer to a smooth, flat surface before the next tracking layer is added. With manufacturers using this process, high-end silicon chips could support up to eight tracking layers by 2002.

Relatively large areas of aluminum called *pads* are constructed at the edges of each integrated circuit for testing and connection purposes. Some of the pads are used to supply power to the device, while the rest are used to provide input and output signals (Figure 14-12).

The pads can be connected to the internal components using the diffusion, polysilicon, or metalization layers. In a step known as *overglassing*, the entire surface of the wafer is coated with a final *barrier* layer (or *passivation* layer) of silicon dioxide or silicon nitride, which provides physical protection for the underlying circuits from moisture and other contaminants. One more lithographic step is required to pattern holes in the barrier layer to allow connections to be made to the pads. In some cases, additional metalization may be deposited on the pads to raise them fractionally above the level of the barrier layer. Augmenting the pads in this way is known as *silicon bumping*.

The entire fabrication process requires numerous lithographic steps, each involving an individual mask and layer of resist to selectively expose different parts of the wafer.

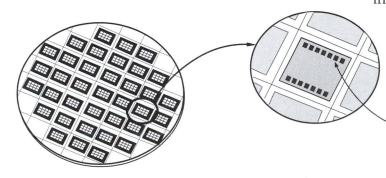

Pads

Figure 14-12. Power and signal pads

The Packaging Process

The individual integrated circuits are tested while they are still part of the wafer in a process known as *wafer probing*. An automated tester places probes on the device's pads, applies power to the power pads, injects a series of signals into the input pads, and monitors the corresponding signals returned from the output pads. Each integrated circuit is tested in turn, and any device that fails the tests is automatically tagged with a splash of dye for subsequent rejection. The *yield* is the number of devices that pass the tests as a percentage of the total number fabricated on that wafer.

The completed circuits, known as *die*,[8] are separated by marking the wafer with a diamond scribe and fracturing it along the scribed lines (much like cutting a sheet of glass or breaking up a Kit Kat® bar) (Figure 14-13).

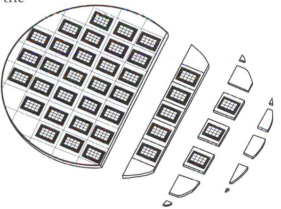

Figure 14-13. Die separation

Following separation, the majority of the die are packaged individually. Since there are almost as many packaging technologies as there are device manufacturers, we will initially restrain ourselves to a relatively traditional process. First, the die is attached to a metallic lead frame using an adhesive (Figure 14-14).

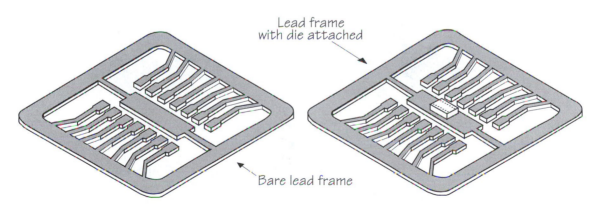

Lead frame
with die attached

Bare lead frame

Figure 14-14. Die attached to lead frame

---

[8] The plural of die is also die (in much the same way that "a shoal of herring" is the plural of "herring").

One of the criteria used when selecting the adhesive is its ability to conduct heat away from the die when the device is operating. An automatic wire bonding tool connects the pads on the die to the leads on the lead frame with wire bonds finer than a human hair.[9] The whole assembly is then encapsulated in a block of plastic or epoxy (Figure 14-15).

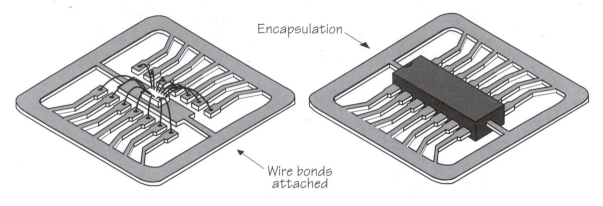

Encapsulation

Wire bonds
attached

Figure 14-15. Wire bonding and encapsulation

A dimple or notch is formed at one end of the package so that the users will know which end is which. The unused parts of the lead frame are cut away and the device's leads, or pins, are shaped as required; these operations are usually performed at the same time (Figure 14-16).

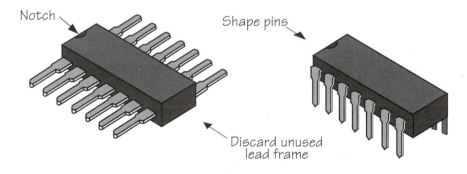

Notch

Shape pins

Discard unused
lead frame

Figure 14-16. Discard unused lead frame and shape pins

---

[9] Human hairs range in thickness from around 0.07 mm to 0.1 mm. A hair from a typical blond lady's head is approximately 0.075 mm (three quarters of one tenth of a millimeter) in diameter. By comparison, integrated circuit bonding wires are typically one-third this diameter, and they can be even thinner.

An individually packaged integrated circuit consists of the die and its connections to the external leads, all encapsulated in the protective package. The package protects the silicon from moisture and other impurities and helps to conduct heat away from the die when the device is operating.

There is tremendous variety in the size and shape of packages. A rectangular device with pins on two sides, as illustrated here, is called a *dual in-line (DIL) package.* A standard 14-pin packaged device is approximately 18 mm long by 6.5 mm wide by 2.5 mm deep, and has 2.5 mm spaces between pins. An equivalent *small outline package (SOP)* could be as small as 4 mm long by 2 mm wide by 0.75 mm deep, and have 0.5 mm spaces between pins. Other packages can be square and have pins on all four sides, and some have an array of pins protruding from the base.

The shapes into which the pins are bent depend on the way the device is intended to be mounted on a circuit board. The package described above has pins that are intended to go all the way through the circuit board using a mounting technique called *lead through hole (LTH).* By comparison, the packages associated with a technique called *surface mount technology (SMT)* have pins that are bent out flat, and which attach to one side (surface) of the circuit board (an example of this is shown in Chapter 18).

It's important to note that the example shown above reflects a very simple packaging strategy for a device with very few pins. By 2002, some integrated circuits had as many as 1,000 pins (with 2,000- and 4,000-pin devices on the horizon). This multiplicity of pins requires a very different approach. In one technique known as *solder bump bonding,* for example, the pads on the die are not restricted to its periphery, but are instead located over the entire face of the die. A minute ball of solder is then attached to each pad, and the die is flipped over and attached to the package substrate (this is referred to as a "flip-chip" technique). Each pad on the die has a corresponding pad on the package substrate, and the package-die combo is heated so as to melt the solder balls and form good electrical connections between the die and the substrate (Figure 14-17).

Eventually, the die will be encapsulated in some manner to protect it from the outside world. The package's substrate itself may be made out of the same material as a printed circuit board, or out of ceramic, or out of some even more esoteric material. Whatever its composition, the substrate will contain multiple

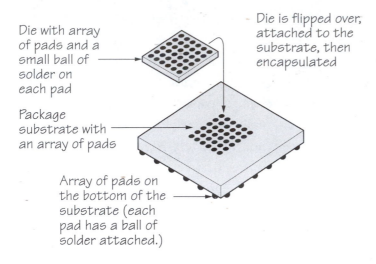

Die with array of pads and a small ball of solder on each pad

Die is flipped over, attached to the substrate, then encapsulated

Package substrate with an array of pads

Array of pads on the bottom of the substrate (each pad has a ball of solder attached.)

Figure 14-17. A solder bump bonded ball grid array packaging technique

internal wiring layers that connect the pads on the upper surface with pads (or pins) on the lower surface. The pads (or pins) on the lower surface (the side that actually connects to the circuit board) will be spaced much father apart—relatively speaking—than the pads that connect to the die.

At some stage the package will have to be attached to a circuit board. In one technique known as a *ball grid array* (BGA), the package has an array of pads on its bottom surface, and a small ball of solder is attached to each of these pads. Each pad on the package will have a corresponding pad on the circuit board, and heat is used to melt the solder balls and form good electrical connections between the package and the board.

Modern packaging technologies are extremely sophisticated. For example, by 2002, some ball grid arrays had pins spaced 0.3 mm (one third of a millimeter) apart! In the case of *chip-scale packages* (CSP), the package is barely larger than the die itself. In the early 1990s, some specialist applications began to employ a technique known as *die stacking*, in which several bare die are stacked on top of each other to form a sandwich. The die are connected together and then packaged as a single entity.

As was previously noted, there are a wide variety of integrated packaging styles. There are also many different ways in which the die can be connected to the package. We will introduce a few more of these techniques in future chapters.[10]

---

[10] Additional packaging styles and alternative mounting strategies are presented in the discussions on circuit boards (Chapter 18), hybrids (Chapter 19), and multichip modules (Chapter 20).

Integrated Circuits versus Discrete Components

The tracks linking components inside an integrated circuit have widths measured in fractions of a millionth of a meter and lengths measured in millimeters. By comparison, the tracks linking components on a circuit board are orders of magnitude wider and have lengths measured in tens of centimeters. Thus, the transistors used to drive tracks inside an integrated circuit can be much smaller than those used to drive their circuit board equivalents, and smaller transistors use less power. Additionally, signals take a finite time to propagate down a track, so the shorter the track, the faster the signal.

A single integrated circuit can contain tens (sometimes hundreds) of millions of transistors. A similar design based on discrete components would be tremendously more expensive in terms of price, size, operating speed, power requirements, and the time and effort required to design and manufacture the system. Additionally, every solder joint on a circuit board is a potential source of failure, which affects the reliability of the design. Integrated circuits reduce the number of solder joints and hence improve the reliability of the system.

In the past, an electronic system was typically composed of a number of integrated circuits, each with its own particular function (say a microprocessor, a communications function, some memory devices, etc.). For many of today's high-end applications, however, electronics engineers are combining all of these functions on a single device, which may be referred to as a *system-on-chip* (*SoC*).

Different Types of ICs

The first integrated circuit—a simple phase shift oscillator—was constructed in 1958.[11] Since that time, a plethora of different device types have appeared on the scene. There are far too many different integrated circuit types for us to cover in this book, but some of the main categories—along with their approximate dates of introduction—are shown in Figure 14-18.[12]

---

[11] The first integrated circuits typically contained around six transistors. By the latter half of the 1960s, devices containing around 100 transistors were reasonably typical.

[12] The white portions of the timeline bars in this figure indicate that although early incarnations of these technologies may have been available, they perhaps hadn't been enthusiastically received during this period. For example, Xilinx introduced the world's first FPGA as early as 1984, but many design engineers didn't really become interested in these little rapscallions until the late 1980s.

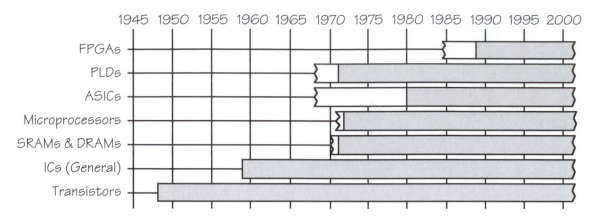

Figure 14-18. Timeline of device introductions (dates are approximate)

Memory devices (in particular SRAMs and DRAMs) are introduced in Chapter 15; programmable integrated circuits (PLDs and FPGAs) are presented in Chapter 16; and application-specific integrated circuits (ASICs) are discussed in Chapter 17.

Technology Considerations

Transistors are available in a variety of flavors called *families* or *technologies*. One of the first to be invented was the *bipolar junction transistor (BJT)*, which was the mainstay of the industry for many years. If bipolar transistors are connected together in a certain way, the resulting logic gates are classed as *transistor-transistor logic (TTL)*. An alternative method of connecting the same transistors results in logic gates classed as *emitter-coupled logic (ECL)*. Another family called *metal-oxide semiconductor field-effect transistors (MOSFETs)* were invented some time after bipolar junction transistors. *Complementary metal-oxide semiconductor (CMOS)* logic gates are based on NMOS and PMOS MOFSETs connected together in a complementary manner.

Logic gates constructed in TTL are fast and have strong drive capability, but consume a relatively large amount of power. Logic gates implemented in CMOS are a little slower than their TTL equivalents and have weaker drive capability, but their static (non-switching) power consumption is extremely low. Technology improvements continue to yield lower-power TTL devices and higher-speed CMOS devices. Logic gates built in ECL are substantially faster than their TTL counterparts, but consume correspondingly more power.

Finally, gates fabricated using the gallium arsenide (GaAs) semiconductor as a substrate are approximately eight times faster than their silicon equivalents, but they are expensive to produce, and so are used for specialist applications only.

If an integrated circuit containing millions of transistors were constructed entirely from high-power transistors, it could literally consume enough power to incinerate itself (or at least melt itself down into an undistinguished puddle of gunk). As a compromise, some integrated circuits use a combination of technologies. For example, the bulk of the logic gates in a device may be implemented in low-power CMOS, but the gates driving the output pins may be constructed from high-drive TTL. A more extreme example is that of *BiCMOS (Bipolar CMOS)*, in which the function of every primitive logic gate is implemented in low-power CMOS, but the output stage of each gate uses high-drive bipolar transistors.

Supply Voltages

Towards the end of the 1980s and the beginning of the 1990s, the majority of circuits using TTL, CMOS, and BiCMOS devices were based on a 5.0-volt supply. However, increasing usage of portable personal electronics such as notebook computers and cellular telephones began to drive the requirement for devices that consume and dissipate less power. One way to reduce power consumption is to lower the supply voltage, so by the mid-to-late 1990s, the most common supplies were 3.3 volts for portable computers and 3.0 volts for communication systems. By 2002, some specialist applications had plunged to 1.8 volts, with even lower supplies on the horizon. Unfortunately, lowering the supply voltage can drastically affect the speed of traditional technologies and greatly lessens any speed advantages of BiCMOS over CMOS. A relatively new low-voltage contender that appeared in the latter half of the 1990s was BiNMOS, in which complex combinations of bipolar and NMOS transistors are used to form sophisticated output stages providing both high speed and low static power dissipation.

Equivalent Gates

One common metric used to categorize an integrated circuit is the number of logic gates it contains. However, difficulties may arise when comparing devices, as each type of logic function requires a different number of transistors.

This leads to the concept of an *equivalent gate*, whereby each type of logic function is assigned an equivalent gate value, and the relative complexity of an integrated circuit is judged by summing its equivalent gates. Unfortunately, the definition of an equivalent gate can vary, depending on whom one is talking to. A reasonably common convention is for a 2-input NAND to represent one equivalent gate. A more esoteric convention defines an ECL equivalent gate as being *"one-eleventh the minimum logic required to implement a single-bit full-adder,"* while some vendors define an equivalent gate as being equal to an arbitrary number of transistors based on their own particular technology. The best policy is to establish a common frame of reference before releasing a firm grip on your hard-earned lucre.

The acronyms SSI, MSI, LSI, VLSI, and ULSI represent *Small-*, *Medium-*, *Large-*, *Very-Large-*, and *Ultra-Large-Scale Integration*, respectively. By one convention, the number of gates represented by these terms are: SSI (1-12), MSI (13-99), LSI (100-999), VLSI (1,000-999,999), and ULSI (1,000,000 or more).

Device Geometries

Integrated circuits are also categorized by their *geometries*, meaning the size of the structures created on the substrate. For example, a 1 µm[13] CMOS device has structures that measure one-millionth of a meter. The structures typically embraced by this description are the width of the tracks and the length of the channel between the source and drain diffusion regions; the dimensions of other features are derived as ratios of these structures.

Geometries are continuously shrinking as fabrication processes improve. In 1990, devices with 1 µm geometries were considered to be state of the art, and many observers feared that the industry was approaching the limits of manufacturing technology, but geometries continued to shrink regardless:

Anything below 0.5 µm is referred to as *deep-submicron (DSM)*, and at some point that isn't particularly well defined (or is defined differently by different people) we move into the realm of *ultra-deep-submicron (UDSM)*.

---

[13] The µ symbol stands for "micro" from the Greek *micros*, meaning "small" (hence the use of µP as an abbreviation for microprocessor). In the metric system, µ stands for "one millionth part of," so 1 µm means one millionth of a meter.

Year	Geometry
1990	1.00 μm
1992	0.80 μm
1994	0.50 μm
1996	0.35 μm
1997	0.25 μm
1999	0.18 μm
2000	0.13 μm
2001	0.10 μm
2002	0.09 μm

With devices whose geometries were 1 μm and higher, it was relatively easy to talk about them in conversation. For example, one might say *"I'm working with a one micron technology."* But things started to get a little awkward when we dropped below 1 μm, because it's a bit of a pain to have to keep on saying things like *"zero point one-three microns."* For this reason, it's become common to talk in terms of "nano," where one nano (short for "nanometer") equates to one thousandth of a micron—that is, one thousandth of one millionth of a meter. Thus, when referring to a 0.13 μm technology, instead of mumbling *"zero point one-three microns,"* you would now proclaim *"one hundred and thirty nano."* Of course both of these mean exactly the same thing, but if you want to talk about this sort of stuff, it's best to use the vernacular of the day and present yourself as hip and trendy as opposed to an old fuddy-duddy from the last millennium.

While smaller geometries result in lower power consumption and higher operating speeds, these benefits do not come without a price. Submicron logic gates exhibit extremely complex timing effects, which make corresponding demands on designers and design systems. Additionally, all materials are naturally radioactive to some extent, and the materials used to package integrated circuits can spontaneously release alpha particles. Devices with smaller geometries are more susceptible to the effects of noise, and the alpha decay in packages can cause corruption of the data being processed by deep-submicron logic gates. Deep-submicron technologies also suffer from a phenomenon known as *subatomic erosion* or, more correctly, *electromigration*, in which the structures in the silicon are eroded by the flow of electrons in much the same way as land is eroded by a river.

What Comes After Optical Lithography?

Although new techniques are constantly evolving, technologists can foresee the limits of miniaturization that can be practically achieved using optical lithography. These limits are ultimately dictated by the wavelength of ultraviolet radiation. The technology has now passed from using standard *ultraviolet (UV)* to *extreme ultraviolet (EUV)*, which is just this side of soft X-rays in the electromagnetic spectrum. One potential alternative is true X-ray lithography, but this requires an intense X-ray source and is considerably more expensive than optical lithography. Another possibility is *electron beam lithography*, in which fine electron beams are used to draw extremely high-resolution patterns directly into the resist without a mask. Electron beam lithography is sometimes used for custom and prototype devices, but it is much slower and more expensive than optical lithography. Thus, for the present, it would appear that optical lithography—in the form of *extreme ultraviolet lithography (EUVL)*—will continue to be the mainstay of mass-produced integrated circuits.

How Many Transistors?

The geometry table presented on the previous page reflects commercially available processes (although widespread adoption typically takes some time), but experimental processes in the laboratories are much further advanced. For example, in December 2000, Intel announced that they had constructed an incredibly small, unbelievably fast CMOS transistor only 0.03 µm (30 nano) in size.

In the first half of 2002, Intel announced its McKinley microprocessor— an integrated circuit based on a 0.13 µm (130 nano) process containing more than 200 million transistors! And by the summer of 2002, Intel had announced a test chip based on a 0.09 µm (90 nano) process that contained 330 million transistors. Some observers are predicting that, using Intel's 0.03 µm (30 nano) process, by 2005, we could have 500 million transistors on a single chip running at 10 GHz[14] with a supply voltage of less than one volt. And by 2010, we could be looking at more than 1.8 billion transistors on a chip! At this level of processing power, we will soon have the capabilities required to create Star Trek–style products like a universal real-time translator.[15,16]

---

[14] The unit of frequency is the Hertz (Hz). One gigahertz (1 GHz) means *"one thousand million cycles per second."*

[15] Speaking of Star Trek, a company called Time Domain (www.timedomain.com) based in Huntsville, Alabama, USA is using ultra wideband wireless technology to create an incredibly low power, Tricorder-like handheld radar. This will allow police, firefighters, and rescue workers to "see through walls" and, for example, detect earthquake victims trapped under piles of rubble.

[16] There are predictions of geometries as low as 0.009 µm (9 nano) becoming available somewhere between 2024 and 2028, at which point the author's mind boggles!

Moore's Law

In 1965, Gordon Moore (who was to co-found Intel Corporation in 1968) was preparing a speech that included a graph illustrating the growth in the performance of memory ICs. While plotting this graph, Moore realized that new generations of memory devices were released approximately every 18 months, and that each new generation of devices contained roughly twice the capacity of its predecessor.

This observation subsequently became known as *Moore's Law*, and it has been applied to a wide variety of electronics trends. These include the number of transistors that can be constructed in a certain area of silicon (the number doubles approximately every 18 months), the price per transistor (which follows an inverse Moore's Law curve and halves approximately every 18 months), and the performance of microprocessors (which again doubles approximately every 18 months).

Liquid crystal displays are often used for personal electronic appliances such as notebook computers. These displays are arranged as a matrix of points called *pixels,* each of which can be light or dark, and each of which is controlled by a transistor. Up until now, these displays have typically been fabricated as a sandwich of liquid crystal between two slices of glass. However, the following interesting little snippet was reported in the November 1994 edition of the IEEE Spectrum magazine:

"Researchers at the French National Center for Scientific Research, in Thiais, say that they have developed an all-plastic, no-metal transistor. Using modern printing techniques, the group built up thin layers of various materials to construct paper-thin field-effect transistors, then employed plastic-like organic polymers and graphite inks to carry electricity. This development could lead to flexible computer screens that roll up like window shades."

Personally the author has enough problems with his window shades spontaneously rolling up (which can offer one of life's unforgettable moments when parading in full glory in front of the bathroom mirror) without his computer screen doing the same thing. However, this opens the door to yet more science fiction predictions coming true in the foreseeable future: for example, comic books, newspapers, and textbooks with animated pictures and graphics. We truly do live in interesting times.

Memory ICs

Memory devices are a special class of integrated circuits that are used to store binary data for later use. There are two main categories of semi-conductor memories: *read-only memory (ROM)* and *read-write memory (RWM)*. Other components in the system can read (extract) data from ROM devices, but cannot write (insert) new data into them. By comparison, data can be read out of RWM devices and, if required, new data can be written back into them. The act of reading data out of a RWM does not affect the master copy of the data stored in the device. For a number of reasons, mainly historical, RWMs are more commonly known as *random-access memories (RAMs)*.

ROM and RAM devices find a large number of diverse applications in electronic designs, but their predominant usage is as memory in computer systems (Figure 15-1).[1]

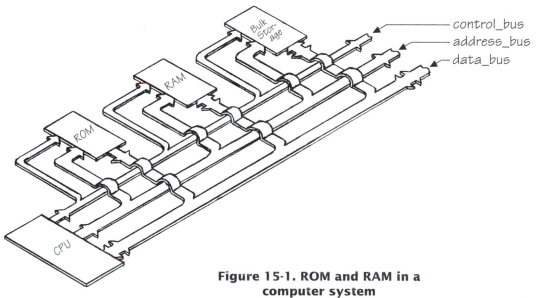

Figure 15-1. ROM and RAM in a computer system

---

[1] In conversation, ROM is pronounced as a single word to rhyme with "bomb," while "RAM" is pronounced to rhyme with "ham."

The brain of the computer is the *central processing unit (CPU)*, which is where all of the number crunching and decision-making are performed. The CPU uses a set of signals called the *address bus* to point to the memory location in which it is interested. The address bus is said to be *unidirectional* because it conveys information in only one direction: from the CPU to the memory. By means of control signals, the CPU either reads data from, or writes data to, the selected memory location. The data is transferred on a set of signals called the *data bus*, which is said to be *bidirectional* because it can convey information in two directions: from the CPU to the memory and vice versa.

The bulk storage is often based on magnetic media such as a disk or tape,[2] which can be used to store a large quantity of information relatively cheaply. Because magnetic media maintains its data when power is removed from the system it is said to be *nonvolatile*.

A general-purpose computer is a system that can accept information from the outside world (from a keyboard, mouse, sensor, or some other input device), process that information, make decisions based on the results of this processing, and return the information to the outside world in its new form (using a computer screen, a robot arm, or some other output device).

The concept of a general-purpose computer on a single chip (or on a group of related chips called a *chipset*) became known as a *microprocessor*. (A *microcontroller* may be considered to be a microprocessor augmented with special-purpose inputs, outputs, and control logic like counter-timers). The written abbreviations for microprocessors and microcontrollers are μP and μC, respectively, where "μ", pronounced "mu" for "micro" comes from the Greek *micros*, meaning "small" (in conversation you always use the full names for these devices and never try to pronounce their abbreviations).

Digital signal processing refers to the branch of electronics concerned with the representation and manipulation of signals in digital form. These signals include voice and video, and this form of processing appears in such things as compressing and decompressing audio and video data, telecommunications, radar, and image processing (including medical imaging). A *digital signal processor (DSP)* is a special form of microprocessor that has been designed to perform a specific processing task on a specific type of digital data much faster and more efficiently than can be achieved using a general-purpose device. (In conversation, DSP is always spelled out as "D-S-P").

---

[2] The use of tape is becoming increasingly rare for almost any form of mainstream application.

One of the major disadvantages of currently available bulk storage units is their relatively slow speed.[3] The CPU can process data at a much higher rate than the bulk storage can supply or store it. Semiconductor memories are significantly more expensive than bulk storage, but they are also a great deal faster.

ROM devices are said to be *mask-programmable* because the data they contain is hard-coded into them during their construction (using photo-masks as was discussed in the previous chapter). ROMs are also classed as being *nonvolatile*, because their data remains when power is removed from the system. By comparison, RAM devices initialize containing random logic 0 or logic 1 values when power is first applied to a system. Thus, any meaningful data stored inside a RAM must be written into it by other components in the system after it has been powered-up. Additionally, RAMs are said to be *volatile*, because any data they contain is lost when power is removed from the system.

When a computer system is first powered up, it doesn't know much about anything. The CPU is hard-wired so that the first thing it does is read an instruction from a specific memory address: for example, address zero. The components forming the system are connected together in such a way that this hard-wired address points to the first location in a block of ROM.[4] The ROM contains a sequence of instructions that are used by the CPU to initialize both itself and other parts of the system. This initialization is known as *boot-strapping*, which is derived from the phrase *"pulling yourself up by your boot-straps."* At an appropriate point in the initialization sequence, instructions in the ROM cause the CPU to copy a set of master programs, known collectively as the *operating system* (OS), from the bulk storage into the RAM. Finally, the instructions in the ROM direct the CPU to transfer its attention to the operating system instructions in the RAM, at which point the computer is ready for the user to enter the game.

---

[3] Note the use of the qualifier "relatively." Modern bulk storage is actually amazingly fast, but not as fast as the rest of the system.

[4] Actually, if the truth be told, these days the block of memory pointed to by the CPU on power-up is typically formed from another form of memory like FLASH, which is non-volatile like ROM, but which can be re-programmed (if necessary) like RAM (FLASH is introduced later in this chapter).

Underlying RAM and ROM Architectures

The smallest unit of memory, called a *cell*, can be used to store a single bit of data: that is, a logic 0 or a logic 1. A number of cells physically grouped together are classed as a *word*, and all the cells in a word are typically written to, or read from, at the same time. The core of a memory device is made up of a number of words arranged as an array (Figure 15-2).

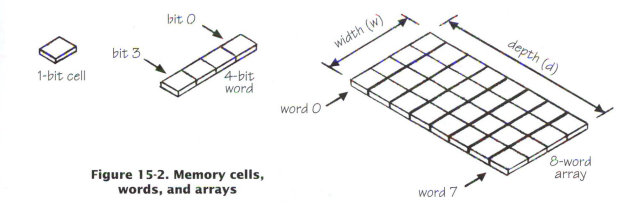

Figure 15-2. Memory cells, words, and arrays

The width (w) of a memory is the number of bits used to form a word, where the bits are usually numbered from 0 to ($w - 1$).[5] Similarly, the depth (d) of a memory is the number of words used to form the array, where the words are usually numbered from 0 to ($d - 1$). The following examples assume a memory array that is four bits wide and eight words deep—real devices can be much wider and deeper.

For the purposes of this discussion, it is convenient to visualize a ROM as containing an array of hard-coded cells. In reality, the physical implementation of a ROM is similar to that of a PROM, which is discussed in more detail in Chapter 16.

In the case of *Dynamic RAMs (DRAMs)*,[6] each cell is formed from a transistor-capacitor pair. The term "dynamic" is applied because a capacitor loses its charge over time and each cell must be periodically recharged to retain its data.

---

[5] Note that there is no official definition as to the width of a word: this is always system-dependent.

[6] In conversation, DRAM is pronounced as "D-RAM." That is, spelling out the "D" and following it with "RAM" to rhyme with "ham."

This operation, known as *refreshing*, requires the contents of each cell to be read out and then rewritten. Some types of DRAM require external circuitry to supervise the refresh process, in which case a special independent controller device is employed to manage a group of DRAMs. In other cases, a DRAM may contain its own internal self-refresh circuitry.

In the case of *Static RAMs (SRAMs)*,[7] each cell is formed from four or six transistors configured as a latch or a flip-flop. The term "static" is applied because, once a value has been loaded into an SRAM cell, it will remain unchanged until it is explicitly altered or until power is removed from the device.

For external components to reference a particular word in the memory, they must specify that word's address by placing appropriate values onto the address bus. The address is decoded inside the device, and the contents of the selected word are made available as outputs from the array (Figure 15-3).

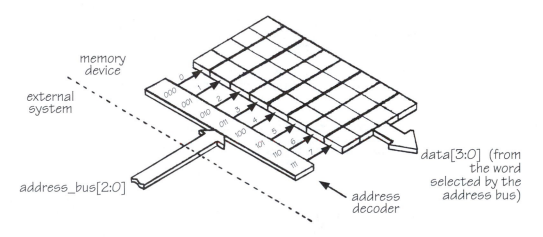

Figure 15-3. Address bus decoding

Standard memory devices are constrained to have a depth of 2^n words, where *n* is the number of bits used to form the address bus. For example, the 3-bit address bus illustrated in Figure 15-3 can be decoded to select one of eight words ($2^3 = 8$) using a 3:8 decoder.[8]

---

[7] In conversation, SRAM is pronounced as "S-RAM." That is, spelling out the "S" and following it with "RAM" to rhyme with "ham."

[8] Decoders were introduced in Chapter 11.

This leads to an interesting quirk when referencing the size of a memory device. In SI units,[9] the qualifier *k* (kilo)[10] represents one thousand (1,000), but the closest power of two to one thousand is 2^{10}, which equals 1,024. Therefore, a 1 kilobit (1 kb or 1 Kb) memory actually refers to a device containing 1,024 bits.

Similarly, the qualifier M (mega)[11] is generally taken to represent one million (1,000,000), but the closest power of two to one million is 2^{20}, which equals 1,048,576. Therefore, a 1 megabit (1 Mb) memory actually refers to a device containing 1,048,576 bits. In the case of the qualifier G (giga),[12] which is now generally taken to represent one billion (1,000,000,000),[13] the closest power of two is 2^{30}, which equals 1,073,741,824. Therefore, a 1 gigaabit (1 Gb) memory actually refers to a device containing 1,073,741,824 bits.

If the width of the memory is equal to a byte or a multiple of bytes, then the size of the memory may also be referenced in terms of bytes. For example, a memory containing 1,024 words, each 8 bits wide, may be referred to as being either an 8 kilobit (8 Kb) or a 1 kilobyte (1 KB) device (note the use of "b" and "B" to represent *bit* and *byte*, respectively).

Because multiple memory devices are usually connected to a single data bus, the data coming out of the internal array is typically buffered from the external system by means of tri-state gates.[14] Enabling the tri-state gates allows the device to drive data onto the data bus, while disabling them allows other devices to drive the data bus.

In addition to its address and data buses, a ROM requires a number of control signals, the two most common being ~chip_select and ~read.

---

[9] The metric system of measurement was developed during the French Revolution and its use was legalized in the U.S. in 1866. The International System of Units (SI) is a modernized version of the metric system.

[10] The term *kilo* comes from the Greek *khiloi*, meaning "thousand" (strangely enough, this is the only prefix with an actual numerical meaning).

[11] The term *mega* comes from the Greek *mega*, meaning "great" (hence the fact that Alexander the Great was known as *Alexandros Megos* in those days).

[12] The term *giga* comes from the Latin *gigas*, meaning "giant."

[13] See the discussions in Chapter 3 on why we now take "one billion" to represent *one thousand million* rather than *one million million*.

[14] Tri-state gates were introduced in Chapter 11.

(The ~read control is sometimes called ~output_enable. Alternatively, some devices have both ~read and ~output_enable controls.) These control signals are commonly *active-low*; that is, they are considered to be ON when a logic 0 is applied.[15] The ~chip_select signal indicates to the ROM that its attention is required, and it is combined with ~read to form an internal signal ~rd which is used to control the tri-state gates (Figure 15-4).

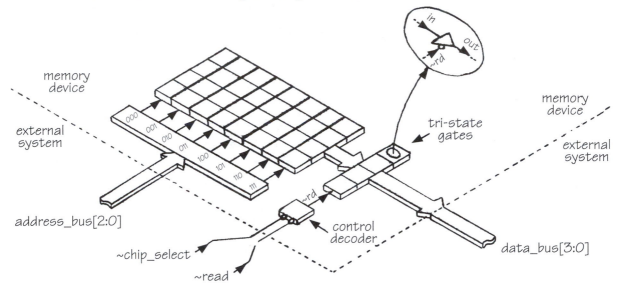

Figure 15-4. ROM control decoding and data bus

When ~rd is active, the tri-state gates are enabled and the data stored in the word selected by the address bus is driven out onto the data bus. When ~rd is inactive, the tri-state gates are disabled and the outputs from the device are placed into high-impedance Z states.

In addition to the control signals used by ROMs, RAMs require a mechanism to control the writing of data *into* the device. Some RAMs employ separate buses for reading and writing called data_out and data_in, respectively. These components usually have an additional control signal called ~write which is also active-low. Once again, the ~chip_select signal indicates to the RAM that its attention is required, and it is combined with ~read to form an internal ~rd

---

[15] Tilde '~' characters prefixing signal names are used to indicate that these signals are active-low. The use of tilde characters is discussed in detail in Appendix A.

signal which is used to control the tri-state gates. Additionally, ~chip_select is combined with ~write to form an internal ~wr signal (Figure 15-5).

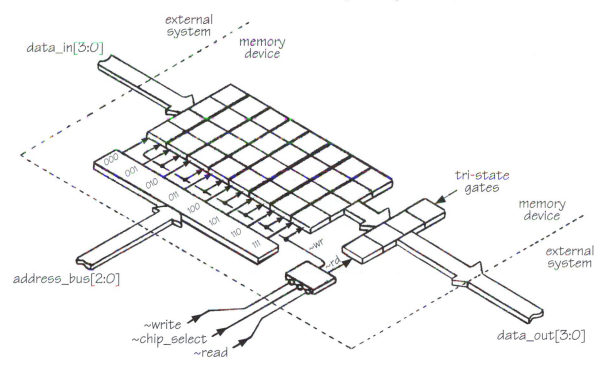

Figure 15-5. RAM with separate data in and data out busses

When ~wr is active, the data present on the data_in bus is written into the word selected by the address_bus. The contents of the word pointed to by the address_bus are always available at the output of the array, irrespective of the value on ~wr. Therefore, devices of this type may be written to and read from simultaneously.

In contrast to those devices with separate data_in and data_out buses, the majority of RAMs use a common bus for both writing and reading. In this case, the ~read and ~write signals are usually combined into a single control input called something like read~write. The name read~write indicates that a logic 1 on this signal is associated with a read operation, while a logic 0 is associated with a write. When the ~chip_select signal indicates to the RAM that its attention is required, the value on read~write is used to determine the type of operation to be performed. If read~write carries a logic 1, the internal ~rd signal is made active and a read operation is initiated; if *read~write* carries a logic *O*, the internal *~wr* signal is made active and a write operation is executed (Figure 15-6).

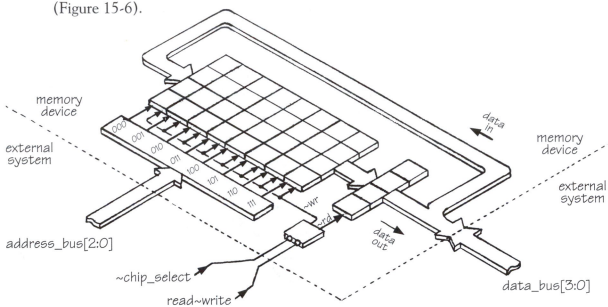

Figure 15-6. RAM with single bi-directional databus

When *~rd* is active, the tri-state gates are enabled and the data from the word selected by the *address_bus* is driven onto the *data_bus*. When *~wr* is active (and therefore *~rd* is inactive), the tri-state gates are disabled. This allows external devices to place data onto the *data_bus* to be written into the word selected by the *address_bus*.

If the value on the *address_bus* changes while *~rd* is active, the data associated with the newly selected word will appear on the *data_bus*. However, it is not permissible for the value on the *address_bus* to change while *~wr* is active because the contents of multiple locations may be corrupted.

Increasing Width and Depth

Individual memory devices can be connected together to increase the width and depth of the total memory as seen by the rest of the system. For example, two 1,024-word devices, each 8 bits wide, can be connected so as to appear to be a single 1,024-word device with a width of 16 bits. An *address_bus* containing 10 bits is required to select between the 1,024 words (Figure 15-7).

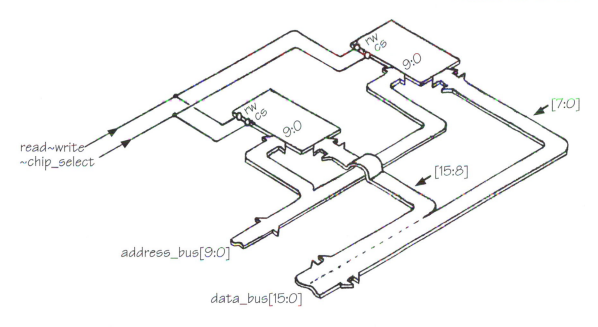

read~write
~chip_select

rw
cs
9:0

rw
cs
9:0

[7:0]

[15:8]

address_bus[9:0]

data_bus[15:0]

Figure 15-7. Connecting memory devices to increase the width

In this case, the address_bus and control signals are common to both memories, but each device handles a subset of the signals forming the data_bus. Additional devices can be added to further increase the total width of the data_bus as required. Alternatively, two 1,024-word devices, each 8 bits wide, can be connected so as to appear to be a single device with a width of 8 bits and a depth of 2,048 words. An address_bus containing 11 bits is required to select between the 2,048 words (Figure 15-8).

In this case, the data_bus, the ten least-significant bits of the address_bus, and the read~write control are common to both memories. However, the most significant bit (MSB) of the address_bus is decoded to generate the ~chip_select signals. A logic 0 on the most-significant bit selects the first device and deselects the second, while a logic 1 deselects the first device and selects the second.

Additional address bits can be used to further increase the total depth as required. If the address_bus in the previous example had contained 12 bits, the two most-significant bits could be passed through a 2:4 decoder to generate four chip select signals. These signals could be used to control four 1,024-word devices, making them appear to be a single memory with a width of 8 bits and a depth of 4,096 words.

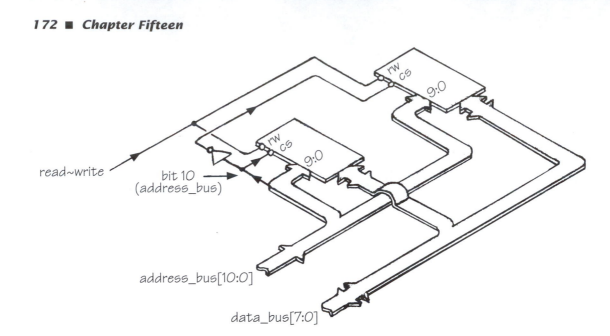

read~write

bit 10
(address_bus)

address_bus[10:0]

data_bus[7:0]

Figure 15-8. Connecting memory devices to increase the depth

Alternative Technologies

SRAMs are faster than DRAMs, but each SRAM cell requires significantly more silicon real estate and consumes much more power than a corresponding DRAM cell. The first SRAM (256-bit) and DRAM (1,024-bit) devices were both created in 1970. For the next few decades, both types of memory quadrupled their capacities approximately every three years, but by the beginning of the twenty-first century this had slowed to a doubling every two to three years.

By mid 2002, the highest density SRAMs contained 16 megabits. These were typically based on 0.18 micron processes with a roadmap in place to shrink to 0.15 and 0.13 microns. By comparison, by the end of 2002, mainstream DRAM devices produced using 0.15 micron processes will contain 256 megabits. DRAM vendors will move to volume production using 0.13 micron processes in 2003, and 512 megabit devices based on these processes will become mainstream in 2004/2005. These will be succeeded by 1 gigabit devices based on 0.09 micron technologies in the second half of the decade.[16]

---

[16] One problem with DRAM memory cells is that there is a limit to the amount by which the size of their capacitors can be reduced before they become too small to retain any data. This problem will hopefully be addressed by MRAM devices, which are introduced later in this chapter.

Due to their regular structures and the numbers of components produced, memory devices used to lead semiconductor technology by approximately two years. For example, DRAMs based on 0.35 micron processes became available as early 1994, but other devices with these geometries didn't hit the mainstream until 1996. By 2002, however, high-end general-purpose logic processes had become just as advanced as memory processes, and both memory and logic foundries routinely offered 0.13 micron technologies.

PROMs

Because ROMs are constructed containing pre-defined data, they are an expensive design option unless used for large production runs. As an alternative, *programmable read-only memories (PROMs)*[17] were developed in the mid-1970s. PROMs are manufactured as standard devices that may be electrically programmed by the designer and are non-volatile. PROMs are *one-time programmable (OTP)* and slightly slower than their ROM equivalents, but are significantly cheaper for small- to medium-sized production runs.

EPROMs

To satisfy the designers' desire to combine the advantages of DRAMs (programmable) and ROMs (nonvolatile), *erasable programmable read-only memories (EPROMs)*[18] were developed in the late 1970s. EPROMs are electrically programmable by the designer and are non-volatile, but can be erased and reprogrammed should the designer so desire. An EPROM device is presented in a ceramic package with a

Until the latter half of the 1990s, DRAM-based computer memories were asynchronous, which means they weren't synchronized to the system clock. Every new generation of computers used a new trick to boost speed, and even the engineers started to become confused by the plethora of names, such as *fast page mode (FPM)*, *extended data out (EDO)*, and *burst EDO (BEDO)*. In reality, these were all based on core DRAM concepts; the differences are largely in how you wire the chips together on the circuit board (OK, sometimes there is a bit of tweaking on the chips also).

Over time, the industry migrated to synchronous DRAM (SDRAM), which is synchronized to the system clock and makes everyone's lives much easier. Once again, however, SDRAM is based on core DRAM concepts—it's all in the way you tweak the chips and connect them together.

[17] In conversation, PROM is pronounced just like the high school dance of the same name.

small quartz window mounted in the top. The quartz window is usually covered by a piece of opaque tape. To erase the device, the tape is removed and the EPROM is exposed to ultraviolet radiation for approximately 20 minutes. EPROMs are relatively expensive due to their complicated packaging requirements, but are ideal for prototyping and other applications that require regular changes to the stored data. EPROMs are often used during the development of a design and are subsequently replaced by equivalent PROM or ROM devices in cheaper plastic packages after the design has stabilized. (The actual construction of EPROMs and some of the other types of memory devices introduced during the remainder of this chapter are discussed in greater detail in Chapter 16.)

By 2002, there were a variety of different types of SDRAM on the market (way too many to cover here). Perhaps the most well known are the ones used in general-purpose personal computers, which are called PC100 and PC133, which operate at 100 MHz and 133 MHz, respectively.

In the future, memory manufacturers may move to faster speeds (say PC166 of higher), but this may cause problems, because the original SDRAM interface was not designed with these high frequencies in mind.

However, the original SDRAM specification was based on using only one of the clock edges (say the rising edge) to read/write data out-of/into the memory as shown below:

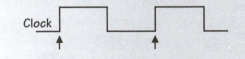

An alternative known as *double data rate (DDR)* is to design the memory in such a way that data can be read/written on both edges of the clock:

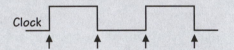

This effectively doubles the amount of data that can be pushed through the system without increasing the clock frequency (it sounds simple if you say it fast, but making this work is trickier than it may at first appear).

Furthermore, a *quad data rate (QDR)* memory is one with separate data in and data out busses (as was illustrated in Figure 15-5), both of which can be used on both edges of the clock.

---

[18] In conversation, EPROM is pronounced as "E-PROM." That is, spelling out the "E" and following it with "prom."

EEPROMs

A further technology called *electrically-erasable read-only memory (EEPROM or E²PROM)*[19] was developed towards the end of the 1970s. E²PROMs are electrically programmable by the designer and are nonvolatile, but can be electrically erased and reprogrammed should the designer so desire. Additionally, by means of additional circuitry, an E²PROM can be erased and reprogrammed while remaining resident on the circuit board, in which case it may be referred to as *in-system programmable (ISP)*.

An EPROM cell is based on a single transistor and is therefore very efficient in terms of silicon real estate, while an equivalent E²PROM cell is based on two transistors and requires more area on the silicon. However, an EPROM device must be erased and reprogrammed in its entirety, while an E²PROM device can be erased and reprogrammed on a word-by-word basis.

---

[19] In conversation, some people pronounce EEPROM as "E-E-PROM" (spelling out "E-E" followed with "prom"), while others say "E-squared-prom."

PC100 and PC133 memory modules are 64-bits wide (although dual configurations have been used to provide 128 bit wide busses in really high-end computers).

Thus, assuming a standard 64-bit (8-byte) wide PC100 (100 MHz) memory module, the peak data throughput will be $8 \times 100 = 800$ megabytes per second. If a DDR version becomes available as discussed earlier, this would rise to 1,600 megabytes per second. Furthermore, a 128-bit wide DDR (using dual 64-bit modules in parallel) would have a peak bandwidth of 3,200 megabytes per second.

Towards the end of the 1990s, an alternative memory concept called *Rambus* began to gain attention (as usual this is based on core DRAM concepts presented in a cunning manner). By around 2000, personal computers equipped with memory modules formed from Rambus DRAM (RDRAM) devices were available. These memory modules were 16-bits (2-bytes) wide, and were accessed on both edges of a 400 MHz clock, thereby providing a peak data throughput of $2 \times 2 \times 400 = 1,600$ megabytes per second. (Just to be confusing, these devices are actually referenced as *"800 MHz Rambus DRAM,"* but this is just marketing spin trying to make using both edges of a 400 MHz clock sound more impressive.)

In the summer of 2002, a new flavor of Rambus was announced with a 32-bit wide bus using both edges of a 533 MHz clock (the official announcements quote *"1,066 MHz Rambus DRAM,"* but this is the same marketing spin as noted above). Whatever, this will provide a peak bandwidth of 4,200 megabytes per second.

FLASH

Yet another technology called FLASH is generally regarded as an evolutionary step that combines the best features from EPROM and E²PROM. The name FLASH is derived from its fast reprogramming time compared to EPROM. FLASH has been under development since the end of the 1970s, and was officially described in 1985, but the technology did not initially receive a great deal of interest. Towards the end of the 1980s, however, the demand for portable computer and communication systems increased dramatically, and FLASH began to attract the attention of designers.

All variants of FLASH are electrically erasable like E²PROMs. Some devices are based on a single transistor cell, which provides a greater capacity than an E²PROM, but which must be erased and reprogrammed on a device-wide basis similar to an EPROM. Other devices are based on a dual transistor cell and can be erased and reprogrammed on a word-by-word basis (see Chapter 16 for more details).

> Computer systems are very complicated and there's always the chance that an error will occur when reading or writing to the memory (a stray pulse of "noise" may flip a logic 0 to a logic 1 while your back is turned). Thus, serious computers use *error-correcting code (ECC)* memory, which includes extra bits and special circuitry that tests the accuracy of data as it passes in and out of memory and corrects any (simple) errors.

FLASH is considered to be of particular value when the designer requires the ability to reprogram a system in the field or via a communications link while the devices remain resident on the circuit board.

MRAMs

A technology that is attracting a great deal of interest for the future is *magnetic random access memory (MRAM)*,[20] which may be able to store more data, read and write data faster, and use less power than any of the current memory technologies. In fact, the seeds of MRAM were laid as far back as 1974, when IBM developed a component called a *magnetic tunnel junction (MTJ)*, which comprises a sandwich of two ferromagnetic layers separated by a

---

[20] In conversation, MRAM is pronounced "M-RAM." That is, by spelling out the "M" and following it with "RAM" to rhyme with "ham."

Whatever flavor of DRAM you are using, a single device can only contain a limited amount of data, so a number of DRAMs are gathered together onto a small circuit board called a *memory module*.

Each memory module has a line of gold-plated pads on both sides of one edge of the board. These pads plug into a corresponding connector on the main computer board.

A *single in-line memory module (SIMM)* has the same electrical signal on corresponding pads on the front and back of the board (that is, the pads on opposite sides of the board are "tied together").

By comparison, in the case of a *dual in-line memory module (DIMM)*, the pads on opposite sides of the board are electrically isolated from each other and form two separate contacts.

Last but not least, we have the RIMM, which really doesn't stand for anything per se, but which is the trademarked name for a Rambus memory module (RIMMs are similar in concept to DIMMs, but have a different pin count and configuration).

thin insulating layer. A memory cell is created by the intersection of two wires (say a "row" line and a "column" line) with an MJT sandwiched between them. MRAMs have the potential to combine the high speed of SRAM, the storage capacity of DRAM, and the non-volatility of FLASH, while consuming very little power. It's taken close to 30 years, but MRAMs test chips are predicted to become available around 2003, with volume production in 2004.

nvRAMs

Finally, a class of devices known as *nonvolatile RAMs (nvRAMs)*[21] may be used in cases where high speed is required and cost is not an overriding issue. These devices are generally formed from an SRAM die mounted in a package with a very small battery, or as a mixture of SRAM and E^2PROM cells fabricated on the same die.

---

[21] In conversation, nvRAM is pronounced as "n-v-RAM." That is, spelling out the "n" and the "v" and following them with "RAM" to rhyme with "ham."

Programmable ICs

Programmable integrated circuits are standard devices constructed in such a way that a designer can configure them to perform a specific function. These devices, which were introduced in the mid-1970s, are generically known as *programmable logic devices (PLDs)*,[1,2] and can range in complexity from a few hundred to tens of thousands of equivalent gates. Traditional PLDs are based on logic gates formed from diodes (Figure 16-1).

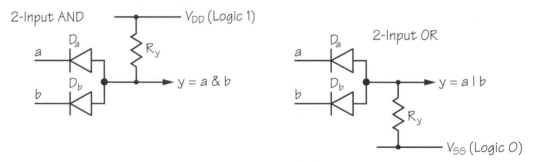

Figure 16-1. Diode implementations of AND and OR gates

In the case of the 2-input AND, the pull-up resistor R_y attempts to hold the output y at V_{DD} (logic 1). When both of the inputs are at logic 1, there is no difference in potential across the diodes and y remains at logic 1. However, if input a is connected to V_{SS} (logic 0), diode D_a will conduct and y will be pulled down to logic 0; similarly for input b and diode D_b. Additional inputs can be formed by adding more diodes.

In the case of the 2-input OR, the pull-down resistor R_y attempts to hold the output y at V_{SS} (logic 0). When both of the inputs are at logic 0, there is no difference in potential across the diodes and y remains at logic 0. However,

---

[1] In conversation, PLD is pronounced by spelling it out as "P-L-D."

[2] Originally all PLDs contained a modest number of equivalent logic gates and were fairly simple. As more *complex PLDs (CPLDs)* arrived on the scene, however, it became common to refer to their simpler cousins as *simple PLDs (SPLDs)*. (Note that CPLDs are introduced in more detail later in this chapter.)

if input *a* is connected to V_{DD} (logic 1), diode D_a will conduct and y will be pulled up to logic 1; similarly for input *b* and diode D_b. Once again, additional inputs can be formed by adding more diodes.

Fusible-link Technologies

In the most common PLD technology, which is known as *fusible-link*, each diode has an associated link known as a *fuse*. The designer can individually remove unwanted fuses by applying pulses of relatively high voltage and current, in the order of 20 volts and 300 milliamps, respectively, to the device's inputs. These pulses are sufficient to effectively vaporize the fuse (Figure 16-2).

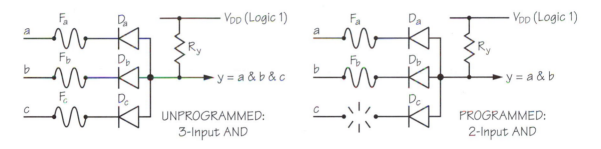

Figure 16-2. Removing unwanted fusible links

This process is typically known as *programming the device*, but may also be referred to as *blowing the fuses* or *burning the device*. Devices based on fusible links are said to be *one-time programmable (OTP)*,[3] because once a fuse has been blown it cannot be replaced.

Antifuse Technologies

As an alternative to fusible links, some PLDs employ *antifuse* technologies in which each diode has an associated *antifuse*. Unlike fusible links—which will conduct unless they are forcibly removed—antifuse links will not conduct in the device's unprogrammed state. The user can individually enable desired antifuse links by applying signals of relatively high voltage and current to the device's inputs (Figure 16-3).

---

[3] In conversation, OTP is pronounced by spelling it out as "O-T-P."

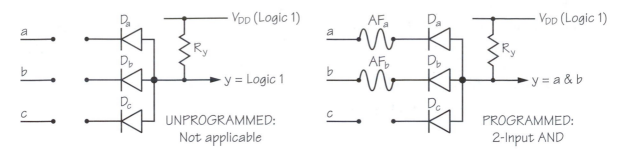

Figure 16-3. Adding desired antifuse links

Antifuse links are formed by creating a connection—called a *via*—of amorphous (noncrystalline) silicon between two layers of metalization. In its unprogrammed state, the amorphous silicon is an insulator with a very high resistance in excess of one billion ohms. The programming signal effectively grows a link by changing the insulating amorphous silicon into conducting polysilicon (Figure 16-4).

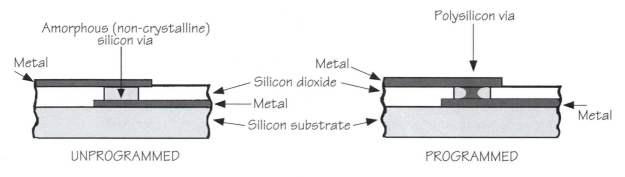

Figure 16-4. Growing an antifuse link

The programmed links in the original antifuse technologies had resistance values in the order of 1,000 ohms, but modern processes have reduced this value to 50 ohms or lower. Devices based on antifuses are also *one-time programmable* (OTP), because once an antifuse link has been grown it cannot be removed.

Special PLD Notation

Due to the fact that PLD structures are very regular, and also because they differ from standard logic gates, a special form of notation has been adopted. This notation is applicable to both fusible-link and antifuse PLDs. Consider a device delivered with four diode-link pairs forming a 4-input AND function (Figure 16-5).

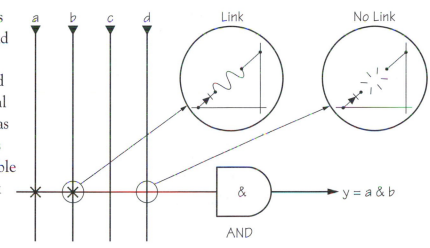

Figure 16-5. Special PLD notation

The AND symbol indicates the function and encompasses the pull-up resistor to V_{DD}. The diode-fuse pairs are represented by the presence or absence of crosses. In a fusible-link device, a cross indicates that the corresponding link has been left intact, while the absence of a cross indicates that the link has been blown away. Thus, the diagram for an *unprogrammed* fusible-link device will show crosses at each intersection of the matrix. By comparison, in antifuse technology, a cross indicates that a link has been grown, while the absence of a cross indicates that no link is present. Thus, the diagram for an *unprogrammed* antifuse device will not show any crosses.

The majority of this book follows the convention laid down in Appendix A, which is that active-low and complementary signals are indicated by prefixing their names with tilde "~" characters.

However, rules are made for the guidance of wise men and the blind obedience of fools, and the purposes of the discussions in this chapter are better served by representing inverted PLD input signals by means of lines, or *bars*, over their names.

Actually, this serves to illustrate an important point: although mixing conventions is not usually recommended, engineers are "switch hitters" who will happily use whatever tool is most appropriate and best suited to get the job done.

Generic PLD Structures

To increase the versatility of PLDs, their inputs are inverted inside the device and both true and inverse versions of the inputs are presented to an array (Figure 16-6).

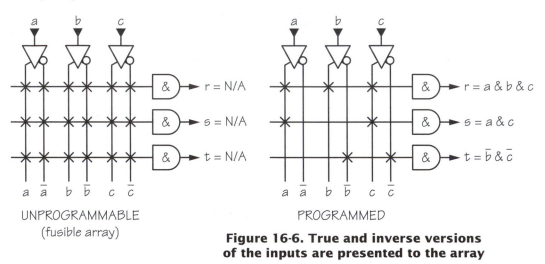

UNPROGRAMMABLE
(fusible array)

PROGRAMMED

Figure 16-6. True and inverse versions of the inputs are presented to the array

The number of AND functions is independent of the number of inputs; additional ANDs can be formed by introducing more rows into the array. Similar techniques can be used to create an array of OR functions, and PLDs typically contain an AND array feeding into an OR array (Figure 16-7).

The number of OR functions is independent of both the number of inputs and the number of AND functions; additional ORs can be formed by introducing more columns into the OR array.

PLDs are not obliged to have AND input arrays feeding OR output arrays; NOR output arrays are also available. However, while it would be possible to create other structures such as OR-AND, NAND-OR, or NAND-NOR, these alternatives tend to be relatively rare or simply not provided. The reason it isn't necessary for the vendor to supply all possible variations is because AND-OR (and AND-NOR) structures directly map onto the *sum-of-products* representations most often used to specify equations.[4] Other equation formats can be mapped to these structures using standard Boolean algebraic techniques —for example, DeMorgan Transformations.[5]

---

[4] Sum-of-products representations were introduced in Chapter 9.

[5] DeMorgan Transformations were introduced in Chapter 9.

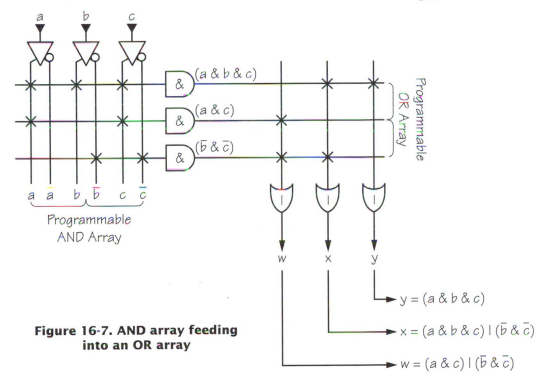

Figure 16-7. AND array feeding into an OR array

$y = (a \& b \& c)$

$x = (a \& b \& c) \,|\, (\overline{b} \& \overline{c})$

$w = (a \& c) \,|\, (\overline{b} \& \overline{c})$

Although core PLD concepts are relatively simple, the wider arena may prove to be a morass of confusion to the unwary. There are a multiplicity of PLD alternatives, most of which seem to have acronyms formed from different combinations of the same three or four letters.[6] This may be, as some suggest, a strategy to separate the *priests* from the *acolytes*, or it may be that the inventors of the devices had no creative energy left to dream up meaningful names for them. Whatever the case, the more common PLD variants are introduced below.

Programmable Logic Arrays (PLAs)

The most user-configurable of the traditional PLDs are *programmable logic arrays (PLAs)*,[7] because both the AND and OR arrays are programmable (Figure 16-8).

---

[6] A common engineering joke is to refer to TLAs, which is an acronym that stands for "three letter acronyms" (in conversation, TLA is pronounced by spelling it out as "T-L-A").

[7] In conversation, PLA is spelled out as "P-L-A."

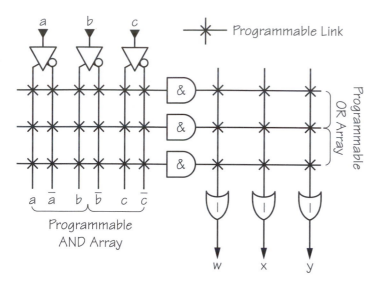

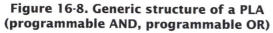

**Figure 16-8. Generic structure of a PLA
(programmable AND, programmable OR)**

This diagram indicates a fusible link technology, because all of the links are present when the device is in its unprogrammed state. Similarly, the following examples are based on fusible link devices unless otherwise stated.

Programmable Array Logic (PAL)

Many applications do not require both the AND and OR arrays to be programmable. In the case of *programmable array logic (PAL)*[8,9] devices, only the AND array is programmable while the OR array is predefined (Figure 16-9).

Although PALs are less flexible than PLAs, they operate somewhat faster because hard-wired connections take less time to switch than their programmable equivalents.[10]

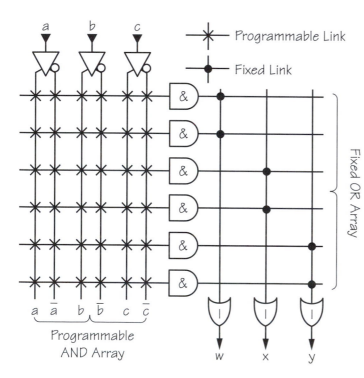

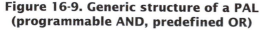

**Figure 16-9. Generic structure of a PAL
(programmable AND, predefined OR)**

---

8 In conversation, PAL is pronounced as a single word to rhyme with "gal."

9 PAL is a registered trademark of Monolithic Memories Inc.

10 There are also variations on PALs called *generic array logic (GAL)* devices from Lattice Semiconductor Corporation (www.latticesemi.com).

Programmable Read-Only Memories (PROMs)

In the case of *programmable read-only memory (PROM)* devices, the AND array is predefined and only the OR array is programmable (Figure 16-10).[11]

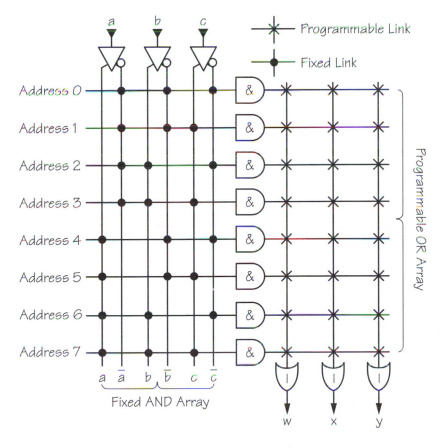

**Figure 16-10. Generic structure of a PROM
(predefined AND, programmable OR)**

PROMs are useful for equations requiring a large number of product terms, but tend to have fewer inputs because every input combination is decoded and used. The internal structure of a PROM is similar to that of a ROM, the main difference being that a ROM's contents are defined during its construction. PROMs are generally considered to be memory devices, in which each address applied to the inputs returns a value programmed into the device.

---

[11] PROMs were introduced in Chapter 15.

Additional Programmable Options

If you thought the above was cunning, just wait until you see some of the really clever tricks that can be used to augment PLDs and make them even more useful.

Tri-stateable Outputs

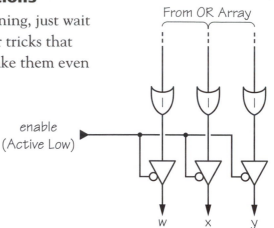

Many PLDs have tri-state buffers on their outputs.[12] All of the buffers share a common *enable* control, which therefore requires only a single input pin on the device (Figure 16-11).

Figure 16-11. PLD with tri-stateable outputs

Some devices may contain additional programmable links associated with the outputs. By means of these additional links (which were omitted from Figure 16-11 for reasons of clarity), each output can be individually programmed to be either permanently enabled or to remain under the control of the *enable* input as required (Figure 16-12).

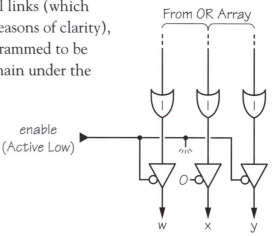

In this example, the tri-state buffer driving the x output has been disconnected from the *enable* control and connected to a constant logic 0 (as these tri-state buffers have active-low control inputs—as is indicated by the bobbles on their symbols—a logic 0

Figure 16-12. PLD with individually tri-stateable outputs

enables them while a logic 1 disables them). As a result, the x output is permanently enabled, irrespective of the value on the *enable* input.

---

12 Tri-state buffers were introduced in Chapter 11.

Using Outputs as Inputs

To further increase the fun, some devices are constructed in such a way that their tri-state outputs are fed back into the AND array (Figure 16-13).

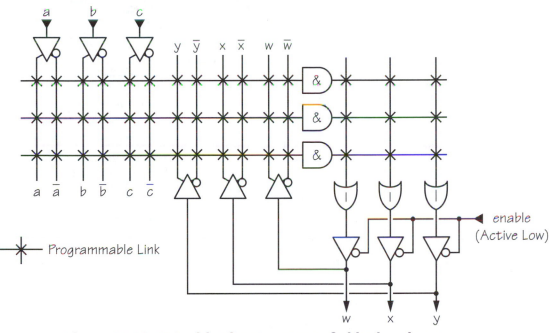

Figure 16-13. PLA with tri-state outputs fed back as inputs

Once again, there are additional programmable links that have been omitted from Figure 16-13 for reasons of clarity. By means of these links, each output's ri-state buffer can be individually configured so as to always drive a high-impedance Z state. Pins that are configured in this way can be used as additional inputs (Figure 16-14).

In this example, the unprogrammed device began with three inputs and three outputs, but the user actually required a device with four inputs and only two outputs. By means of the appropriate links, the tri-state buffer on the y output was disconnected from the *enable* control and connected to a constant logic 1. Because w and x are still required to function as outputs, all of their associated links in the AND array must be blown away to ensure that these pins will not have any effect as inputs. The ability to configure pins as outputs or inputs provides a great amount of flexibility; instead of purchasing myriad devices with every conceivable combination of inputs and outputs, the designer can configure a single device type as required.

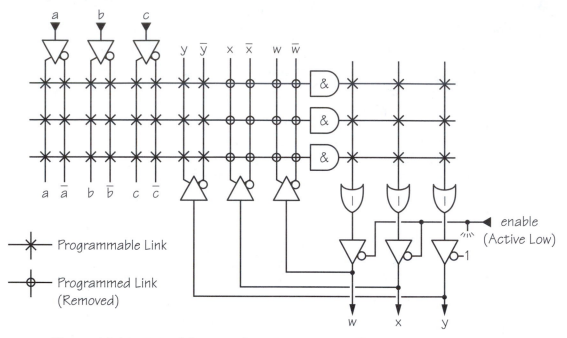

Figure 16-14. PLA with one tri-state output configured as an input

Registered Outputs

Certain PLDs are equipped with registers on the outputs, and others with latches. Depending on the particular device, the registers (or latches) may be provided on all of the outputs or on a subset of the outputs. Registered devices are particularly useful for implementing finite state machines (FSMs).[13] All of the registers share a common clock signal (this will be an enable signal in the case of latches) which therefore requires only a single input pin on the device (Figure 16-15).

In this example, the outputs are shown as being registered with D-type flip-

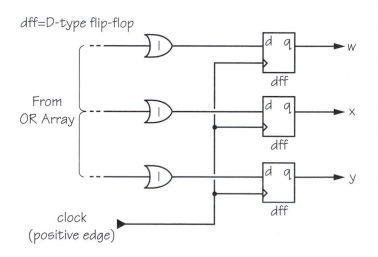

Figure 16-15. PLD with registered outputs

---

[13] FSMs were introduced in Chapter 12.

flops, but alternative register types such as JK flip-flops or T-type flip-flops may be more suitable for certain applications. It can be inconvenient to support a dedicated device for each type of register. As a solution, some devices have configurable register elements whose types can be selected by programming appropriate fuses.

Registered (or latched) outputs may also incorporate by-pass multiplexers (Figure 16-16).

By means of appropriate fuses, the control inputs to the multiplexers can be individually configured to select either the *non-registered* data or its *registered* equivalent.

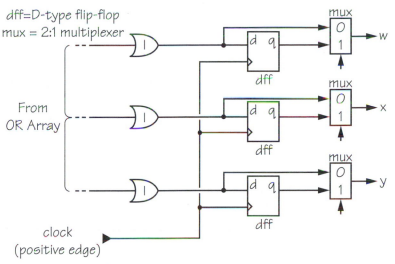

Figure 16-16. PLD with registered outputs and by-pass multiplexers

Other common programmable options are the ability to select *true* or *complemented* outputs and TTL- or CMOS-compatible output drivers. An individual PLD typically only provides a subset of the above capabilities, but these may be combined in a variety of ways; for example, registered outputs may be followed by tri-state buffers.

Programming PLDs

Programming a traditional PLD is relatively painless because there are computer programs and associated tools dedicated to the task. The user first creates a computer file known as a *PLD source file* containing a textual description of the required functionality (Figure 16-17).

In addition to Boolean equations, the PLD source file may also support truth tables, state tables, and other constructs, all in textual format. The exclamation marks (called shrieks) shown in these equations provide a textual way of indicating inversions. Additional statements allow the user to specify which outputs are to be tri-statable, which are to be registered, and any of the other programmable options associated with PLDs.

```
INPUT     =    a , b , c ;
OUTPUT    =    w , x , y ;

EQUATION w   =    (a & b) ;
EQUATION x   =    (a & b) | (a & !b & !c) ;
EQUATION y   =    (a & b) | (b & c) | (a & !c) ;
```

Figure 16-17. Very simple example PLD source file

A special computer program is used to process the PLD source file. This program makes use of a knowledge database that contains details about the internal construction of all of the different types of PLDs. After the user has instructed the program as to which type of device they wish to use, it analyzes the equations in the source file and performs algebraic minimization to ensure optimal utilization of the device's resources. The program accesses the knowledge database for details about the designated device and evaluates which fuses need to be blown to implement the desired functionality. The program then generates a textual output file comprising 0 and 1 characters, which represent the fuses to be blown; this file is in an industry-standard format known as JEDEC (Figure 16-18).

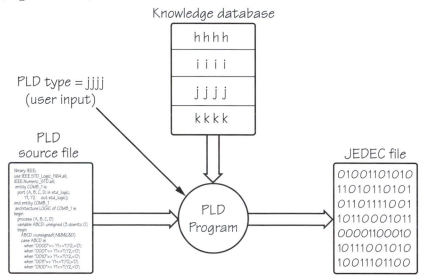

Figure 16-18. Using the textual source file to create a fuse file

As an alternative to the user specifying a particular device, the program can be instructed to automatically select the best device for the task. The program can base its selection on a variety of criteria, such as the speed, cost, and power consumption of the devices. The program may also be used to partition a large design across several devices, in which case it will output a separate JEDEC file for each device.

Finally, the designer takes a virgin device of the appropriate type and places it in a socket on a special tool, which may be referred to as a *programmer*, *blower*, or *burner*. The main computer passes the JEDEC file to the programmer, which uses the contents of the file to determine which fuses to blow. The designer presses the GO button, the burner applies the appropriate signals to the device's inputs, and a new device is born (Figure 16-19).

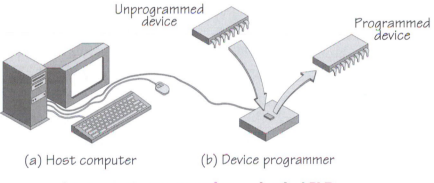

(a) Host computer   (b) Device programmer

Figure 16-19. Programming a physical PLD

Reprogrammable PLDs

One consideration with fusible link and antifuse technologies is that, once they have been programmed, there is no going back. This may be of particular concern with PROMs as the data they store is prone to change.

EPROMs

One alternative is a technology known as *erasable programmable read-only memory (EPROM)*. An EPROM transistor has the same basic structure as a standard MOS transistor, but with the addition of a second polysilicon *floating gate* isolated by layers of oxide (Figure 16-20).

In its unprogrammed state, the floating gate is uncharged and doesn't affect

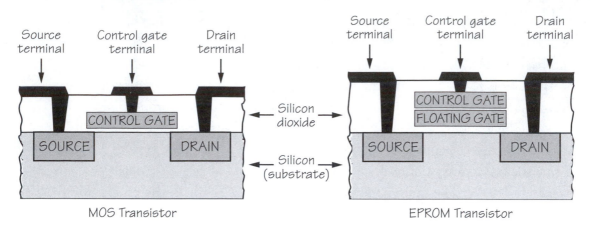

Figure 16-20. Standard MOS transistor versus EPROM transistor

the normal operation of the control gate. To program the transistor, a relatively high voltage in the order of 12V is applied between the control gate and drain terminals. This causes the transistor to be turned hard on, and energetic electrons force their way through the oxide into the floating gate in a process known as *hot* (high energy) *electron injection*. When the programming signal is removed, a negative charge remains on the floating gate. This charge is very stable and will not dissipate for more than a decade under normal operating conditions. The stored charge on the floating gate inhibits the normal operation of the control gate, and thus distinguishes those cells that have been programmed from those which have not.

EPROM cells are efficient in terms of silicon real estate, being half the size of DRAM cells and an order of magnitude smaller than fusible links. However, their main claim to fame is that they can be erased and reprogrammed. An EPROM cell is erased by discharging the electrons on the floating gate. The energy required to discharge the electrons is provided by a source of ultraviolet (UV) radiation. An EPROM device is delivered in a ceramic package with a small quartz window in the top; this window is usually covered with a piece of opaque sticky tape. For the device to be erased, the tape is first removed from the circuit board, the quartz window is uncovered, and the package is placed in an enclosed container with an intense ultraviolet source.

The main problems with EPROM devices are their expensive ceramic packages with quartz windows and the time it takes to erase them, which is in the order of 20 minutes. A foreseeable problem with future generations of these devices is paradoxically related to improvements in process technologies that allow transistors to be made increasingly smaller. The problem is one of scaling, because not all of the structures on a silicon chip are shrinking at the same rate. Most notably, the transistors are shrinking faster than the metal interconnections and, as the features become smaller, a larger percentage of the surface of the die is covered by metal. This makes it difficult for the EPROM cells to absorb the ultraviolet and increases the required exposure time.

EEPROMs

A somewhat related technology is that of *electrically-erasable programmable read-only memory (EEPROM or E²PROM)*. An E^2PROM cell is approximately 2.5 times larger than an EPROM cell because it contains two transistors. One of the transistors is similar to that of an EPROM transistor in that it contains a floating gate, but the insulating oxide layers surrounding the floating gate are very much thinner. The second transistor can be used to erase the cell electrically, and E^2PROM devices can typically be erased and reprogrammed on a word-by-word basis.

FLASH

Finally, a development known as FLASH can trace its ancestry to both EPROM and E^2PROM technologies. The name FLASH was originally coined to reflect the technology's rapid erasure times compared to EPROM. Components based on FLASH can employ a variety of architectures. Some have a single floating gate transistor cell with the same area as an EPROM cell, but with the thinner oxide layers characteristic of an E^2PROM component. These devices can be electrically erased, but only by erasing the whole device or a large portion of it. Other architectures have a two-transistor cell—which is very similar to that of an E^2PROM cell—allowing them to be erased and reprogrammed on a word-by-word basis.

Initial versions of FLASH could only store a single bit of data per cell. By 2002, however, technologists were experimenting with a number of different ways of increasing this capacity. One technique involves storing distinct levels of charge in the FLASH transistor's floating gate to represent two bits per cell. An alternative approach involves creating two distinct storage nodes in a layer below the gate, thereby supporting two bits per cell.

In-System Programmable

One advantage of E^{2}PROM and FLASH devices is that they can be reprogrammed while still on the circuit board, in which case they may be referred to as *in-system programmable (ISP)*.[14] A large proportion of FLASH and E^{2}PROM devices use hot (high energy) electron injection similar to EPROM and require 12V signals for programming. Some designers see this as an advantage, because these devices cannot be accidentally erased by means of the 5.0V (or lower, such as 3.3V, 3.0V, etc.) signals used during the circuit's normal operation. However, an increasing number of devices use *cold* (low energy) *Fowler-Nordheim electron tunneling* and can be programmed using the standard signal voltage (5.0V, 3.3V, 3.0V, etc.).

In addition to their use as memories, EPROM, E^{2}PROM, and FLASH technologies are also employed in other PLD types, which are generically known as EPLDs, E^{2}PLDs, and FLASH PLDs. For example, each fusible link in a standard PLA is replaced by an EPROM cell in an EPLA, an E^{2}PROM cell in an E^{2}PLA, or a FLASH cell in a FLASH PLA. Regardless of the technology used—from fusible link to FLASH—all of the components are created on the surface of a single piece of silicon substrate. However, it may be useful to visualize the device as comprising two distinct strata, with the AND and OR arrays on the lower level and the links "floating" above them (Figure 16-21).

In yet another alternative, some PLDs are constructed with each link being replaced with an SRAM cell. Unlike the other reprogrammable options, PLDs

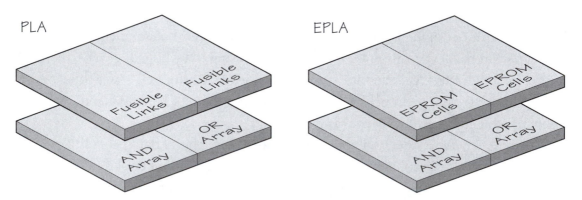

Figure 16-21. A fusible-link PLA versus an EPLA

---

[14] In conversation, ISP is spelled out as "I-S-P."

based on SRAM cells are volatile and lose their data when power is removed from the system. However, these devices can be dynamically reprogrammed while the circuit is in use. Some SRAM-based devices employ a double-buffering scheme using two SRAM cells per link. In this case, the device can be reprogrammed with a new pattern while it is still operating with an old pattern. At the appropriate time, the device can be instructed to switch from one bank of SRAM cells to the other.

Reprogrammable devices also convey advantages over fusible link and antifuse devices in that they can be more rigorously tested at the factory. For example, reprogrammable devices typically undergo one or more program and erase cycles before being shipped to the end user.

Complex PLDs (CPLDs)

The term *complex PLD (CPLD)* is generally taken to refer to a class of devices that contain a number of simple PLA or PAL functions—generically referred to as *simple PLDs (SPLDs)*—sharing a common programmable interconnection matrix (Figure 16-22).

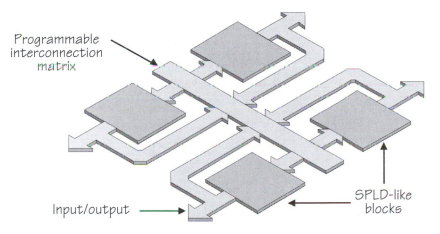

Figure 16-22. Generic CPLD structure

In addition to programming the individual PLD blocks, the connections between the blocks can be configured by means of the *programmable interconnection matrix*. Additional flexibility may be provided with a CPLD whose links are based on SRAM cells. In this case, it may be possible to configure each PLD block to either act in its traditional role or to function as a block of SRAM.

Field-Programmable Gate Arrays (FPGAs)

SPLDs and CPLDs are tremendously useful for a wide variety of tasks, but they are somewhat limited by the structures of their programmable AND and OR planes. At the other end of spectrum are full-blown *application-specific integrated circuits (ASICs).*[15]

Around the middle of the 1980s, a new breed of devices called *Field-Programmable Gate Arrays (FPGAs)* began to appear on the scene. These devices combined many aspects traditionally associated with ASICS (such as high-density) with characteristics traditionally associated with SPLDs and CPLDs (such as the ability to program them in the field).

FPGAs are somewhat difficult to categorize, because each FPGA vendor fields a proprietary architecture; however, a generic architecture that illustrates the sophistication of FPGAs could be represented as shown in Figure 16-23.

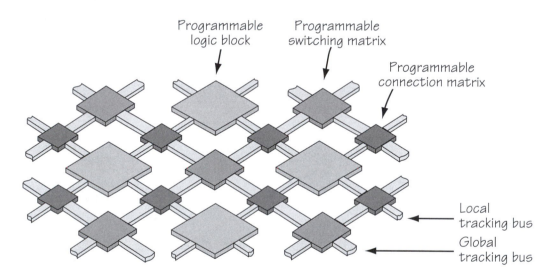

Figure 16-23. Generic field-programmable gate array (FPGA) structure

The device consists of a number of programmable logic blocks, each connected to a number of programmable connection matrices, which are in turn connected to a number of programmable switching matrices. Each programmable logic block may contain a selection of primitive logic gates and register elements. By programming the appropriate links, each logic block can

---

[15] ASICs are introduced in detail in Chapter 17.

be individually configured to provide a variety of combinational and/or sequential logic functions. The programmable connection matrices are used to establish links to the inputs and outputs of the relevant logic blocks, while the programmable switch matrices are used to route signals between the various connection matrices. In short, by programming the appropriate links in the connection and switch matrices, the inputs and outputs of any logic block can be connected to the inputs and outputs of any other logic block.

The majority of FPGAs are based on SRAM programmable switches (some are based on antifuse technology, but these are in the minority). The two predominant architectural variants are to use *look-up tables (LUTs)* or multiplexers. First consider the LUT approach, of which there are two main sub-variants (Figure 16-24)

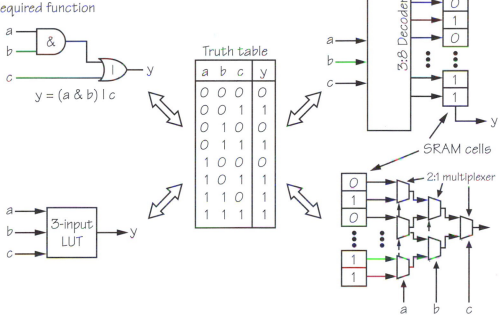

Figure 16-24: Look-up table (LUT)–based FPGAs

In Figure 16-24—which assumes 3-input LUTs[16]—the required function (upper left) is converted into an LUT (lower left) whose contents can be

---

[16] FPGA vendors can spend countless hours debating the advantages and disadvantages of 3-input versus 4-input versus 5-input LUTs (you can only imagine how much fun it is to be cornered by one of their salespersons at a party).

represented using a truth table (center). One technique involves using a 3:8 decoder (upper right) to select one of eight SRAM cells that contain the required truth table output values. Alternatively, the SRAM cells can be used to drive a "pyramid" structure of multiplexers (lower right), where the multiplexers are themselves controlled by the values on the input variables and "funnel down" to generate the output.

As opposed to LUTs, some FPGA architectures are based almost exclusively on multiplexers (Figure 16-25).

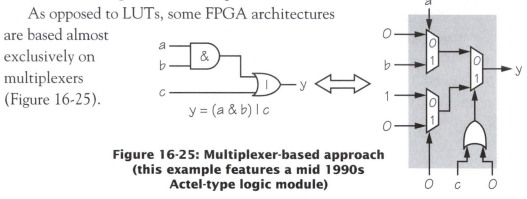

$$y = (a \& b) \mid c$$

Figure 16-25: Multiplexer-based approach (this example features a mid 1990s Actel-type logic module)

There are several considerations to bear in mind when using FPGAs. First, interconnect delays are not as predictable as they are with SPLDs and CPLDs. Second, each vendor employs special fitting software to map designs into their devices, which makes it pretty darn tricky to migrate a design from one vendor to another while maintaining anything like the same propagation delays. Third, the majority of synthesis tools are geared towards fine-grained ASIC architectures, so they output gate-level netlists, but FPGA fitting tools usually do a less than superb job of placement or packing from these netlists. Having said this, there have been tremendous strides in FPGA design tools over the years, and modern versions of these tools work very hard to address these issues.

The original FPGAs contained a significant number of logic gates, but nothing like the little rascals you see these days. For example, by 2002, platform FPGAs from Xilinx (www.xilinx.com) were available with upwards of 10 million equivalent logic gates, which is pretty impressive whichever way you look at it. Furthermore, FPGAs of this caliber don't simply sport simple programmable logic, but may include such things as embedded microprocessor cores, incredibly fast communications functions, large blocks of RAM, large numbers of special-purpose macro-cells such as multipliers and adders, and . . . the list goes on.

Why Use Programmable ICs?

A single programmable device can be used to replace a large number of simpler devices, thereby reducing board space requirements, optimizing system performance by increasing speed and decreasing power consumption, and reducing the cost of the final system. Additionally, replacing multiple devices with one or more programmable devices increases the reliability of the design. Every integrated circuit in a design has to be connected to the circuit board by means of soldered joints, and every joint is a potential failure mechanism.[17] Thus, using programmable devices reduces the number of components, which reduces the number of joints, thereby reducing the chances of a bad or intermittent connection.

Electronic designs typically undergo a number of changes during the course of their development. A design modification may require the addition or removal of a number of logic functions and corresponding changes to the tracks connecting them. However, when a programmable device is used to replace a number of simpler devices, the effects of any design modifications are minimized.

Reprogrammable EPLDs and E^2PLDs are often employed during the initial stages of a design. Once the design has stabilized, these devices may be replaced by cheaper, one-time programmable PLD equivalents. In certain cases involving large production runs, both reprogrammable and one-time programmable devices may eventually be replaced by mask-programmed equivalents, in which the patterns are hard-coded during the construction of the component.

Traditionally, the majority of designers have not availed themselves of the capabilities for in-circuit reprogrammability offered by devices based on the E^2PROM technology. However, the current interest in *design-for-test (DFT)* and the increasing requirement for portable computer and telecommunication equipment are driving an interest in in-circuit reprogrammability. This is especially true in the case of components based on FLASH and SRAM technologies, which allow these devices to be easily and quickly programmed, erased, and reprogrammed while remaining embedded in the middle of a system.

To facilitate the testing of a system, these highly reprogrammable devices can first be configured so as to make internal signals on the circuit board

---

[17] Circuit boards are introduced in more detail in Chapter 18.

available to the test system. After the circuit has been tested, the devices can be erased and reprogrammed with the patterns required to make the board perform its normal function. In some cases, it may make sense to create a single design that can be used to replace a number of different variants. The designers may create one circuit board type that can be reprogrammed with different features for diverse applications.

Portable electronic equipment such as computers can be reprogrammed in the field or remotely by means of a telephone link. Of most interest is the concept of *adaptive hardware*, which can be reprogrammed on-the-fly. For example, state machines may be reprogrammed to follow different sequences, pattern detectors may be reprogrammed to search for different patterns, and computational devices may be reprogrammed with different algorithms. For all of the reasons discussed above, programmable integrated circuits continue to form an extremely significant market segment.

Application-Specific Integrated Circuits (ASICs)

As opposed to a standard "off-the-shelf" integrated circuit whose function is specified by the manufacturer, an *application-specific integrated circuit (ASIC)*[1] is a device whose function is determined by design engineers to satisfy the requirements of a particular application. (The programmable logic devices introduced in the previous chapter fall into both camps; in one respect they are standard devices in that they are mass produced, but they may also be classed as subspecies of ASIC because of the design engineers' ability to customize their functionality.)

In 1967, Fairchild Semiconductor introduced a device called the *Micromosaic*, which contained a few hundred transistors. The key feature of the Micromosaic was that the transistors were not initially connected together. Design engineers used a computer program to specify the function they wished the device to perform. This program determined the necessary interconnections required to link the transistors and generated the masks required to complete the device. Although relatively simple, the Micromosaic is credited as being the forerunner of the modern ASIC and the first real application of *computer-aided design (CAD)*.

A formal classification of ASICs tends to become a bit fluffy around the edges. However, in addition to the plethora of programmable logic devices introduced in Chapter 16, there are generally taken to be three other major categories: *gate array* (including sea-of-gates), *standard cell* (including compiled cells), and *full custom* (which also includes compiled cells) (Figure 17-1).

---

[1] In conversation, ASIC is pronounced "A-SIC," that is, by spelling out the "A" to rhyme with "hay" followed by "SIC" to rhyme with "tick."

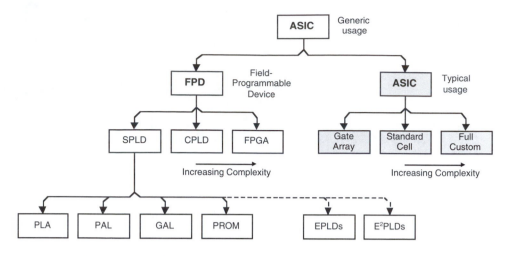

Figure 17-1. Generic and typical usage of the term ASIC

Gate Array Devices

A silicon chip may be considered to consist of two major facets: components such as transistors and resistors, and the tracks connecting them together. In the case of *gate arrays*, the ASIC vendor prefabricates wafers of silicon chips containing unconnected transistors and resistors.[2] Gate arrays are based on the concept of a *basic cell*, which consists of a selection of components; each ASIC vendor determines the numbers and types of components provided in their particular basic cell (Figure 17-2).

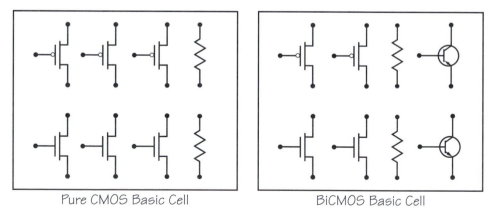

Pure CMOS Basic Cell   BiCMOS Basic Cell

Figure 17-2. Examples of gate array basic cells

---

[2] In the early days, gate arrays were also known as *uncommitted logic arrays (ULAs)*, but this term has largely fallen into disuse.

Channeled Gate Arrays

In the case of *channeled gate arrays*, the basic cells are typically presented as either single-column or dual-column arrays; the free areas between the arrays are known as the *channels* (Figure 17-3). (Note that, although these diagrams feature individual silicon chips, at this stage in the process these chips are still embedded in the middle of a wafer.)

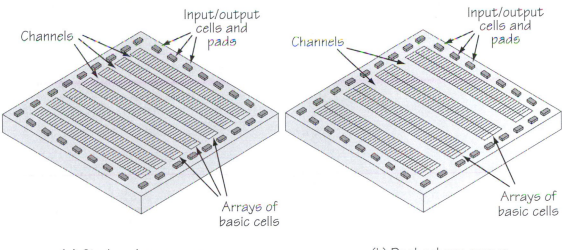

(a) Single-column arrays   (b) Dual-column arrays

Figure 17-3. Channeled gate array architectures

Channel-less Gate Arrays

In the case of *channel-less* or *channel-free* devices, the basic cells are presented as a single large array. The surface of the device is covered in a "sea" of basic cells, and there are no dedicated channels for the interconnections. Thus, these devices are popularly referred to as *sea-of-gates* or *sea-of-cells* (Figure 17-4).

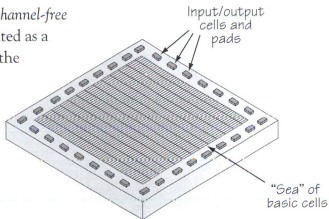

Figure 17-4. Channel-less gate array architecture

Cell Libraries

Although the transistors from one or more basic cells could be connected together to implement practically any digital function, design engineers using gate arrays do not work at the transistor level. Instead, each ASIC vendor selects a set of logic functions such as primitive gates, multiplexers, and registers that they wish to make available to the engineers. The vendor also determines how each of these functions can be implemented using the transistors from one or more basic cells.

A primitive logic gate may only require the transistors from a single basic cell, while a more complex function like a D-type flip-flop or a multiplexer may require the use of several basic cells. Each of these "building block" functions is referred to as a *cell*—not to be confused with a *basic cell*—and the set of functions provided by the ASIC vendor are known collectively as the *cell library*. The number of functions in a typical cell library can range from 50 to 250 or more. The ASIC vendor also provides the design engineers with a cell library data book containing details about each function, including its truth table and timing attributes.

A High-level View of the Design Flow

Aided by a suite of design tools (see also the main *ASIC Design Flow* discussions later in this chapter), design engineers select the cells and the connections between them that are necessary for the gate array to perform its desired function. In the classical ASIC design flow, this information is passed to the ASIC vendor in a computer-readable form, and the vendor uses another suite of tools to complete the device. One of these tools—called the *placer*—assigns the cells selected by the engineers to basic cells on the silicon (Figure 17-5).

Another tool—called the *router*—determines the optimal way to connect the cells together (these tools operate hand-in-hand and are referred to collectively as *place-and-route*). Further tools are used to create the masks required to implement the final metalization layers. These layers are used to connect the transistors in the basic cells to form logic gates, and to connect the logic gates to form the complete device. The channels in channeled devices are used for the tracks that connect the logic gates together. By comparison, in channel-less devices, the connections between logic gates have to be deposited over the top of other basic cells. In the case of early processes based on two layers of

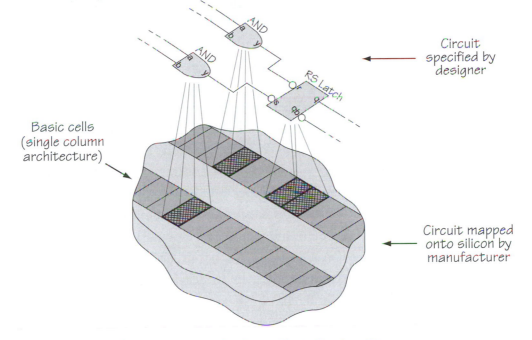

Figure 17-5. Assigning cells to basic cells

metalization, any basic cell overlaid by a track was no longer available to the user. More recent processes, which can potentially have up to eight metalization layers, overcome this problem.

Gate arrays can contain from thousands to millions of equivalent gates,[3] but there is a difference between the number of *available gates* and the number of *usable gates*. Due to limited space in the routing channels of channeled devices, the design engineers may find that they can only actually use between 70% and 90% of the total number of available gates. Sea-of-gates architectures provide significantly more available gates than those of channeled devices because they don't contain any dedicated wiring channels. However, in practice, only about 40% of the gates were usable in devices with two layers of metalization, rising to 60% or 70% in devices with three or four metalization layers, and even higher as the number of metalization layers increases.

Functions in gate array cell libraries are generally fairly simple, ranging from primitive logic gates to the level of registers. The ASIC vendor may also provide libraries of more complex logical elements called *hard-macros* (or *macro-cells*)

---

[3] The concept of *equivalent gates* was introduced in Chapter 14.

and *soft-macros* (or *macro-functions*).[4] In the case of gate arrays, the functions represented by the hard-macros and soft-macros are usually at the MSI and LSI level of complexity, such as comparators, shift-registers, and counters. Hard-macros and soft-macros are both constructed using cells from the cell library. In the case of a hard-macro, the ASIC vendor predetermines how the cells forming the macro will be assigned to the basic cells and how the connections between the basic cells will be realized. By comparison, in the case of a soft-macro, the assignment of cells to basic cells is performed at the same time, and by the same tool, as for the simple cells specified by the design engineers.

Gate arrays are classed as *semi-custom* devices. The definition of a semi-custom device is that it has one or more customizable mask layers, but not all the layers are customizable. Additionally, the design engineers can only utilize the predefined logical functions provided by the ASIC vendor in the cell and macro libraries.

Standard Cell Devices

Standard cell devices bear many similarities to gate arrays. Once again, each ASIC vendor decides which logic functions they will make available to the design engineers. Some vendors supply both gate array and standard cell devices, in which case the majority of the logic functions in the cell libraries will be identical and the main differences will be in their timing attributes.

Standard cell vendors also supply hard-macro and soft-macro libraries, which include elements at the LSI and VLSI level of complexity, such as processors, controllers, and communication functions. Additionally, these macro libraries typically include a selection of RAM and ROM functions, which cannot be implemented efficiently in gate array devices.[5] Last but not least, the design engineers may decide to reuse previously designed functions and/or to purchase blocks of *intellectual property (IP)*[6] (see sidebar).

---

[4] The term *macro* was inherited from the software guys. In software terms, a macro is like a subroutine. So a single computer instruction (to call the macro) initiates a series of additional instructions for the computer to perform. Similarly, in hardware terms, a single macro represents a large number of primitive logic gates.

[5] In the case of a design that contains large functions connected together by relatively small amounts of simple logic, this interfacing logic is referred to as "glue logic."

[6] In conversation, IP is pronounced by spelling it out as "I-P."

A High-level View of the Design Flow

Once again, aided by a suite of design tools, the design engineers determine which elements they wish to use from the cell and macro libraries, and how they require these elements to be connected together (see also the main *ASIC Design Flow* discussions later in this chapter). As before, in the classical ASIC design flow, the design engineers pass this information to the ASIC vendor in a computer-readable form, and the vendor uses another suite of tools to complete the device.

Unlike gate arrays, however, standard cell devices do not use the concept of a basic cell, and no components are prefabricated on the silicon chip. The ASIC vendor creates custom masks for every stage of the device's fabrication. This allows each logic function to be created using the minimum number of transistors required to implement that function with no redundant components. Additionally, the cells and macro functions can be located anywhere on the chip, there are no dedicated interconnection areas, and the functions are placed so as to facilitate any connections made between them. Standard cell devices therefore provide a closer-to-optimal utilization of the silicon than do gate arrays (Figure 17-6).

> There are several different flavors of intellectual property (IP). In the generic sense, if you have a capriciously cunning idea, then that idea is your intellectual property and no one else can use it unless they pay you for it.
>
> When a team of electronics engineers is tasked with designing a complex integrated circuit, rather than "reinvent the wheel," they may decide to purchase the plans for one or more functional blocks that have already been created by someone else. The plans for these functional blocks are known as *intellectual property (IP)*.
>
> IP blocks can range all the way up to sophisticated communications functions and microprocessors. The more complex functions—like microprocessors—may be referred to as "cores."

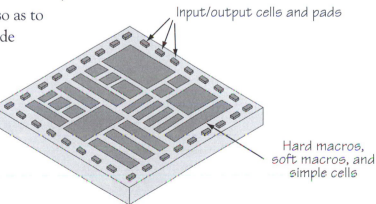

Input/output cells and pads

Hard macros, soft macros, and simple cells

Figure 17-6. Standard cell architecture

Compiled Cell Technology

Design engineers working with standard cell devices may also make use of *compiled cell technology*. In this case, the engineers specify a desired function in such terms as Boolean equations or state tables, and then employ a tool called a *silicon compiler* to generate the masks required to implement the transistors and interconnections. These tools can generate highly efficient structures, because they have the ability to vary the size of individual transistors to achieve the optimum tradeoff between speed requirements and power consumption. Silicon compilers are also useful in the creation of selected macro functions such as highly structured RAMs and ROMs. The user simply specifies the required widths and depths of such functions, leaving the silicon compiler to determine an optimal implementation.

Compiled cell technology was originally intended to provide a mechanism for the creation of entire devices. This promise was never fully realized, and the technology is now considered to be an adjunct to standard cell devices, allowing the design engineers to create custom cells and macros to augment the libraries supplied by the ASIC vendor.

From the ASIC vendor's point of view, standard cell devices may be considered to be *full custom* on the basis that all the mask layers are customized. From the design engineer's point of view, however, these devices are still considered to be *semi-custom* because, with the exception of compiled cells, the engineers can only utilize the predefined functions provided by the ASIC vendor in the cell and macro libraries.

Full Custom Devices

In the case of *full custom* devices, the design engineers have complete control over every mask layer used to fabricate the silicon chip. The ASIC vendor does not prefabricate any components on the silicon and does not provide any cell libraries. With the aid of appropriate design tools, engineers can "handcraft" the dimensions of individual transistors, and then create higher-level functions based on these components. For example, if the engineers require a slightly faster logic gate, they can alter the dimensions of the transistors used to build that gate. The design tools used for full custom devices are often created in-house.

Full custom devices may employ compiled cell technology and may also include analog circuitry such as comparators, amplifiers, filters, and digital-to-analog and analog-to-digital converters. The design of full custom devices is highly complex and time consuming, but the resulting chips contain the maximum amount of logic with minimal waste of silicon real estate.

Input/Output Cells and Pads

Around the periphery of the silicon chip are power and signal pads used to interface the device to the outside world. The cell library data books for gate array and standard cell devices include a set of functions known as *input/output (I/O) cells.* The signal pads contain a selection of transistors, resistors, and diodes necessary to implement *input*, *output*, or *bi-directional* buffers, and the design engineers can decide how each pad will be configured. The masks and metalization used to interconnect the internal logic are also used to configure the components in the input/output cells and to connect the internal logic to these cells (Figure 17-7).

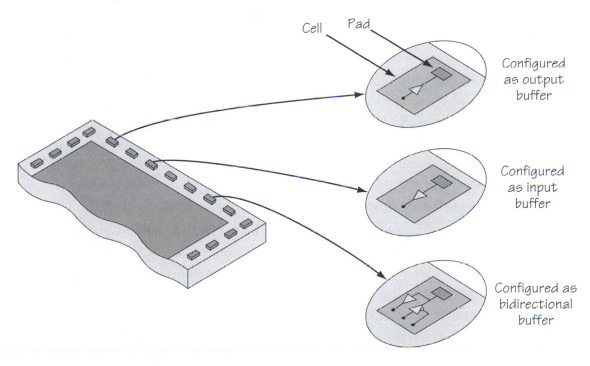

Figure 17-7. Input/output (I/O) cells and pads

The ASIC vendor may permit the design engineers to individually specify whether each input/output cell should present CMOS, TTL, or ECL characteristics to the outside world. The input/output cells also contain any circuitry required to provide protection against *electrostatic discharge* (ESD). After the metalization layers have been added, the chips are separated and packaged using the same techniques as for standard integrated circuits.

Who Are All the Players?

In a little while we're going to look at the ASIC design flow, but before that we need to have some understanding as to who all of the players are in this complicated game. As you will soon come to realize, understanding the technology is the easy part—it's when you try to work out who does what that things start to get hairy.

First of all we have the folks who create the tools (software programs) that the engineers use to design integrated circuits, circuit boards, and electronic systems. Prior to the 1970s, electronic circuits were handcrafted. Circuit diagrams (known as *schematics*) showing symbols for the components to be used and the connections between them were drawn using pen, paper, and stencils. Similarly, the copper tracks on a circuit board were drawn using red and blue pencils to represent the top and bottom of the board. Any form of analysis (for example, *"What frequency will this oscillator run at if I use this capacitor and this resistor?"*) was performed with pencil, paper, and a slide rule (or a mechanical calculator if you were lucky). Not surprisingly, this style of design was time-consuming, expensive, and prone to error.

Computer-aided Design (CAD)

As electronic designs and devices grew more complex, it became necessary to develop automated techniques to aid in the design process. In the early 1970s, companies like Calma, ComputerVision, and Applicon created special computer programs that helped personnel in the drafting department[7] capture hand-drawn designs in digital form using large-scale digitizing tables.

Over time, these early computer-aided drafting tools evolved into interactive programs that performed integrated circuit layout (that is, they could be

---

[7] The drafting department is referred to as the "drawing office" in the UK.

used to describe the locations of the transistors forming the integrated circuit and the connections between them). Other companies like Racal-Redac, SCI-Cards, and Telesis created equivalent layout programs for printed circuit boards. These integrated circuit and circuit board layout programs became known as *computer-aided design (CAD)* tools.[8,9]

Computer-aided Engineering (CAE)

Also in the late 1960s and early 1970s, a number of universities and commercial companies started to develop computer programs known as *simulators.* These programs allowed students and engineers to emulate the operation of an electronic circuit without actually having to build it first. Perhaps the most famous of the early simulators was the *simulation program with integrated circuit emphasis (SPICE).*[10] This was developed by the University of California in Berkeley and was made available for widespread use around the beginning of the 1970s. SPICE was designed to simulate the behavior of analog circuits—other programs called logic simulators were developed to simulate the behavior of digital circuits.

Around the beginning of the 1980s, companies like Daisy, Mentor, and Valid spawned computer programs that allowed engineers to capture schematic (circuit) diagrams on the computer screen. These tools could then be used to generate textual representations of the circuits called *netlists* that described the components to be used and the connections between them. In turn, these netlists could be used to drive analog and digital simulators (and eventually layout tools).

The companies promoting front-end tools for schematic capture and simulation classed them as *computer-aided engineering (CAE).*[11] This was based on the fact that these tools were targeted toward design engineers, and the CAE companies wished to distinguish their products from the CAD tools that were originally used by the drafting department.

---

[8] In conversation, CAD is pronounced as a single word to rhyme with "bad."

[9] The term CAD is also used to refer to computer-aided design tools intended for a variety of other engineering disciplines, such as mechanical and architectural design.

[10] In conversation, SPICE is pronounced like the seasoning to rhyme with "mice."

[11] In conversation, CAE is spelled out as "C-A-E".

Designers versus Engineers

If you say things the wrong way when talking to someone in the industry, you immediately brand yourself as an outsider (one of "them" instead of one of "us"). For historical reasons that are based on the origins of the terms CAD and CAE, the term *layout designer* or simply *designer* is typically used to refer to someone who lays out a circuit board or integrated circuit (determines the locations of the components and the routes of the tracks connecting them together). By comparison, the term *design engineer* or simply *engineer* is typically used to refer to someone who conceives and describes the functionality of an integrated circuit, printed circuit board, or electronic system (what it does and how it does it).

Electronic Design Automation (EDA)

Sometime during the 1980s, all of the CAE and CAD tools used to help design electronic components and systems came to be referred to by the "umbrella" name of *electronic design automation (EDA)*,[12] and everyone was happy (apart from the ones who weren't, but they don't count).

Some of the Key Players

This is where things start to get interesting. Let's start with the *system house*, A block in the middle of Figure 17-8. These are the folks who design and build the system-level products (from cell phones to televisions to computers) that eventually wend their way into our hands.

When these system house folks are creating a new product, they may decide it requires one or more ASICs, which they also design. Once they've created the design for an ASIC, they may pass it over to an ASIC vendor to be implemented and fabricated (see also the *ASIC Design Flow* discussions below). In this case, the ASIC vendor is in charge of creating the masks and constructing and packaging the final devices.

As opposed to a full-line ASIC vendor, a *fabless semiconductor company* is one that designs and sells integrated circuits, but does not have the ability to manufacture them. By comparison, a *foundry* is a company that manufactures integrated circuits, but doesn't actually do any designs of their own (these are also known as "fabs" because they fabricate the integrated circuits).

---

[12] In conversation, EDA is spelled out as "E-D-A".

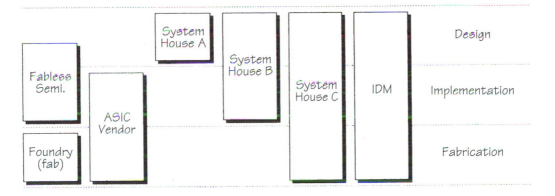

Figure 17-8. Some of the key players

Initially we described a system house as passing its ASIC designs over to an ASIC vendor for implementation and fabrication. However, some system houses (B) perform both the design and implementation, and then hand the masks over to a foundry for fabrication. Alternatively, a system house (C) may have its own fab capability, in which case it will perform the entire process—design, implementation, and fabrication—"in-house."

And just to confuse the issue even further, we have *integrated device manufacturers (IDMs)*.[13] These are companies that are very similar to system houses, except that IDMs focus on designing, manufacturing, and selling integrated circuits as opposed to complete electronic systems.

The ASIC Design Flow

Before we start this portion of our discussions, we should note that some integrated circuits contain only digital functions, others are analog in nature, and some contain both analog and digital elements (these latter components are referred to as *mixed-signal* devices). For the purpose of these discussions, we'll concentrate on a digital ASIC design flow. The full-up design flow is quite complex and is beyond the scope of this book, so we'll only consider a much-simplified version of the flow here (Figure 17-9).[14]

---

[13] In conversation, IDM is spelled out as "I-D-M".

[14] The book *EDA: Where Electronics Begins* (www.techbites.com/eda), which was co-authored by the author of this book, introduces design flows and tools (*capture, simulation, synthesis, verification, layout*, etc.), for integrated circuits, printed circuit boards, and electronic systems. Don't worry; *EDA: Where Electronics Begins* is presented at a nontechnical level that won't make your brains leak out of your ears.

Verilog started life as a commercial (proprietary) simulation language. However, in the mid 1990s it became an open standard under the auspices of the Open Verilog International (OVI) organization.

Similarly, VHDL, which came out of the American Department of Defense (DoD), has also evolved into an open standard. VHDL is an acronym for *VHSIC HDL* (where VHSIC is itself an acronym for *very high speed integrated circuit*).

VHDL was originally intended to provide a way of documenting designs in such a way that the design intent could be preserved for a long time. This was important for military projects with life spans in terms of decades. However, when VHDL was first conceived, little thought was given to using it to drive synthesis and simulation tools (which may explain why doing so can be such a painful experience).

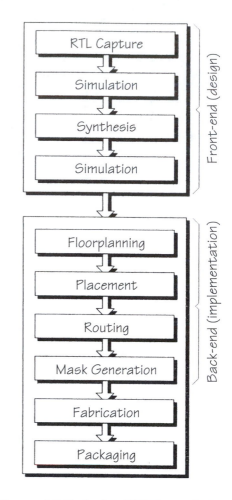

Figure 17-9. A simplified version of the ASIC design flow

Capturing the Design

First of all, the design engineers have to decide exactly what logic gates and functions they wish to use in their integrated circuit and how they are to be connected together. In the early days of integrated circuit design, schematic (circuit) diagrams were hand-drawn by the engineers using pencil and paper. Over time, schematic capture programs were developed that allowed engineers to place symbols representing components on a computer screen and then draw the wires connecting the symbols together.

However, today's digital integrated circuits can contain tens of millions of logic gates, and it simply isn't possible to capture and manage designs of this complexity at the schematic level. Thus, as opposed to using schematics, the functionality of a high-end integrated circuit is now captured using a textual *hardware description language (HDL)*.[15] The two most popular HDLs for digital designs are Verilog and VHDL.[16,17]

Initially, the design engineers describe the device's functionality at a high level of abstraction called *register transfer level (RTL)*. For example, assuming that we had already declared an input signal called *clock* and three 8-bit registers called a[7:0], b[7:0], and y[7:0], an RTL statement might look something like the following:

WHEN clock RISES
    y[7:0] = a[7:0] & b[7:0];
END;

When capturing the design's functionality, the design engineers may decide to reuse RTL descriptions of functional blocks from previous designs. Also, they may decide to purchase *intellectual property (IP)* blocks from a third-party design house.

Simulation and Synthesis

Once the RTL has been captured, the engineers use a computer program called a *simulator*, which reads in the RTL and creates a virtual representation of the integrated circuit in the computer's memory. The engineers use the simulator to apply a sequence of test patterns to the simulation model's virtual inputs and check the responses on its virtual outputs to ensure that the device will function as planned (this sequence of patterns is called a *testbench*). The simulator models how signals propagate through—and are processed by—the high-level logic functions forming the integrated circuit.

Next, a *logic synthesis* program is used to automatically convert the high-level RTL representation into equivalent Boolean Equations (like the ones we discussed in Chapter 9). The synthesis tool automatically performs simplifications and minimizations, and eventually outputs a gate-level netlist (Figure 17-10).

---

[15] In conversation, HDL pronounced by spelling it out as "H-D-L."

[16] In conversation, VHDL pronounced by spelling it out as "V-H-D-L."

[17] In the 1980s, engineers in Japan came up with their own HDL called UDL/I. This was a very useful language that was designed with synthesis and simulation in mind. However, Verilog and VHDL already held the high ground, and interest in UDL/I is now almost nonexistent even in Japan.

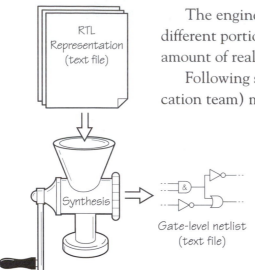

The engineers can direct the synthesis tool to optimize different portions of designs for area (to use the smallest amount of real-estate on the silicon) or for speed.

Following synthesis, the engineers (or a separate verification team) may re-simulate the design at the gate level using the same testbench as before. This simulation is performed to ensure that the synthesis tool didn't inadvertently change the functionality of the design. As opposed to simulation—or in addition to it—the engineers may also run a *formal verification* program that compares the RTL and gate-level views of the design to ensure that they are functionally equivalent (synthesis tools have been known to make mistakes).

Figure 17-10. A synthesis program converts an RTL description into a gate-level netlist

Floorplanning

Traditionally, once the design engineers had generated a gate-level netlist, they handed it over to the layout designers to perform *floorplanning*, *placement*, and *routing*. This may still be the case for some design flows. In the case of the majority of today's high-performance digital integrated circuits, however, the design engineers typically perform these functions also. For the purposes of these discussions, we'll assume that the design engineers undertake all of the tasks associated with designing the integrated circuit.

Floorplanning[18] involves creating a high-level representation showing where the major functional blocks will be placed on the surface of the silicon chip. Depending on the particular design flow, the floorplan may be created by hand, interactively, or automatically. The creator of the floorplan tries to ensure that blocks that communicate with each other are positioned as close to each other as possible. Also, it may be preferable to place certain blocks near to the periphery of the chip, while other blocks may be better situated toward the center.

---

[18] The term "floorplan" comes from the architectural and construction industries, because the graphical floorplan for an integrated circuit looks similar to a floorplan showing the rooms and corridors in a building.

A gate-level netlist is a textual representation that describes the logic gates to be used and the connections between them. For example, consider a very simple circuit called "test" comprising just two logic gates:

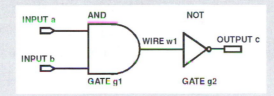

A simple textual netlist representation of this circuit could be as follows:

    START CIRCUIT NAME=test;
    INPUTS = a, b;
    OUTPUTS = c;
    INTERNAL WIRES = w1;
    GATE=g1, TYPE=AND, INPUT=a, INPUT=b, OUTPUT=w1;
    GATE=g2, TYPE=NOT, INPUT=w1, OUTPUT=c;
    END CIRCUIT NAME=test;

Such a netlist can be used to drive simulation and layout tools. There are a variety of different netlist formats, such as the industry-standard *electronic design interchange format (EDIF)*. However, integrated circuit designers often work with netlists in *VHDL* or *Verilog*.

Placement and Routing

Next, the design engineer uses a placement program to determine the optimum locations for the individual logic gates forming the major functions. This program tries to ensure that gates that communicate with each other are placed as close to each other as possible. Also, the program may decide that it's preferable to place certain gates near to the periphery of a block, while other gates may be better situated toward its center.

Once all of the logic gates have been placed, the engineer uses a routing program to determine the optimum paths for the tracks linking the logic gates together. Somewhere in the flow, the engineer will also use special programs to design and route the chip's clock signals and power supplies.

Mask Generation

Throughout the entire process, the engineers will use a variety of complex verification and timing analysis programs to ensure that the device will work as planned. Eventually, when everyone is happy, the computer will take the placed and routed design and generate all of the masks that will be required to actually fabricate the device. These masks, which are represented in a machine-readable format known as GDSII,[19,20] will be used by the foundry to fabricate the device.

ASIC, ASSP, and COT

This is where things start to get really "fluffy" around the edges when you're trying to explain them to someone, but we'll take a stab at trying to put everything into some sort of context. Throughout this chapter we've referred to ASICs as being all-encompassing. In reality, an ASIC is generally considered to be a component that is designed by and/or used by a single company in a specific system (for example, a telecommunications system house may design an ASIC for use in their new cell phone). By comparison, an *application-specific standard part (ASSP)*[21] is a more general-purpose device targeted at a specific market (say the communications market) and intended for use by multiple design houses.

It's important to note that there are a number of points at which the team designing the integrated circuit may decide to hand over control to the ASIC vendor or foundry. In some (rare) cases, the design team may create only the high-level RTL description and then pass this to the ASIC vendor to take the design through synthesis, floorplanning, and layout (place-and-route) to a physical implementation. In this case, the ASIC vendor will charge a lot of money for its services.

A traditional ASIC design flow has the design engineers performing RTL capture, synthesis, and simulation (this may be referred to as the *front-end*

---

[19] In conversation, GDSII is pronounced by spelling it out as "G-D-S-2."

[20] Officially, GDSII doesn't actually stand for anything in particular. Unofficially, it's rumored to be an acronym for *Graphics Design Station Two*, because GDSII was the output format from the second generation of design stations from Calma.

[21] In conversation, ASSP is pronounced by spelling it out as "A-S-S-P."

portion of the flow). The design engineers then hand over the resulting gate-level netlist to the ASIC vendor's layout designers, who perform floorplanning and layout and progress the design to its final implementation (this may be referred to as the *back-end* portion of the flow).

In contrast, in the case of today's high-end ASIC and ASSP devices, the design engineers are typically in charge of capturing the RTL and performing simulation, synthesis, floorplanning, layout, etc. The resulting placed-and-routed design is then passed to the ASIC vendor, whose task is largely reduced to generating the masks used to create the chips and then manufacturing the chips themselves.

There is also the *customer-owned tooling (COT)*[22] business model, in which the design engineers may take the design all the way through to generating the masks used to create the chips. In this case, the only task remaining is for a foundry to actually fabricate the chips.

Non-recurring engineering (NRE)

One important concept that people in the electronics industry often talk about is that of *non-recurring engineering (NRE)*,[23] which refers to the costs associated with developing an ASIC or ASSP. The NRE depends on a number of factors, including the complexity of the design, the style of packaging, and who does what in the design flow (that is, how the various tasks are divided between the design house and the semiconductor vendor).

Summary

Electronic designs typically contain a selection of highly complex functions—such as processors, controllers, and memory—interfaced to each other by a plethora of simple functions such as primitive logic gates. The simple interfacing functions are often referred to as the *glue logic*. Both complex and simple functions are available as standard integrated circuits. Devices containing small numbers of simple logic functions (for example, four 2-input AND gates) are sometimes referred to as "jellybean" parts. When glue logic is implemented

---

[22] In conversation, COT is pronounced by spelling it out as "C-O-T" (people in the know do NOT pronounce COT to rhyme with "hot").

[23] In conversation, NRE is pronounced by spelling it out as "N-R-E."

using jellybean parts, these components may require a disproportionate amount of the circuit board's real estate. A few complex devices containing the vast majority of the design's logic gates may occupy a small area, while a large number of jellybean devices containing relatively few logic gates can occupy the bulk of the board's surface.

Early ASICs contained only a few hundred logic gates. These were mainly used to implement glue logic, and it was possible to replace fifty or more jellybean devices with a single ASIC. This greatly reduced the size of circuit boards, increased speed and reliability, and reduced power consumption. As time progressed, application-specific devices with tens of millions of gates and thousands of pins became available. Thus, today's ASICs can be used to implement the most complex functions.

Both gate array and standard cell devices essentially consist of building blocks designed and characterized by the manufacturer and connected together by the designer. Gate arrays are mask programmable with a predefined number of transistors, and different designs are essentially just changes in the inter-connect. This means that gate arrays require the customization of fewer layers than standard cell devices. Gate arrays are therefore faster to implement than their standard cell equivalents, but the latter can contain significantly more logic gates.

In certain respects gate arrays are moving towards having similar capabilities to those of standard cells, and some support complex functions such as memory and processor cores. The design engineers generally know the memory and processing requirements in advance of the rest of the logic, and the gate array manufacturer may supply devices with pre-built processor and memory functions surrounded by arrays of basic cells. Some gate array devices support simple analog cells in addition to the digital cells, and standard cell devices can be constructed with complex analog functions if required.

Circuit Boards

Electronic components are rarely useful in isolation, and it is usually necessary to connect a number of them together in order to achieve a desired effect. Early electronic circuits were constructed using discrete (individually packaged) components such as transistors, resistors, capacitors, and diodes. These were mounted on a non-conducting board and connected using individual pieces of insulated copper wire. The thankless task of wiring the boards by hand was time-consuming, boring, prone to errors, and expensive.

The First Circuit Boards

The great American inventor Thomas Alva Edison had some ideas about connecting electronic circuits together. In a note to Frank Sprague, founder of Sprague Electric, Edison outlined several concepts for printing additive traces on an insulating base. He even talked about the possibility of using conductive inks (it was many decades before this technology—which is introduced later in this chapter—came to fruition).

In 1903, Albert Hanson (a Berliner living in London) obtained a British patent for a number of processes for forming electrical conductors on an insulating base material. One of these described a technique for cutting or stamping traces out of copper foil and then sticking them to the base. Hanson also came up with the idea of double-sided boards and through-holes (which were selectively connected by wires).

In 1913, Arthur Berry filed a British patent for covering a substrate with a layer of copper and selectively etching parts of it away to leave tracks. In another British patent issued in 1925, Charles Ducas described etching, plating up, and even multi-layer circuit boards (including the means of interconnecting the layers). For the next few decades, however, it was easier and cheaper to wire boards manually. The real push into circuit boards only came with the invention of the transistor and later the integrated circuit.

PCBs and PWBs

By the 1950s, the interconnection technology now known as *printed wire boards (PWBs)* or *printed circuit boards (PCBs)* had gained commercial acceptance. Both terms are synonymous, but the former is more commonly used in America, while the latter is predominantly used in Europe and Asia. These circuit boards are often referred to as *laminates* because they are constructed from thin layers or sheets. In the case of the simpler boards, an insulating base layer has conducting tracks formed on one or both sides. The base layer may technically be referred to as the *substrate*, but this term is rarely used in the circuit board world.[1]

The original board material was *Bakelite*, but modern boards are predominantly made from woven glass fibers which are bonded together with an epoxy. The board is cured using a combination of temperature and pressure, which causes the glass fibers to melt and bond together, thereby giving the board strength and rigidity. These materials may also be referred to as *organic substrates*, because epoxies are based on carbon compounds as are living creatures. The most commonly used board material of this type is known as FR4, where the first two characters stand for flame retardant, and you can count the number of people who know what the "4" stands for on the fingers of one hand.

To provide a sense of scale, a fairly representative board might be 15 cm × 20 cm in area and in the region of 1.5 mm to 2.0 mm thick, but they can range in size from 2 cm × 2 cm or smaller (and thinner) to 50 cm × 50 cm or larger (and thicker).[2]

Subtractive Processes

In a subtractive process, a thin layer of copper foil in the order of 0.02 mm thick is bonded to the surface of the board. The copper's surface is coated with an *organic resist*, which is cured in an oven to form an impervious layer (Figure 18-1).

---

[1] See also the glossary definition of "substrate."

[2] These dimensions are not intended to imply that circuit boards must be square or even rectangular. In fact, circuit boards may be constructed with whatever outline is necessary to meet the requirements of a particular enclosure: for example, the shape of a car dashboard.

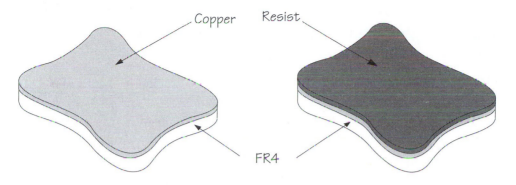

**Figure 18-1. Subtractive process:
applying resist to copper-clad board**

Next, an optical mask is created with areas that are either transparent or opaque to ultraviolet light. The mask is usually the same size as the board, and can contain hundreds of thousands of fine lines and geometric shapes.

The mask is placed over the surface of the board, which is then exposed to ultraviolet (UV) light. This ionizing radiation passes through the transparent areas of the mask to break down the molecular structure of the resist. After the board has been exposed, it is bathed in an organic solvent, which dissolves the degraded resist. Thus, the pattern on the mask has been transferred to a corresponding pattern in the resist (Figure 18-2).

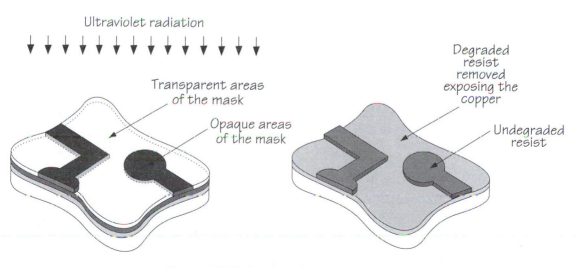

**Figure 18-2. Subtractive process:
degrading the resist and exposing the copper**

A process in which ultraviolet light passing through the transparent areas of the mask causes the resist to be degraded is known as a *positive-resist* process; *negative-resist* processes are also available.[3] The following discussions assume positive-resist processes unless otherwise noted.

After the unwanted resist has been removed, the board is placed in a bath containing a cocktail based on sulfuric acid, which is agitated and aerated to make it more active. The sulfuric acid dissolves any exposed copper not protected by the resist in a process known as *etching*. The board is then washed to remove the remaining resist. Thus, the pattern in the mask has now been transferred to a corresponding pattern in the copper (Figure 18-3).

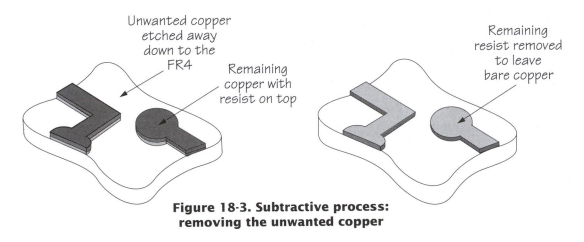

Figure 18-3. Subtractive process: removing the unwanted copper

This type of process is classed as *subtractive* because the board is first covered with the conductor and any unwanted material is then removed, or subtracted. As a point of interest, much of this core technology predates the modern electronics industry. The process of copper etching was well known by the printing industry in the 1800s, and opto-lithographic techniques involving organic resists were used to create printing plates as early as the 1920s. These existing processes were readily adopted by the fledgling electronics industry. As process technologies improved, it became possible to achieve ever-finer features. By 2002, a large proportion of boards were using *lines* and *spaces* of

---

[3] In a *negative-resist* process the ultraviolet radiation passing through the transparent areas of the mask is used to cure the resist. The remaining uncured areas are then removed using an appropriate solvent. Thus, a mask used in a negative-resist process is the photographic negative of one used in a positive-resist process to achieve the same effect; that is, the transparent areas are now opaque and vice versa.

5 mils or 4 mils, with some as small as 3 mils (one mil is one thousandth of an inch). In this context, the term "lines" refers to the widths of the tracks, while "spaces" refers to the gaps between adjacent tracks.

Additive Processes

An additive process does not involve any copper foil being bonded to the board. Instead, the coating of organic resist is applied directly to the board's surface (Figure 18-4).

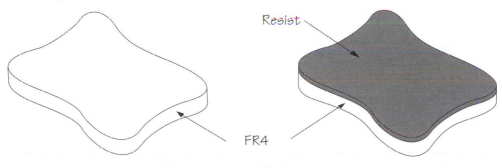

**Figure 18-4. Additive process:
applying resist to bare board**

Once again the resist is cured, an optical mask is created, ultraviolet light is passed through the mask to break down the resist, and the board is bathed in an organic solvent to dissolve the degraded resist (Figure 18-5).[4]

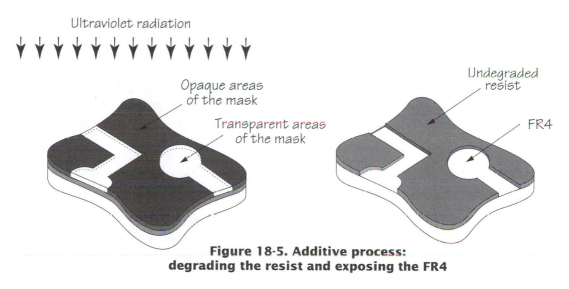

**Figure 18-5. Additive process:
degrading the resist and exposing the FR4**

---

[4] Note that a mask used in an additive process is the photographic negative of one used in a subtractive process to achieve the same effect; that is, the transparent areas are now opaque and vice versa.

After the unwanted resist has been removed, the board is placed in a bath containing a cocktail based on copper sulfate where it undergoes a process known as *electroless plating*. Tiny crystals of copper grow on the exposed areas of the board to form copper tracks. The board is then washed in an appropriate solvent to remove the remaining resist (Figure 18-6).

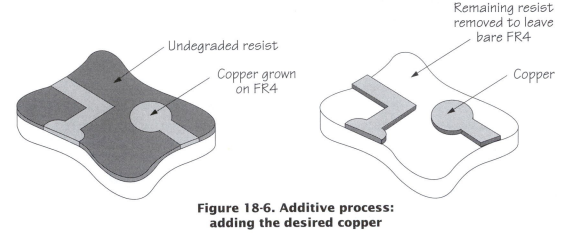

**Figure 18-6. Additive process:
adding the desired copper**

A process of this type is classed as *additive* because the conducting material is only grown on, or added to, specific areas of the board. Additive processes are increasing in popularity because they require less processing and result in less wasted material. Additionally, fine tracks can be grown more accurately in additive processes than they can be etched in their subtractive counterparts. These processes are of particular interest for high-speed designs and microwave applications, in which conductor thicknesses and controlled impedances are critical. Groups of tracks, individual tracks, or portions of tracks can be built up to precise thicknesses by iterating the process multiple times with selective masking.

Single-sided Boards

It probably comes as no great surprise to find that *single-sided* boards have tracks on only one side. These tracks, which may be created using either subtractive or additive processes, are terminated with areas of copper known as *pads*. The shape of the pads and other features are, to some extent, dictated by the method used to attach components to the board. Initially, these discussions will assume that the components are to be attached using a technique known as *through-hole*, which is described in more detail below. The pads associated with the through-hole technique are typically circular, and holes are drilled through both the pads and the board using a computer-controlled drilling machine (Figure 18-7).[5]

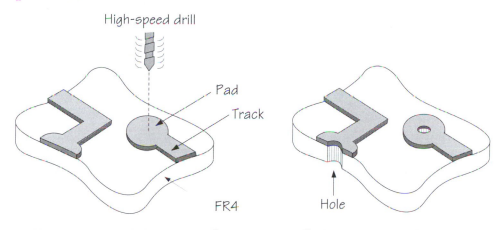

**Figure 18-7. Single-sided boards:
drilling the holes**

Once the holes have been drilled, an electroless plating process referred to as *tinning* is used to coat the tracks and pads with a layer of *tin-lead* alloy. This alloy is used to prevent the copper from oxidizing and provides protection against contamination. The deposited alloy has a rough surface composed of vertical crystals called *spicules*, which, when viewed under a microscope, resemble a bed of nails. To prevent oxygen from reaching the copper through pinholes in the alloy, the board is placed in a *reflow oven* where it is heated by either *infrared (IR)* radiation or hot air. The reflow oven causes the alloy to melt and form a smooth surface (Figure 18-8).

After the board has cooled, a layer known as the *solder mask* is applied to the surface carrying the tracks (the purpose of this layer is discussed in the next section). One common technique is for the solder mask to be screen printed onto the board. In this case, the screen used to print the mask has patterns that leave the areas around the pads exposed, and the mask is then cured in an

---

[5] This is usually referred to as an *NC drilling machine*, where NC stands for "Numerically Controlled."

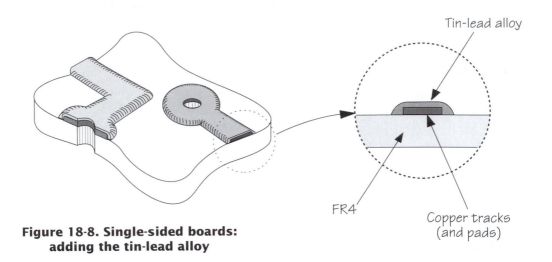

**Figure 18-8. Single-sided boards:
adding the tin-lead alloy**

oven. In an alternative technique, the solder mask is applied across the entire surface of the board as a film with an adhesive on one side. In this case, a further opto-lithographic stage is used to cure the film with ultraviolet light. The optical mask used in this process contains opaque patterns, which prevent the areas around the pads from being cured; these areas are then removed using an appropriate solvent (Figure 18-9).

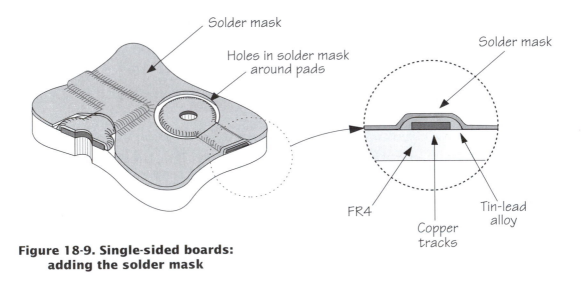

**Figure 18-9. Single-sided boards:
adding the solder mask**

Beware! Although there are relatively few core processes used to manufacture circuit boards, there are almost endless variations and techniques. For example, the tracks and pads may be created before the holes are drilled or vice versa. Similarly, the tin-lead alloy may be applied before the solder mask or vice versa. This latter case, known as *solder mask over bare copper (SMOBC)*, prevents solder from leaking under the mask when the tin-lead alloy melts during the process of attaching components to the board. Thus, the tin-lead alloy is applied only to any areas of copper that are left exposed by the solder mask, such as the pads at the end of the tracks. As there are so many variations and techniques, these discussions can only hope to offer an overview of the main concepts.

Lead Through-Hole (LTH)

Prior to the early 1980s, the majority of integrated circuits were supplied in packages that were attached to a circuit board by inserting their leads through holes drilled in the board. This technique, which is still widely used, is known as *lead through-hole (LTH)*, *plated through-hole (PTH)*, or, more concisely, *through-hole*. In the case of a single-sided board, any components that are attached to the board in this fashion are mounted on the opposite side to the tracks. This means that any masks used to form the tracks are actually created as mirror-images of the required patterns (Figure 18-10).

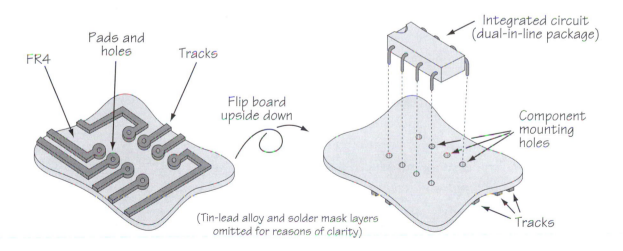

Figure 18-10. Lead through-hole (LTH)

The act of attaching components is known as *populating* the board, and the area of the board occupied by a component is known as its *footprint*. Early manufacturing processes required the boards to be populated by hand, but modern processes make use of automatic insertion machines. Similarly, component leads used to be hand-soldered to the pads, but most modern processes employ automatic *wave-soldering* machines (Figure 18-11).

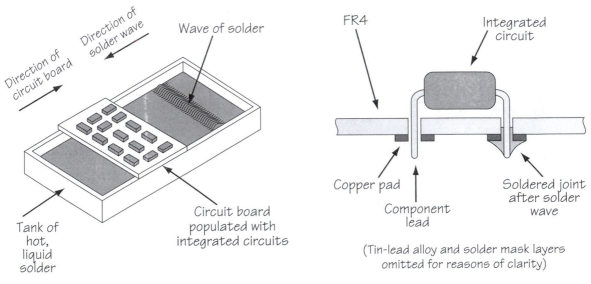

Figure 18-11. Wave-soldering

A wave-soldering machine is based on a tank containing hot, liquid *solder*.[6] The machine creates a wave (actually a large ripple) of solder which travels across the surface of the tank. Circuit boards populated using the through-hole technique are passed over the machine on a conveyer belt. The system is carefully controlled and synchronized such that the solder wave brushes across the bottom of the board only once.

The solder mask introduced in the previous section prevents the solder from sticking to anything except the exposed pads and component leads. Because the solder is restricted to the area of the pads, surface tension causes it to form

---

[6] An alloy of tin and lead with a relatively low melting point.

good joints between the pads and the component leads. Additionally, capillary action causes the solder to be drawn up the hole, thereby forming reliable, low-resistance connections. If the solder mask were omitted, the solder would run down the tracks away from the component leads. In addition to forming bad joints, the amount of heat absorbed by the tracks would cause them to separate from the board (this is not considered to be a good thing to happen).

Surface Mount Technology (SMT)

In the early 1980s, new techniques for packaging integrated circuits and populating boards began to appear. One of the more popular is *surface mount technology (SMT)*, in which component leads are attached directly to pads on the surface of the board. Components with packages and lead shapes suitable for this technology are known as *surface mount devices (SMDs)*. One example of a package that achieves a high lead count in a small area is the *quad flat pack (QFP)*, in which leads are present on all four sides of a thin square package.[7]

Boards populated with surface mount devices are fabricated in much the same way as their through-hole equivalents (except that the pads used for attaching the components are typically square or rectangular and do not have any holes drilled through them).[8] However, the processes begin to diverge after the solder mask and tin-lead plating layers have been applied. A layer of solder paste is screen-printed onto the pads, and the board is populated by an automatic *pick-and-place machine*, which pushes the component leads into the paste (Figure 18-12).

Thus, in the case of a single-sided board, the components are mounted on the same side as the tracks. When all of the components have been attached, the solder paste is melted to form good conducting bonds between their leads and the board's pads. The solder paste can be melted by placing the board in a *reflow oven* where it is heated by infrared (IR) radiation or hot air. Alternatively, the solder may be melted using *vapor-phase soldering*, in which the board is lowered into the vapor-cloud above a tank containing boiling hydrocarbons.

---

[7] Other packaging styles such as *pad grid arrays (PGAs)* and *ball grid arrays (BGAs)*—which are also suitable for surface mount technology—are introduced in more detail in Chapter 20.

[8] Actually, the pads may have holes in the case of the microvia technologies discussed later in this chapter.

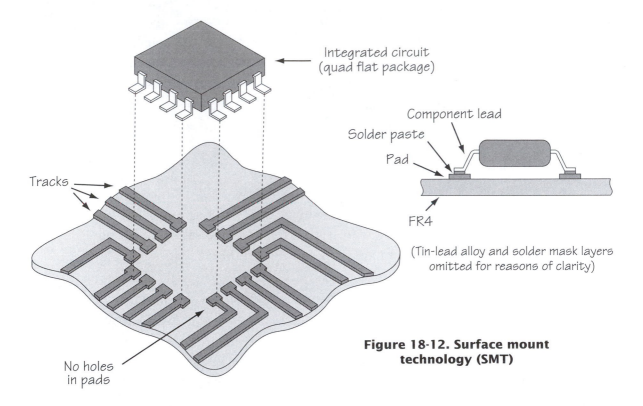

Integrated circuit
(quad flat package)

Component lead

Solder paste

Pad

FR4

(Tin-lead alloy and solder mask layers
omitted for reasons of clarity)

Tracks

No holes
in pads

**Figure 18-12. Surface mount
technology (SMT)**

However, vapor-phase soldering is becoming increasingly less popular due to environmental concerns.

Surface mount technology is well suited to automated processes. Due to the fact that the components are attached directly to the surface of the board, they can be constructed with leads that are finer and more closely spaced than their through-hole equivalents. The result is smaller and lighter packages which can be mounted closer together and occupy less of the board's surface area, which is referred to as *real estate*. This, in turn, results in smaller, lighter, and faster circuit boards. The fact that the components do not require holes for their leads is also advantageous, because drilling holes is a time-consuming and expensive process. Additionally, if any holes are required to make connections through the board (as discussed below), they can be made much smaller because they do not have to accommodate component leads.[9]

---

[9] Typical pad and hole diameters are presented in the *Holes versus Vias* section later in this chapter.

Double-sided Boards

There is a simple game played by children all over the world. The game commences by drawing three circles and three squares on a piece of paper, and then trying to connect each circle to each square without any of the connecting lines crossing each other (Figure 18-13).

Children can devote hours to this game, much to the delight of their parents. Unfortunately, there is no solution,

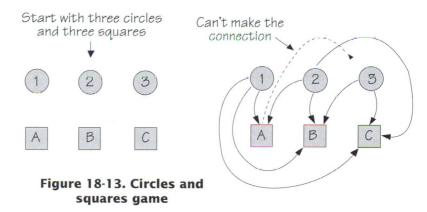

Figure 18-13. Circles and squares game

and one circle-square pair will always remain unconnected. This simple example illustrates a major problem with single-sided boards, which may have to support large numbers of component leads and tracks. If any of the tracks cross, an undesired electrical connection will be made and the circuit will not function as desired. One solution to this dilemma is to use wire links called *jumpers* (Figure 18-14).

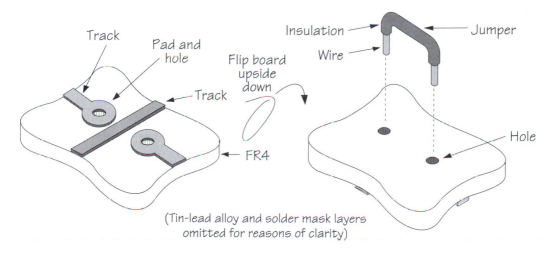

(Tin-lead alloy and solder mask layers omitted for reasons of clarity)

Figure 18-14. Single-sided boards: using jumpers

Unfortunately, the act of inserting a jumper is as expensive as for any other component. At some point it becomes more advantageous to employ a *double-sided* board, which has tracks on both sides.

Initially, the construction of a double-sided board is similar to that for a single-sided board. Assuming a subtractive process, copper foil is bonded to both sides of the board, and then organic resist is applied to both surfaces and cured. Separate masks are created for each side of the board, and ultraviolet light is applied to both sides. The ultraviolet radiation that is allowed to pass through the masks degrades the resist, which is then removed using an organic solvent. Any exposed copper that is not protected by resist is etched, the remaining resist is removed, and holes are drilled. However, a double-sided board now requires an additional step. After the holes have been drilled, a plating process is used to line them with copper (Figure 18-15).

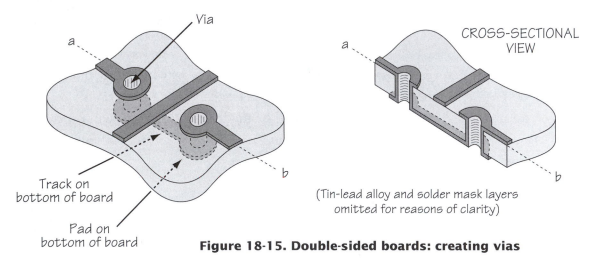

Figure 18-15. Double-sided boards: creating vias

Instead of relying on jumpers, a track can now pass from one side of the board to the other by means of these copper-plated holes, which are known as *vias*.[10,11] The tracks on one side of the board usually favor the y-axis (North-South), while the tracks on the other side favor the x-axis (East-West).

---

[10] The term *via* is taken to mean a conducting path linking two or more conducting layers, but does not include a hole accommodating a component lead (see also the following section entitled *Holes versus Vias*).

[11] There are a number of alternative techniques that may be used to create circuit board vias. By default, however, the term is typically understood to refer to holes plated with copper as described here.

The inside of the vias and the tracks on both sides of the board are plated with tin-lead alloy, and solder masks are applied to both surfaces (or vice versa in the case of the SMOBC-based processes, which were introduced earlier).

Some double-sided boards are populated with through-hole or surface mount devices on only one side. Some boards may be populated with through-hole devices on one side and surface mount devices on the other. And some boards may have surface mount devices attached to both sides. This latter case is of particular interest, because surface mount devices do not require holes to be drilled through the pads used to attach their leads (they have separate fan-out vias as discussed below). Thus, in surface mount technology, it is possible to place two devices directly facing each other on opposite sides of the board without making any connections between them.

Having said this, in certain circumstances it may be advantageous to form connections between surface mount devices directly facing each other on opposite sides of the board. The reason for this is that a through-board connection can be substantially shorter than an equivalent connection between adjacent devices on the same side of the board. Thus, this technique may be of use for applications such as high-speed data buses, because shorter connections result in faster signals.

Holes versus Vias

Manufacturers of circuit boards are very particular about the terminology they use, and woe betide anyone caught mistakenly referring to a *hole* as a *via*, or vice versa. Figure 18-16 should serve to alleviate some of the confusion.

In the case of a single-sided board (as illustrated in Figures 18-7 and 18-14), a hole that is used to accommodate a through-hole component lead is simply referred to as a *hole*. By comparison, in the case of double-sided boards (or multilayer boards as discussed below), a hole that is used to accommodate a through-hole component lead is plated with copper and is referred to as a *plated through-hole*. Additionally, a hole that is only used to link two or more conducting layers, but does not accommodate a component lead, is referred to as a *via* (or, for those purists among us, an *interstitial via*). The qualification attached to this latter case is important, because even if a hole that is used to accommodate a component lead is also used to link two or more conducting layers, it is still referred to as a plated through-hole and not a via (phew!).

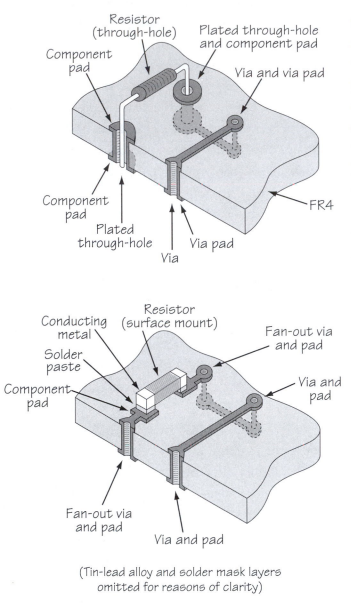

Resistor (through-hole)

Plated through-hole and component pad

Component pad

Via and via pad

Component pad

Plated through-hole

Via pad

Via

FR4

Conducting metal

Resistor (surface mount)

Fan-out via and pad

Solder paste

Via and pad

Component pad

Fan-out via and pad

Via and pad

(Tin-lead alloy and solder mask layers omitted for reasons of clarity)

Figure 18-16. Holes versus vias

Due to the fact that vias do not have to accommodate component leads, they can be created with smaller diameters than plated through-holes, thereby occupying less of the board's real estate. To provide a sense of scale, the diameters of plated through-holes and their associated pads are usually in the order of 24 mils (0.6 mm) and 48 mils (1.2 mm), respectively, while the diameters of vias and via pads are typically 12 mils (0.3 mm) and 24 mils (0.6 mm), respectively.

Finally, in the case of surface mount devices attached to double-sided boards (or multilayer boards as discussed below), each component pad is usually connected by a short length of track to a via, known as a *fan-out via*,[12] which forms a link to other conducting layers (an example of a fan-out via is shown in Figure 18-16). However, if this track exceeds a certain length (it could meander all the way around the board), an otherwise identical via at the end would be simply referred to as a via. Unfortunately,

---

[12] Some engineers attempt to differentiate vias that fall inside the device's footprint (under the body of the device) from vias that fall outside the device's footprint by referring to the former as *fan-in vias*, but this is not a widely used term.

there is no standard length of track that differentiates a fan-out via from a standard via, and any such classification depends solely on the in-house design rules employed by the designer and board manufacturer.

Multilayer Boards

It is not unheard of for a circuit board to support thousands of components and tens of thousands of tracks and vias. Double-sided boards can support a higher population density than single-sided boards, but there quickly comes a point when even a double-sided board reaches its limits. A common limiting factor is lack of space for the necessary number of vias. In order to overcome this limitation, designers may move onwards and upwards to *multilayer* boards.

A multilayer board is constructed from a number of single-sided or double-sided sub-boards.[13] The individual sub-boards can be very thin, and multilayer boards with four or six conducting layers are usually around the same thickness as a standard double-sided board. Multilayer boards may be constructed using a double-sided sub-board at the center with single-sided sub-boards added to each side (Figure 18-17a).[14] Alternatively, they may be constructed using only double-sided sub-boards separated by non-conducting layers of semi-cured FR4 known as *prepreg* (Figure 18-17b).

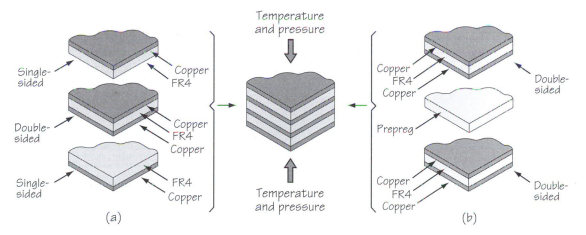

Figure 18-17. Multilayer boards: alternative structures

---

[13] The term "sub-board" is not an industry standard, and it is used in these discussions only to distinguish the individual layers from the completed board.

[14] This technique is usually reserved for boards that carry only four conducting layers.

After all of the layers have been etched to form tracks and pads, the sub-boards and prepreg are bonded together using a combination of temperature and pressure. This process also serves to fully cure the prepreg. Boards with four conducting layers are typical for designs intended for large production runs. The majority of multilayer boards have less than ten conducting layers, but boards with twenty-four conducting layers or more are not outrageously uncommon, and some specialized boards like *backplanes* (as discussed later in this chapter) may have 60 layers or more!

Through-Hole, Blind, and Buried Vias

To overcome the problem of limited space, multilayer boards may make use of *through-hole*, *blind*, and *buried* vias. A through-hole via passes all the way through the board, a blind via is only visible from one side of the board, and a buried via is used to link internal layers and is not visible from either side of the board (Figure 18-18).

Unfortunately, although they help to overcome the problem of limited space, blind and buried vias significantly increase the complexity of the manufacturing process. When these vias are only used to link both sides of a sub-board, that board must be drilled individually and a plating process used to line its vias with copper.

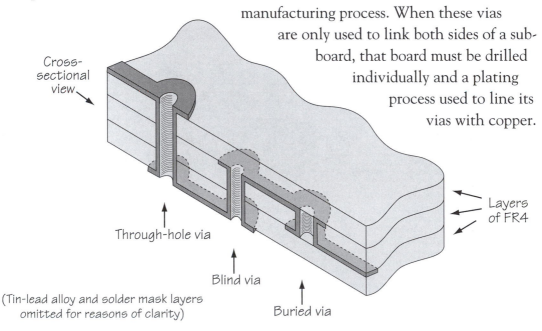

Cross-sectional view

Layers of FR4

Through-hole via

Blind via

Buried via

(Tin-lead alloy and solder mask layers omitted for reasons of clarity)

**Figure 18-18. Multilayer boards:
through-hole, blind, and buried vias**

Similarly, when these vias only pass through a number of sub-boards, those boards must be bonded together, drilled, and plated as a group. Finally, after all of the sub-boards have been bonded together, any holes that are required to form plated through-holes and vias are drilled and plated. Blind and buried vias can greatly increase the number of tracks that a board can support but, in addition to increasing costs and fabrication times, they can also make it an absolute swine to test.

Power and Ground Planes

The layers carrying tracks are known as the *signal layers*. In a multilayer board, the signal layers are typically organized so that each pair of adjacent layers favors the y-axis (North-South) and the x-axis (East-West), respectively. Additionally, two or more conducting layers are typically set aside to be used as *power* and *ground* planes. The power and ground planes usually occupy the central layers, but certain applications have them on the board's outer surfaces. This latter technique introduces a number of problems, but it also increases the board's protection from external sources of noise such as electromagnetic radiation.

Unlike the signal layers, the bulk of the copper on the power and ground planes remains untouched. The copper on these layers is etched away only in those areas where it is not required to make a connection. For example, consider a through-hole device with eight leads. Assume that leads 4 and 8 connect to the ground and power planes respectively, while the remaining leads are connected into various signal layers (Figure 18-19).

For the sake of simplicity, the exploded view in Figure 18-19 only shows the central sub-board carrying the power planes; the sub-boards carrying the signal layers would be bonded to either side. Also note that the holes shown in the prepreg in the exploded view would not be drilled and plated until all of the sub-boards had been bonded together.

In the case of component leads 1, 2, 3, 5, 6, and 7, both the power and ground planes have copper removed around the holes. These etched-away areas, which are referred to as *anti-pads*, are used to prevent connections to the planes when the holes are plated. Similarly, the power plane has an anti-pad associated with lead 4 (the ground lead), and the ground plane has an anti-pad associated with lead 8 (the power lead).

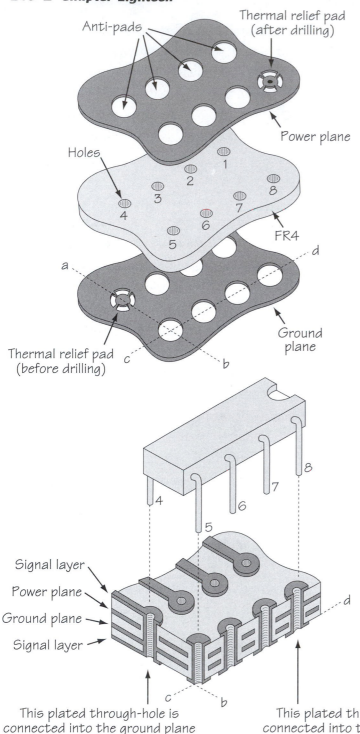

Anti-pads

Thermal relief pad (after drilling)

Holes

Power plane

1
2
3
4
5
6
7
8

FR4

a
d

c
b

Thermal relief pad (before drilling)

Ground plane

The power plane has a special pattern etched around the hole associated with lead 8 (the power lead), and a similar pattern is present on the ground plane around the hole associated with lead 4 (the ground lead). These patterns, which are referred to as *thermal relief* pads,[15] are used to make electrical connections to the power and ground planes. The spokes in the thermal relief pads are large enough to allow sufficient current to flow, but not so large that they will conduct too much heat.

Thermal relief pads are necessary to prevent excessive heat from being absorbed into the ground and power planes when the board is being

---

[15] The pattern of a thermal relief pad is often referred to as a "wagon wheel," because the links to the plated-through hole or via look like the spokes of a wheel. Depending on a number of factors, a thermal relief pad may have anywhere from one to four spokes.

4
6
7
8
5

Signal layer
Power plane
Ground plane
Signal layer

d

c
b

This plated through-hole is connected into the ground plane

This plated through-hole is connected into the power plane

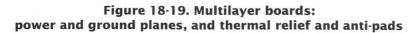

Figure 18-19. Multilayer boards:
power and ground planes, and thermal relief and anti-pads

soldered.[16] When the solder is applied, a surface-tension effect known as capillary action sucks it up the vias and plated through-holes. The solder must be drawn all the way through to form reliable, low-resistance connections. The amount of copper contained in the power and ground planes can cause problems because it causes them to act as *thermal heat sinks*. The use of thermal relief pads ensures good electrical connections, while greatly reducing heat absorption. If the thermal relief pads were not present, the power and ground planes would absorb too much heat too quickly. This would cause the solder to cool and form plugs in the vias resulting in unreliable, high-resistance connections. Additionally, in the case of wave soldering, so much heat would be absorbed by the power and ground planes that all of the layers forming the board could separate in a process known as *delamination*.

A special flavor of multilayer boards known as *Padcap* (or "Pads-Only-Outer-Layers") are sometimes used for high-reliability military applications. Padcap boards are distinguished by the fact that the outer surfaces of the board only carry pads, while any tracks are exclusively created on inner layers and connected to the pads by vias. Padcap technology offers a high degree of protection in hostile environments because all of the tracks are inside the board.

Microvia, HDI, and Build-up Technologies

One exciting recent circuit board development is that of *microvia* technology, which officially refers to any vias and via pads with diameters of 6 mils (0.15 mm) and 12 mils (0.3 mm)—or smaller—respectively. By 2002, boards using 4 mil, 3 mil, and 2 mil diameter microvias were reasonably common, and some folks were even using 1 mil microvias.

In a typical implementation, one or two microvia layers are added to the outer faces of a standard multiplayer board, which is why the term *buildup technology* is also commonly used when referring to microvia boards. Just to make things even more confusing, the term *high density interconnect (HDI)* is also commonly used. In reality, the terms *microvia*, *HDI*, and *buildup technology* are synonymous.

---

[16] For future reference, the term *pad-stack* refers to any pads, anti-pads, and thermal relief pads associated with a particular via or plated-through hole as it passes through the board.

Microvias—which are actually blind vias that just pass through one or more of the buildup layers on the outer faces of the main board—may be created using a variety of techniques. One common method is to use a laser, which can "drill" 20,000+ microvias per second.

The reason microvias are so necessary is largely tied to recent advances in device packaging technologies. It's now possible to get devices with 1,000 pins (or pads or leads or connections or whatever you want to call them), and packages with 2,000 and 4,000 pins are on the way. These pins are presented as an array across the bottom of the device. The pin pitch (the distance between adjacent pins) has shrunk to the extent that it simply isn't possible to connect the package to a board using conventional via diameters and line widths (there just isn't enough space to squeeze in all of the fan-out vias and route all of the tracks). The use of microvia technology alleviates this problem, making it possible to place a microvia in the center of a component pad, thereby eliminating the need for fan-out vias.

As a simple example, consider a simple 8-pin TSOP type package. Using conventional technologies, each of the component pads would be connected to a fan-out via (Figure 18-20a). (This is obviously a much-magnified view, and this simple component would actually be only a few millimeters in size.) Compared to the footprint of the device itself, having these fan-out vias means

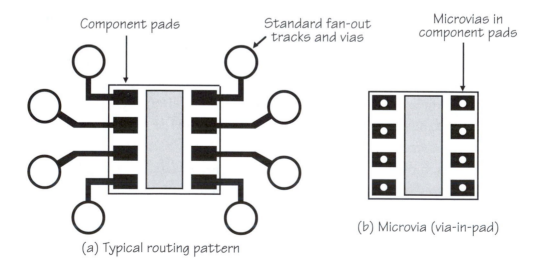

(a) Typical routing pattern

(b) Microvia (via-in-pad)

Figure 18-20. Microvias can save a lot of board real estate

that this component now occupies a substantial amount of the board's real estate. By comparison, placing vias in the component's pads using microvia technology means that the component occupies much less space on the board (Figure 18-20b).

Although using microvia technology isn't cheap per se, it can actually end up being very cost effective. In one example of which the author is aware, the use of microvias enabled an 18-layer board to be reduced to 10 layers, made the board smaller, and halved the total production cost.

Discrete Wire Technology

Discrete wire technology is an interesting discipline that has enjoyed only limited recognition, but its proponents continue to claim that it's poised to emerge as the technology-of-choice for designs that demand the highest signal speeds and component densities. Circuit boards created using this technology are known as *discrete wired boards (DWBs)*.

Multiwire Boards

The earliest form of discrete wire technology, commonly known as *multiwire*,[17,18] was developed in the late 1960s, and had gained both military and commercial acceptance by the late 1970s. The discrete wire process commences with a conventional FR4 base layer with copper foil bonded to both sides to form the power and ground planes.[19] After any thermal relief pads and anti-pads have been etched into the copper using conventional opto-lithographic processes (Figure 18-21), a layer of insulating prepreg is bonded to each side and cured. This is followed by a layer of *adhesive*,[20] or *wiring film*, which is applied using a hot roll laminating machine.

---

[17] Multiwire is a registered trademark of Advanced Interconnection Technology (AIT), Islip, New York, USA.

[18] Many thanks for the wealth of information on discrete wire technology, which was provided by Hitachi Chemical Electro-Products Division (Atlanta, Georgia, USA), I-CON Industries (Euless, Texas, USA), and MW Technologies (Aldershot, Hampshire, England).

[19] These discussions concentrate on a reasonably simple implementation: more complex boards with multiple and/or split power and ground planes are also available.

[20] Manufacturers of discrete wire boards are trying to discard the term "adhesive" on the basis that this layer is not actually "sticky."

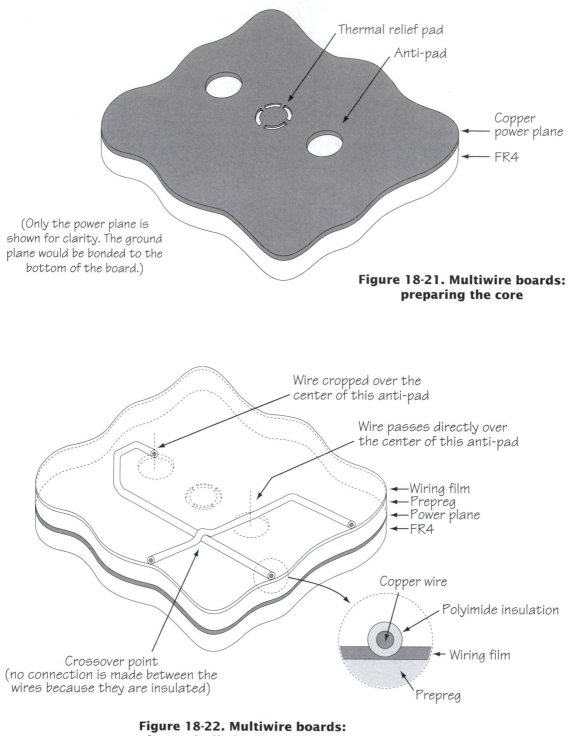

Thermal relief pad

Anti-pad

Copper power plane

FR4

(Only the power plane is shown for clarity. The ground plane would be bonded to the bottom of the board.)

Figure 18-21. Multiwire boards: preparing the core

Wire cropped over the center of this anti-pad

Wire passes directly over the center of this anti-pad

Wiring film
Prepreg
Power plane
FR4

Copper wire

Polyimide insulation

Wiring film

Prepreg

Crossover point (no connection is made between the wires because they are insulated)

Figure 18-22. Multiwire boards: ultrasonically bonding the wire

A special computer-controlled wiring machine[21] is used to ultrasonically bond extremely fine insulated wires[22] into the wiring film. The wire is routed through all the points to which it will eventually be connected. When the last point of a particular net is reached, the wiring machine cuts the wire and positions the wiring head at the next starting point (Figure 18-22).

The wire has an insulating coat of *polyimide* and can be wired *orthogonally* (North-South and East-West), *diagonally*, or as a combination of both. Due to the fact that the wires are insulated, they can cross over each other without making any unwanted electrical connections. After all the wires have been applied to one side of the board, the board is inverted and more wires can be applied to the other side. The majority of such boards have just these two wiring layers, one on each side. If necessary, however, additional layers of prepreg and wiring film can be bonded to the outer surfaces of the board and more wiring layers can be applied. It is not unusual for a very dense multiwire board to support four wiring layers, two on each side. Supporters of this technology claim that an equivalent multilayer board might require twenty or more signal layers to achieve the same effect.

After all of the wiring layers have been created, a final layer of prepreg and copper foil are laminated onto the outer surfaces of each side of the board, and any necessary holes are drilled through the board (Figure 18-23).

The drilling process leaves the wire ends exposed in the holes. The board is now exposed to a polyimide-selective etchant, which etches away the insulation at the end of wires. Although the insulation is only etched away to a depth of approximately 0.05 mm, this is sufficient for the subsequent plating process to form a wrap-around joint, which provides mechanical reliability as opposed to a simple electrical connection.

The holes are plated with copper to form plated through-holes and vias, and then the outer surfaces of the board are etched to form any via pads and component mounting pads (Figure 18-24). (Note that the pads shown in this illustration indicate that the board is to be populated with surface mount devices.)

---

[21] This is usually referred to as an NC wiring machine, where NC stands for "Numerically Controlled."

[22] The wires used in the original process were 0.16 mm in diameter. Later processes made use of additional wire diameters of 0.10 mm and 0.06 mm.

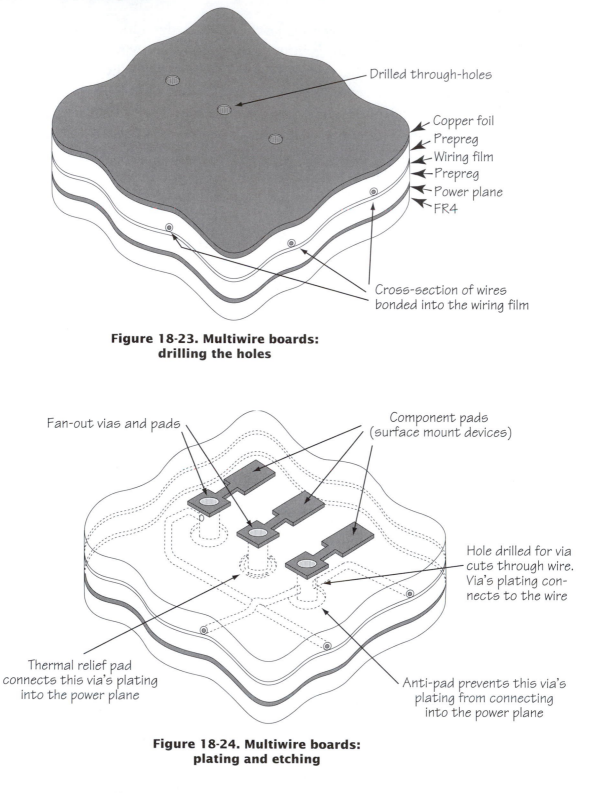

**Figure 18-23. Multiwire boards:
drilling the holes**

**Figure 18-24. Multiwire boards:
plating and etching**

Finally, the board is tinned, solder masks are applied, and components are attached in the same way as for standard printed circuit boards. In fact, a completed multiwire board appears identical to a standard printed circuit board.

Microwire Boards

If multiwire is a niche market, then its younger brother, *microwire*,[23] forms a niche within a niche. Microwire augments the main attributes of multiwire with laser-drilled blind vias, allowing these boards to support the maximum number of wires and components.[24] Due to the large numbers of components they support, microwire boards typically have to handle and dissipate a great deal of heat, so their cores may be formed from sandwiches of copper-Invar-copper[25] or similar materials. This sandwich structure is used because the resulting coefficient of thermal expansion of the core combined with the other materials used in the board is almost equal to that of any components with ceramic packages that may be attached to the board. The coefficient of thermal expansion defines the amount a material expands and contracts due to changes in temperature. If materials with different coefficients are bonded together, changes in temperature will cause shear forces at the interface between them. Engineering the board to have a similar coefficient to any ceramic component packages helps to ensure that a change in temperature will not result in those components leaping off the board.

To provide a sense of scale, in one of the more common microwire implementations each copper-Invar-copper sandwich is 0.15 mm thick. The board's central core is formed from two of these sandwiches separated by a layer of insulating prepreg, where the upper sandwich forms the power plane and the lower sandwich forms the ground plane (Figure 18-25).[26]

---

[23] Microwire is a registered trademark of Advanced Interconnection Technology (AIT), Islip, New York, USA.

[24] Actually, as multiwire boards can also be created with blind vias, and as the diameter of these vias can be as small as their microwire equivalents, someone less trusting than the author might wonder whether the main justification for microwire was the fact that some people simply wanted to play with lasers.

[25] Invar is an alloy similar to bronze.

[26] More complex structures are also possible, including multiple and/or split power and ground planes.

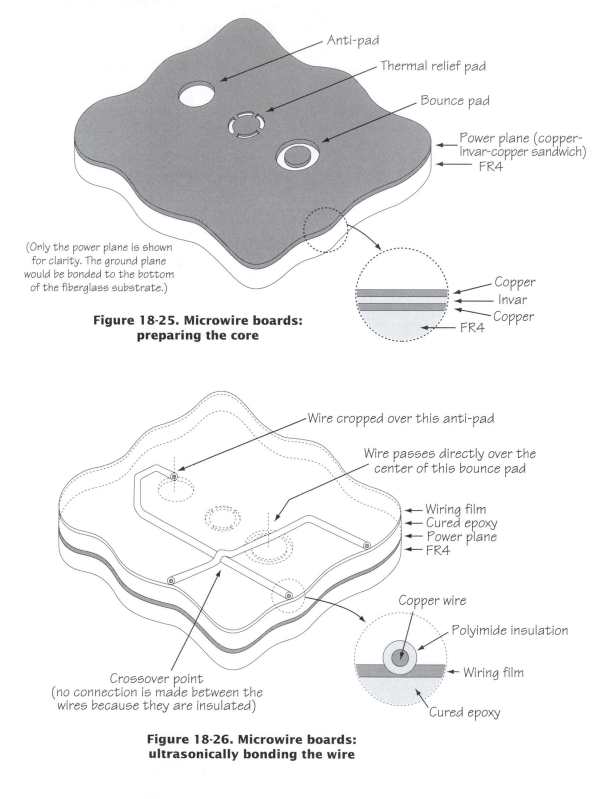

Anti-pad

Thermal relief pad

Bounce pad

Power plane (copper-Invar-copper sandwich)

FR4

(Only the power plane is shown for clarity. The ground plane would be bonded to the bottom of the fiberglass substrate.)

Copper
Invar
Copper
FR4

Figure 18-25. Microwire boards: preparing the core

Wire cropped over this anti-pad

Wire passes directly over the center of this bounce pad

Wiring film
Cured epoxy
Power plane
FR4

Copper wire

Polyimide insulation

Wiring film

Crossover point (no connection is made between the wires because they are insulated)

Cured epoxy

Figure 18-26. Microwire boards: ultrasonically bonding the wire

Thermal relief pads and anti-pads are etched into the upper and lower copper-Invar-copper sandwiches in the same manner as for multiwire boards. The etch is performed through all of the layers forming the sandwich, all the way down to the surface of the prepreg in the center. In addition to the standard thermal relief and anti-pads, microwire boards also contain special pads known as *bounce pads* for use during the laser drilling process discussed below.

Following the etching process, a layer of liquid epoxy is applied to the outer surfaces.[27] After this epoxy has been cured, a layer of wiring film is applied into which wires are ultrasonically bonded as for multiwire boards. However, in the case of microwire, the wires are always 0.06 mm in diameter (Figure 18-26).

As with multiwire, the majority of microwire boards have just two wiring layers, one on each side. However, if necessary, additional layers of liquid epoxy and wiring film can be bonded to the outer surfaces of the board and more wiring layers can be applied.

After all of the wiring layers have been created, a final encapsulation layer of liquid epoxy is applied to the outer surfaces of the board and cured. Now comes the fun part of the process, in which blind vias are drilled using a carbon-dioxide laser (Figure 18-27).

The laser evaporates the organic materials and strips the polyimide coating off the top of wire, but does not affect the wire itself. The laser beam continues down to the bounce pad, which reflects, or bounces, it back up, thereby stripping the polyimide coating off the bottom of the wire. To complete the board, holes for through-hole vias and, possibly, plated through-holes are created using a conventional drilling process. The through-hole vias, laser-drilled blind vias, and plated through-holes are plated with copper, along with any via pads and component pads which are plated onto the board's outer surfaces (Figure 18-28), and standard tinning and solder mask processes are then applied.

---

[27] Liquid epoxy is used to cover the wires because the glass fibers in prepreg would be impervious to the subsequent laser-drilling operations.

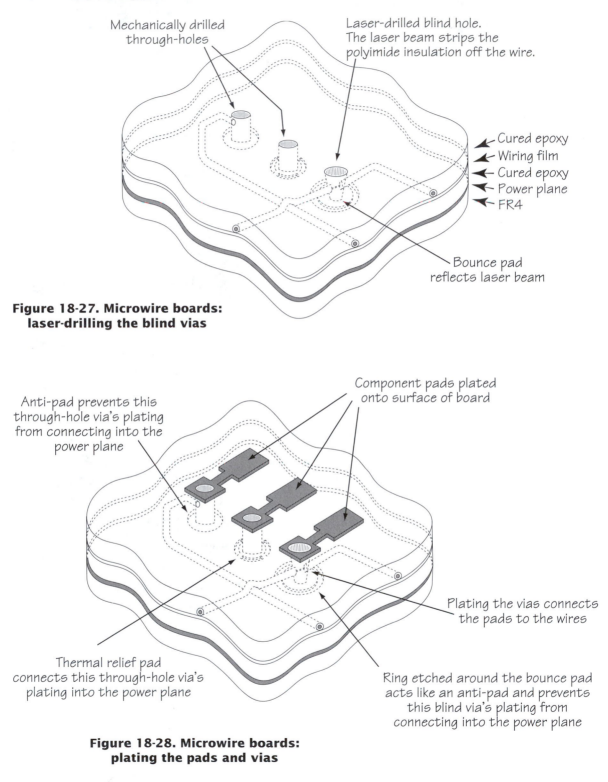

Mechanically drilled through-holes

Laser-drilled blind hole. The laser beam strips the polyimide insulation off the wire.

Cured epoxy
Wiring film
Cured epoxy
Power plane
FR4

Bounce pad reflects laser beam

Figure 18-27. Microwire boards: laser-drilling the blind vias

Anti-pad prevents this through-hole via's plating from connecting into the power plane

Component pads plated onto surface of board

Plating the vias connects the pads to the wires

Thermal relief pad connects this through-hole via's plating into the power plane

Ring etched around the bounce pad acts like an anti-pad and prevents this blind via's plating from connecting into the power plane

Figure 18-28. Microwire boards: plating the pads and vias

The Advantages of Discrete Wire Technology

Since its inception, discrete wire technology has attracted a small number of dedicated users, but has never achieved widespread recognition. Both multiwire and microwire processes are more expensive than their conventional printed wire counterparts, and therefore they are normally reserved for the most dense and electrically sophisticated applications. In fact, this discipline was originally regarded by many designers as simply being a useful prototyping technology, which offered fast turnaround times on design modifications. However, as many of its devotees have known for a long time, discrete wire technology offers a number of advantages over traditional printed wire boards. One obvious advantage is that discrete wired boards with only two wiring layers (one on each side) can provide an equivalent capability to ten or more signal layers in their multilayer counterparts. Thus, discrete wire technology can offer substantially thinner and lighter circuit boards.

Additionally, discrete wire technology is particularly advantageous in the case of high-speed applications. One major requirement of high-speed designs is to control the impedance (capacitance and inductance) of the interconnect, where the impedance is a function of the distance of the tracks from the power and ground planes. In the case of multilayer boards, the tracks keep on changing their distance from the planes as they pass through vias from one layer to another. To compensate for this, additional ground planes must be added between each pair of signal layers, which increases the thickness, weight, and complexity of the boards. By comparison, the wires on discrete wire boards maintain a constant distance from the planes.

Another consideration is the vias themselves. Transmitting a signal down a conducting path may be compared to shouting down a corridor, where any sharp turns in the corridor cause reflections and echoes. Similarly, the vias in a multilayer board have significant capacitive and inductive parasitic effects associated with them, and these effects cause deterioration and corruption of the signals being passed through them. In the case of discrete wire technology, the points at which wires cross over each other have similar effects, but they are orders of magnitude smaller.

Yet another consideration is that, as the frequency of signals increases, electrons start to be conducted only on the outer surface, or skin, of a conductor. This phenomenon is known as the *skin effect*. By the nature of their construction, the etched tracks on multilayer boards have small imperfections at their edges. These uneven surfaces slow the propagation of signals and cause noise. By comparison, the wire used in discrete wire technology is uniform to ± 0.0025 mm, which is smoother than etched tracks by a factor of 10:1.

Last but not least, the optimal cross-section for a high-speed conductor is circular rather than square or rectangular, and a constant cross-section should be maintained throughout the conductor's length. Both of these criteria are met by discrete wire technology, which, in fact, offers interconnections that are close to ideal transmission lines. The end result is that discrete wire boards can operate at frequencies that are simply not possible with their traditional multi-layer counterparts.[28] It is for this reason that discrete wire technology may yet be poised to emerge as the technology-of-choice for designers of high-speed circuit boards.

Backplanes and Motherboards

Another flavor of multilayer boards, known as *backplanes*, have their own unique design constraints and form a subject in their own right. Backplanes usually have a number of connectors into which standard circuit boards are plugged (Figure 18-29).

Backplanes typically do not carry any active components such as integrated circuits, but they often carry passive components such as resistors and capacitors. If a backplane does contain active components, then it is usually referred to as a *motherboard*. In this case, the boards plugged into it are referred to as *daughter boards* or *daughter cards*.

Because of the weight that they have to support, backplanes for large systems can be 1 cm thick or more. Their conducting layers have thicker copper plating than standard boards and the spaces between adjacent tracks are wider to reduce noise caused by inductive effects.

Backplanes may also support a lot of hardware in the form of bolts, earth straps, and power cables. It is not unusual for a backplane to have multiple power planes such as +5 volts, –5 volts, +12 volts, –12 volts, +24 volts, and –24 volts. Each power plane typically has an independent ground plane associated

---

[28] Discrete wire boards can be constructed using microcoaxial cable, and such boards can operate at frequencies that would make your eyes water!

with it to increase noise immunity. So, our hypothetical backplane would require 12 conducting layers for the power and ground planes alone. Additionally, systems containing both analog and digital circuits often require independent power and ground planes for purposes of noise immunity. Backplanes also require excellent thermal tolerance, because some of the more heroic systems can consume upwards of 200 amps and generate more heat than a rampaging herd of large electric radiators.

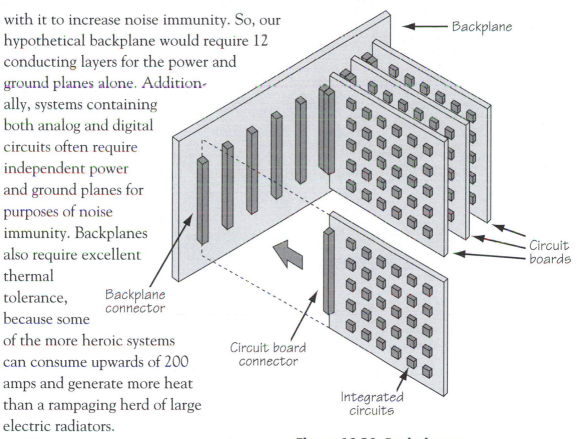

Figure 18-29. Backplanes

Backplanes may be constructed using multilayer techniques or discrete wire technology. In fact, discrete wire technology is starting to see increased use in backplane applications because of its high performance, and also due to the resulting reduction in layers, thickness, and weight.

Conductive Ink Technology

The underlying concept of conductive ink technology is relatively simple. Tracks are screen-printed onto a bare board using a conducting ink, which is then cured in an oven. Next, a dielectric, or insulating, layer is screen-printed over the top of the tracks. The screen used to print the dielectric layer is patterned so as to leave holes over selected pads on the signal layer. After the dielectric layer has been cured, the cycle is repeated to build a number of signal layers separated by dielectric layers. The holes patterned into the dielectric layers are used to form vias between the signal layers (Figure 18-30).

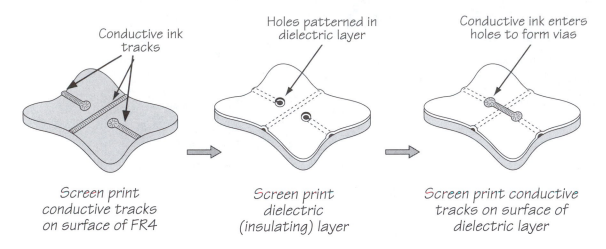

Figure 18-30. Conductive ink technology

Finally, plated through-holes and vias can be created, and components can be mounted, using the standard processes described previously. The apparent simplicity of the conductive ink technique hides an underlying sophistication in materials technology. Early inks were formed from resin pastes loaded with silver or copper powder. These inks required high firing temperatures to boil off the paste and melt the powder to form conducting tracks. Additionally, the end product was not comparable to copper foil for adhesion, conductivity, or solderability.

In the early 1990s, new inks were developed based on pastes containing a mixture of two metal or alloy powders. One powder has a relatively low melting point, while the other has a relatively high melting point. When the board is cured in a reflow oven at temperatures as low as 200°C, a process called *sintering* occurs between the two powders, resulting in an alloy with a high melting point and good conductivity.

Conductive ink technology has not yet achieved track widths as fine as traditional circuit board processes, but it does have a number of attractive features, not the least of which is that it uses commonly available screen-printing equipment. Modern inks have electrical conductivity comparable to copper and they work well with both wave soldering and reflow soldering techniques. Additionally, these processes generate less waste and are more cost-effective and efficient than the plating and etching of copper tracks.

Chip-On-Board (COB)

Chip-on-board (COB) is a relatively modern process that only began to gain widespread recognition in the early 1990s, but which is now accepted as a common and cost-effective die attachment technique. As the name implies, unpackaged integrated circuits are mounted directly onto the surface of the board. The integrated circuits are mechanically and electrically connected using similar *wire bonding, tape-automated bonding,* and *flip-chip* techniques to those used for hybrids and multichip modules.[29] The final step is *encapsulation,* in which the integrated circuits and their connections are covered with "globs" of epoxy resin or plastic, which are then cured to form hard protective covers (Figure 18-31).[30]

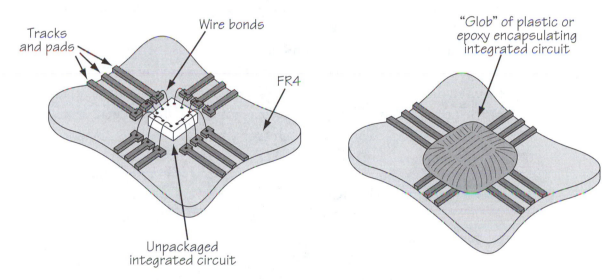

Figure 18-31. Chip-on-board (COB)

There are a number of variations of chip-on-board. For example, the designer may wish to maintain an extremely low profile for applications such as intelligent credit cards. One way to achieve this is to form cavities in the board into which the integrated circuits are inserted. Compared to surface mount technology, and especially to through-hole technology, chip-on-board

---

[29] Hybrids and multichip modules are introduced in Chapters 19 and 20, respectively.

[30] In addition to mechanical and environmental protection, the encapsulating material is also used to block out light.

offers significant reductions in size, area, and weight. Additionally, this technique boosts performance because the chips can be mounted closer together, resulting in shorter tracks and faster signals.

Flexible Printed Circuits (FPCs)

Last, but not least, are *flexible printed circuits (FPCs)*, often abbreviated to *flex*, in which patterns of conducting tracks are printed onto flexible materials. Surprisingly, flexible circuits are not a recent innovation: they can trace their ancestry back to 1904 when conductive inks were printed on linen paper. However, modern flexible circuits are made predominantly from organic materials such as *polyesters* and *polyimides*. These base layers can be thinner than a human hair, yet still withstand temperatures up to approximately 700°C without decomposing.

There are many variants of flexible circuits, not the least being *flexing*, or *dynamic flex*, and *non-flexing*, or *static flex*. Dynamic flex is used in applications that are required to undergo constant flexing such as ribbon cables in printers, while static flex can be manipulated into permanent three-dimensional shapes for applications such as calculators and high-tech cameras requiring efficient use of volume and not just area (Figure 18-32).

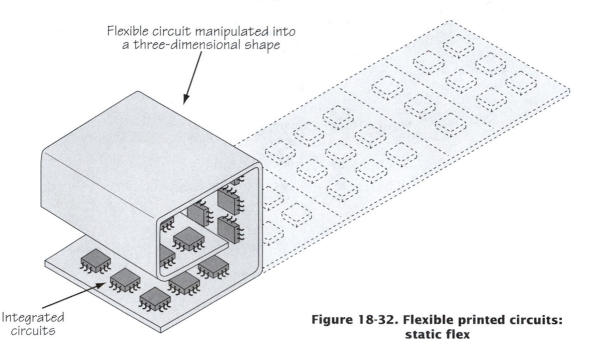

Flexible circuit manipulated into a three-dimensional shape

Integrated circuits

Figure 18-32. Flexible printed circuits: static flex

As well as single-sided flex, there are also double-sided and multilayer variants. Additionally, unpackaged integrated circuits can be mounted directly onto the surface of the flexible circuits in a similar manner to chip-on-board discussed above. However, in this case, the process is referred to as *chip-on-flex* (COF).

A common manifestation of flex technology is found in hybrid constructions known as *rigid flex*, which combine standard rigid circuit boards with flexible printed circuits (Figure 18-33).

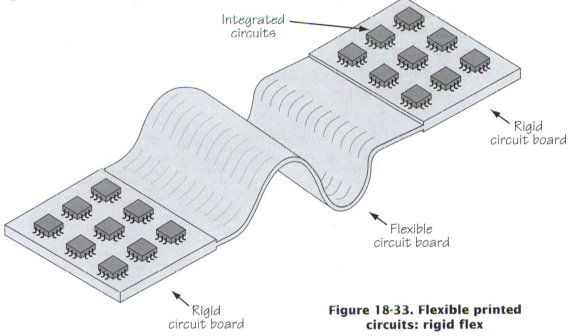

Integrated circuits

Rigid circuit board

Flexible circuit board

Rigid circuit board

Figure 18-33. Flexible printed circuits: rigid flex

In this example, the flexible printed circuit linking two standard (rigid) boards eliminates the need for connectors on the boards (which would have to be linked by cables), thereby reducing the component count, weight, and susceptibility to vibration of the circuit, and greatly increasing its reliability.

While the use of flexible circuits is relatively low, it is beginning to increase for a number of reasons. These include the ongoing development of miniaturized, lightweight, portable electronic systems such as cellular phones, and the maturing of surface mount technology, which has been described as the ideal packaging technology for flexible circuits. Additionally, flexible circuits are amenable to being produced in the form of a continuous roll, which can offer significant manufacturing advantages for large production runs.

Hybrids

The word hybrid is defined as *"the offspring resulting from crossbreeding."* Many would agree that this is an apt description for the species of electronic entities known as *hybrids*, which combine esoteric mixtures of interconnection and packaging technologies. In electronic terms, a hybrid consists of a collection of components mounted on a single insulating base layer called the *substrate*. A typical hybrid may contain a number of packaged or unpackaged integrated circuits and a variety of discrete components such as resistors, capacitors, and inductors, all attached directly to the substrate. Connections between the components are formed on the surface of the substrate; also, some components such as resistors and inductors may be fabricated directly onto the surface of the substrate.

Hybrid Substrates

Hybrid substrates are predominantly formed from alumina (aluminum oxide) or similar ceramic materials. Ceramics have many valuable properties, which have been recognized since the Chinese first created their superb porcelains during the Ming dynasty. In addition to being cheap, light, rugged, and well understood, ceramics have a variety of characteristics that make them particularly well suited to electronic applications. They are nonporous and do not absorb moisture, they can be extremely tough,[1] they have very good lateral thermal conductivity, and their coefficient of thermal expansion is close to that of silicon.

Good lateral thermal conductivity means that heat generated by the components can be conducted horizontally across the substrate and out

---

[1] As examples of their toughness, ceramics can be used to create artificial bone joints and to line the faces of golf clubs. In fact, during the cold war, Glock developed a handgun for espionage purposes that was fabricated almost completely out of ceramics (so it wouldn't trigger metal detectors at airports).

through its leads. The coefficient of thermal expansion defines the amount a material expands and contracts due to changes in temperature. If materials with different coefficients are bonded together, changes in temperature will cause shear forces at the interface between them. Because silicon and ceramic have similar coefficients of thermal expansion, they expand and contract at the same rate. This is particularly relevant when unpackaged silicon chips are bonded directly to the hybrid's substrate, because it helps to ensure that a change in temperature will not result in the chips leaping off the substrate.

Hybrid substrates are usually created by placing ceramic powder in a mold and firing it at high temperatures. The resulting substrates have very smooth surfaces and are flat without any significant curvature. Hybrid substrates typically range in size from 2.5 cm × 2.5 cm to 10 cm × 15 cm and are in the order of 0.8 mm thick.

Some hybrids require numbers of small holes between the top and bottom surfaces of the substrate. These holes will eventually be plated to form conducting paths between the two surfaces, at which time they start to be called *vias*. Additionally, depending on the packaging technology being used, slightly larger holes may be required to accommodate the hybrid's leads. One technique for forming these vias and lead-holes is to introduce tiny pillar structures into the mold. The ceramic powder flows around the pillars and, when the mold (including the pillars) is removed after firing, the substrate contains holes corresponding to the pillars. Until recent times there was no cost-effective method for drilling vias through a ceramic substrate after it had been fired.[2] However, developments in laser technology have made it possible to punch holes through fired substrates using laser beams.

While the majority of hybrid substrates are ceramic, a wide variety of other materials may also be employed. These include glass, small FR4 circuit boards (laminated substrates), and even cardboard. The latter may have appeared among your Christmas presents embedded in a pair of socks that play an annoying tune when you squeeze them.[3]

---

[2] A process not dissimilar to attempting to bore holes through a dinner plate.

[3] Thank you, Auntie Barbara, I wear them all the time.

The Thick-Film Process

The two most common techniques used to create tracks and components on the surface of hybrid substrates are known as the *thick-film* and *thin-film* processes. The thick-film process is based on screen-printing, an ancient art whose invention is usually attributed to the Chinese around 3,000 BC.

Creating Tracks

An optical mask is created carrying a pattern formed by areas that are either transparent or opaque to ultraviolet frequencies. The simple patterns shown in the following diagrams are used for reasons of clarity; in practice, such a mask may contain hundreds or thousands of fine lines and geometric shapes. Next, an extremely fine steel mesh the same size as the hybrid substrate is coated with a layer of photo-resistive emulsion (Figure 19-1).

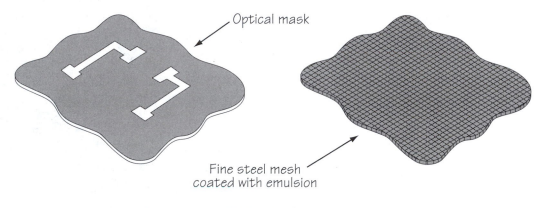

Optical mask

Fine steel mesh
coated with emulsion

**Figure 19-1. Thick-film: optical mask and
emulsion-coated fine steel mesh**

The emulsion-coated mesh is first dried, then baked, and then exposed to ultraviolet radiation passed through the optical mask. The ionizing radiation passes through the transparent areas of the optical mask into the emulsion where it breaks down the molecular structure of the resist. The mesh is then bathed in an appropriate solvent to dissolve the degraded resist. Thus, the pattern on the optical mask has been transferred to a corresponding pattern in the resist. The steel mesh with the patterned resist forms a screen-print mask, through which a paste containing metal and glass particles suspended in a solvent is applied onto the surface of the substrate (Figure 19-2). (Note that these figures show a magnified view of a very small portion of the entire substrate.)

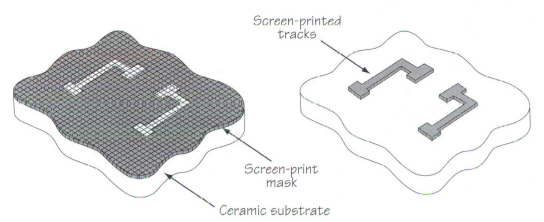

Screen-printed
tracks

Screen-print
mask

Ceramic substrate

**Figure 19-2. Thick-film: tracks screen-
printed onto the substrate**

The metal particles suspended in the paste are usually those of a noble metal such as gold, silver, or platinum, or an alloy of such metals as pladium-silver (platinum and silver). When the substrate is dried, the solvent evaporates, leaving the particles of glass and metal forming tracks on the surface of the substrate. Thick-film tracks are in the order of 0.01 mm thick. The widths of the tracks and the spaces between adjacent tracks are normally in the order of 0.25 mm, but can be as low as 0.1 mm or even finer in a leading-edge process.

Multiple layers of tracks can be printed onto the surface, each requiring the creation of a unique screen-print mask. A pattern of insulating material called a *dielectric layer* must be inserted between each pair of tracking layers to keep them separated. The dielectric patterns are formed from a paste containing only glass particles suspended in a solvent. Each dielectric layer requires its own screen-print mask and is applied using an identical process to that used for the tracking layers. Holes are included in the dielectric patterns where it is required for tracks from adjacent tracking layers to be connected to each other. Typical hybrids employ four tracking layers, commercial applications are usually limited to between seven and nine tracking layers, and a practical limit for current process technologies is around fourteen tracking layers.[4] More layers can be used, but

---

[4] The surface of the substrate becomes increasingly irregular with every layer that is applied. Eventually, the screen-print mask does not make sufficiently good contact across the substrate's surface, and paste "leaks out" under the edges of the patterns.

there is a crossover point where falling yields make the addition of successive layers cost-prohibitive. When all the tracking and insulating layers have been laid down and dried, the substrate is re-fired to approximately 1,000°C.

Creating Resistors

Resistors can be formed from a paste containing carbon compounds suspended in a solvent; the mixture of carbon compounds determines the resistivity of the final component. The resistors require their own screen-print mask and are applied to the substrate using an identical process to that described above (Figure 19-3).

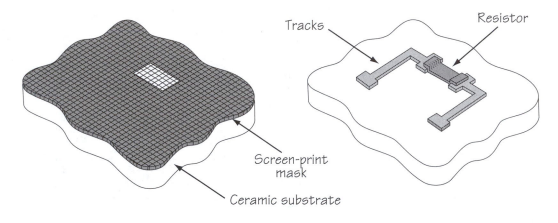

Figure 19-3. Thick-film: resistors screen-printed onto the substrate

When the substrate is dried, the solvent evaporates, leaving the carbon compounds to form resistors on the surface of the substrate. Assuming a constant thickness, each resistor has a resistance defined by its length divided by its width and multiplied by the resistivity of the paste. Thus, multiple resistors with different values can be created in a single screen-print operation by controlling the length and width of each component. However, if a low resistivity paste is used, resistors with large values will occupy too great an area. Similarly, if a high resistivity paste is used, resistors with small values will be difficult to achieve within the required tolerances. To overcome these limitations, the process may be repeated with a series of screen-print masks combined with pastes of different resistivity.

When all of the resistors have been created, the substrate is refired to approximately 600°C. An additional screen-print operation is employed to lay a protective overglaze over the resistors. This protective layer is formed from a paste containing glass particles in a solvent and is fired at approximately 450°C.

Laser Trimming

Unfortunately, creating resistors as described above is not as exact a process as one could wish. In order to compensate for process tolerances and to achieve precise values, the resistors have to be trimmed using a laser beam. There are two types of laser trimming known as *passive trimming* and *active trimming*. Passive trimming is performed before any of the integrated circuits or discrete components are mounted on the substrate. Probes are placed at each end of a resistor to monitor its value while a laser beam is used to cut parts of the resistive material away. There are a variety of different cuts which may be used to modify the resistor, including *plunge cuts*, *double plunge cuts*, or *L-shaped* cuts (Figure 19-4).

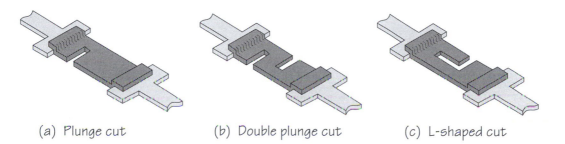

(a) Plunge cut   (b) Double plunge cut   (c) L-shaped cut

Figure 19-4. Thick-film: laser trimming of resistors

The L-shaped cut combines a plunge cut with a second cut at 90°. In this case the plunge cut provides a coarse alteration and the second cut supplies finer modifications. After each resistor has been trimmed, the probes are automatically moved to the next resistor and the process is repeated.

By comparison, active trimming is used to fine-tune analog circuits (such as active filters) and requires the integrated circuits and discrete components to be mounted on the substrate. The whole circuit is powered up and the relevant portion of the circuit is stimulated with suitable signals. A probe is placed at the output of the circuit to monitor characteristics such as amplification and frequency response. A laser is then used to trim the appropriate resistors to achieve the required characteristics while the output of the circuit is being monitored.

Creating Capacitors and Inductors

Capacitors and inductors can also be fabricated directly onto the substrate. However, capacitors created in this way are usually not very accurate, and discrete components are typically used. If inductors are included on the substrate, they are created at the same time and using the same paste as one of the tracking layers. There are two main variations of such inductors: spiral and square spiral (Figure 19-5).

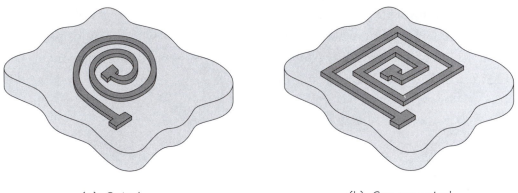

(a) Spiral   (b) Square spiral

**Figure 19-5. Thick-film: inductors
screen-printed onto the substrate**

The connection to the center-tap of the inductor can be made in several ways. A wire link can be connected to the center-tap, arched over the paths forming the spiral, and connected to a pad on the substrate outside the spiral. A somewhat similar solution is to use a track on another tracking layer to connect the center-tap to a point outside the spiral. As usual, an insulating dielectric layer would be used to separate the layer forming the inductor from the tracking layer. In yet another alternative, the center-tap can be connected to tracks on the bottom side of the substrate by means of a via placed at the center of the spiral.

Double-sided Thick-film Hybrids

Thick-film hybrids can support tracking layers on both sides of the substrate, with vias being used to make any necessary connections between the two sides. Components of all types can be mounted on both sides of the substrate as required, but active components such as integrated circuits are usually mounted only on the upper side.

Subtractive Thick-film Technology

As with any branch of electronics, new developments are always appearing on the scene. For example, a company called Silonex (www.silonex.com) has developed a process called *subtractive thick film (STF)*, which they claim is the next generation in substrate technology. STF is the integration of different technologies. In addition to standard thick film conductors, resistors, capacitors, and inductors, STF features photolithography and chemical etching steps used in conjunction with novel materials to produce unprecedented line width density and repeatability. Silonex say that STF expands the capabilities and performance of thick film modules for wireless telecommunication, instrumentation, telemetry and medical devices such as hearing aids—without sacrificing reliability or cost.

The Thin-Film Process

Thin-film processes typically employ either ceramic or glass substrates. The substrate is prepared by spluttering a layer of nichrome (nickel and chromium) alloy across the whole of its upper surface, then electroplating a layer of gold on top of the nichrome. The nichrome sticks to the substrate and the gold sticks to the nichrome. The nichrome and gold layers are each in the order of 5 μm (five-millionths of a meter) thick.

The thin-film process is similar to the opto-lithographic processes used to create integrated circuits. An optical mask is created carrying a pattern formed by areas that are either transparent or opaque to ultraviolet frequencies (Figure 19-6). As usual, the simple patterns shown in the following diagrams are used for reasons of clarity; in practice, such a mask may contain hundreds of thousands of fine lines and geometric shapes.

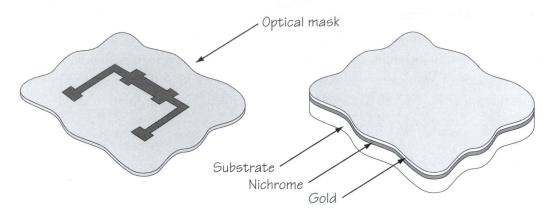

Figure 19-6. Thin-film: optical mask and substrate

The surface of the gold is coated with a layer of photo-resistive emulsion, which is first dried, then baked, and then exposed to ultraviolet radiation passed through the optical mask. The ionizing radiation passes through the transparent areas of the mask to break down the molecular structure of the emulsion. The substrate is bathed in a solvent which dissolves the degraded resist, then etched with a solvent that dissolves both the gold and nichrome from any areas left unprotected. The nichrome and gold remaining after the first mask-and-etch sequence represent a combination of tracks and resistors (Figure 19-7).

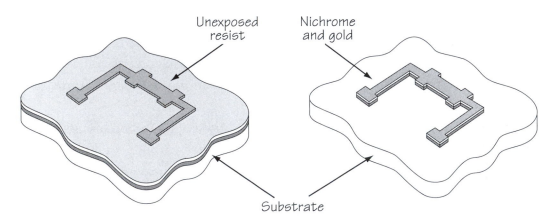

Figure 19-7. Thin-film: combined tracks and resistors

The thin-film tracks are typically in the order of 0.025 mm in width, but can be as narrow as 0.001 mm in a leading-edge process. The substrate is now recoated with a second layer of emulsion and the process is repeated with a different mask. The solvent used in this iteration only dissolves any exposed gold, but does not affect the underlying nichrome. The gold is removed from specific sites to expose the nichrome underneath, and it is these exposed areas of nichrome that form the resistors (Figure 19-8).

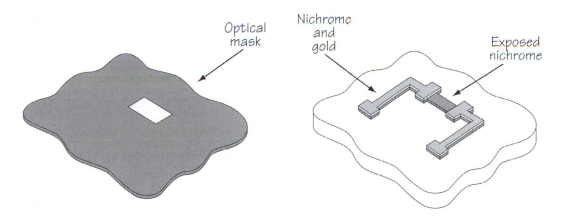

Figure 19-8. Thin-film: separated tracks and resistors

Laser Trimming

In certain respects, thin-film designers have less freedom in their control of resistance values than do their thick-film counterparts. Although the resistivity of the nichrome layer can be varied to some extent from hybrid to hybrid, the resistivity for a single hybrid is constant across the whole surface. Thus, the only way to select the value of an individual resistor is by controlling its length and width. In addition to simple rectangles, thin-film resistors are often constructed in complex concertina shapes with associated trimming blocks. The resistor values can be subsequently modified by laser trimming (Figure 19-9).

Resistors may also be created in ladder structures and their values modified by selectively cutting out rungs from the ladder using the laser. Thin-film hybrids normally employ only one tracking layer on a single side of the substrate; however, multilayer and double-sided variations are available.

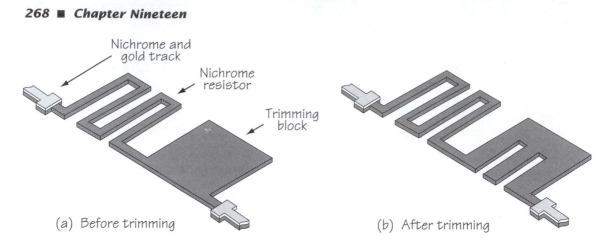

Figure 19-9. Thin-film: resistors may have complex shapes

The Assembly Process

Some hybrids contain only passive components such as resistor-capacitor networks; others contain active devices such as integrated circuits. If integrated circuits are present, they may be individually packaged surface mount devices or unpackaged bare die. Individually packaged integrated circuits are mounted onto the substrate using similar techniques to those used for surface mount technology printed circuit boards.[5] Unpackaged die are mounted directly onto the substrate using the techniques discussed below.

Attaching the Die

The primary method of removing heat from an unpackaged integrated circuit is by conduction through the back of the die into the hybrid's substrate. One technique for attaching die to the substrate is to use an adhesive such as *silver-loaded epoxy*. The epoxy is screen-printed onto the sites where the die are to be located, the die are pushed down into the epoxy by an automated pick-and-place machine, and the epoxy is then cured in an oven at approximately 180°C (Figure 19-10).

The diagrams shown here have been highly simplified for reasons of clarity; in practice, a hybrid may contain many die, each of which could have numerous pads for power and ground connections and hundreds of pads for input and output signals.

---

[5] These techniques were introduced in Chapter 18.

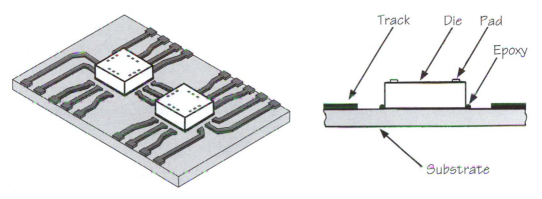

Figure 19-10. Bare die attached to substrate

An alternative method of die attachment is that of a *eutectic bond*, which requires *gold flash* (a layer of gold with a thickness measured on the molecular level) on the back of the die and also at the target site on the substrate. The substrate is heated to approximately 300ºC, and then an automatic machine presses the die against the substrate and vibrates it at ultrasonic frequencies to create a friction weld. The process of vibrating the die ultrasonically is called *scrubbing*. (These eutectic bonds are used mainly for military applications that are intended to withstand high accelerations.)

Wire Bonds

After the die have been physically attached to the substrate, electrical connections are made between the pads on each die and corresponding pads on the substrate. The most commonly used technique is *wire bonding*, in which an automatic machine makes the connections using gold or aluminum wires (Figure 19-11).

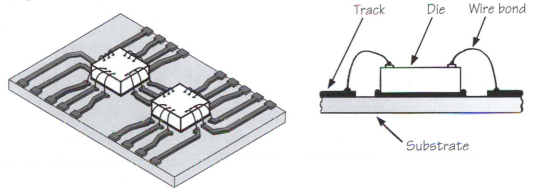

Figure 19-11. Wire bonding

These bonding wires are finer than a human hair, the gold and aluminum wires typically having diameters in the order of 0.05 mm and 0.025 mm, respectively. In one technique known as a *ball bond*, the end of the wire is heated with a hydrogen flame until it melts and forms a ball, which is then brought into contact with the target. As an alternative to ball bonds, aluminum wires are usually attached by pressing the wire against the target and scrubbing it at ultrasonic frequencies to form a friction weld. Similar wire bonds can also be used to form links between tracks on the substrate and to form connections to the hybrid's pins.

Tape-Automated Bonding

Another die attachment process is that of *tape automated bonding (TAB)*, which is mainly used for high-volume production runs or for devices with a large number of pads. In this process, a transparent flexible tape is coated with a thin layer of metal into which lead frames are patterned using standard opto-lithographic techniques. The pads on the bare die are attached to corresponding pads on the tape, which is then stored in a reel (Figure 19-12).

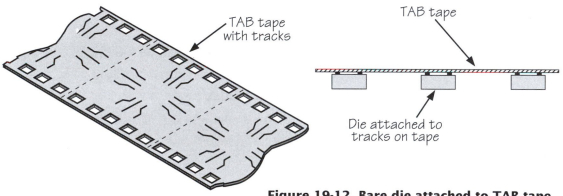

Figure 19-12. Bare die attached to TAB tape

In practice, the lead frame tracks on the tape can be so fine and so close together that they are difficult to distinguish with the human eye. When the time comes to attach the die to the hybrid's substrate, silver-loaded epoxy is screen-printed onto the substrate at the sites where the devices are to be located. This silver-loaded epoxy is also printed onto the substrate's pads to which the TAB leads are to be attached. The reel of TAB tape with attached die is fed through an automatic machine, which pushes the die and the TAB leads down into the epoxy at their target sites. When the silver-loaded epoxy is subsequently cured, it forms electrical connections between the TAB leads and the pads on the substrate (Figure 19-13).

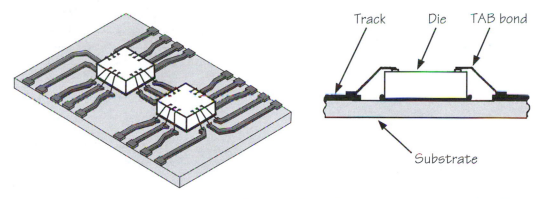

Figure 19-13. Tape-automated bonding (TAB)

Flipped-Chip Techniques

As an alternative to all of the above, there are several attachment techniques grouped under the heading of *flipped chip*. In addition to having the lowest inductance for the connections between the die and the substrate, flipped-chip techniques provide the highest packing densities because each die has a small *footprint* (the amount of area occupied by the die). Thus, flipped-chip techniques may be used where the substrate's real estate is at a premium and the die have to be mounted very closely together.

One flipped chip technique called *solder bumping* requires perfect spheres of solder to be formed on the pads of the die. The die are then "flipped over" to bring their solder bumps into contact with corresponding pads on the substrate (Figure 19-14).

One advantage of this solder bumping technique is that the pads are not obliged to be at the periphery of a die; connections may be made to pads located anywhere on a die's surface. When all of the die have been mounted on the substrate, the solder bumps can be melted in a *reflow oven* using infrared radiation or hot air. Alternatively, the solder may be melted using *vapor-phase soldering*, in which the board is lowered into the vapor-cloud above a tank containing boiling hydrocarbons. However, as for printed circuit boards, vapor-phase soldering is becoming increasingly less popular due to environmental concerns.

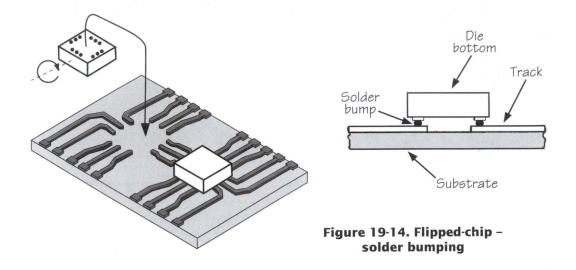

**Figure 19-14. Flipped-chip –
solder bumping**

Another flipped chip alternative is *flipped TAB*, which, like standard TAB,
may be of use for high volume production runs or for devices with a large
number of pads (Figure 19-15).

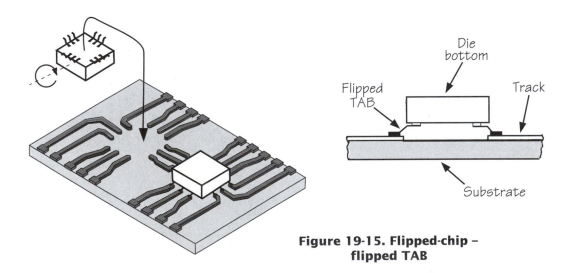

**Figure 19-15. Flipped-chip –
flipped TAB**

Advantages of Using Bare Die

Hybrids in which bare die are attached directly to the substrate are smaller, lighter, and inherently more reliable than circuits using individually packaged integrated circuits. Every solder joint and other form of connection in an assembly is a potential failure mechanism. Individually packaged integrated circuits already have a connection from each pad on the die to the corresponding package lead and require a second connection from this lead to the substrate. By comparison, in the case of a bare die, only a single connection is required between each pad on the die and its corresponding pad on the substrate. Thus, using bare die results in one less level of interconnect, which can be extremely significant when using die with hundreds of pads.

The Packaging Process

Hybrids come in a tremendous variety of shapes, sizes, and packages. Many hybrids are not packaged at all, but simply have their external leads attached directly to the substrate. In this case, any bare die are protected by covering them with blobs of plastic polymer in a solvent; the solvent is evaporated, leaving the plastic to protect the devices (Figure 19-16).

This is very similar to the chip-on-board (COB) techniques introduced in Chapter 18. In fact, if the hybrid's substrate were a small printed circuit, this would fall into the category of chip-on-board.

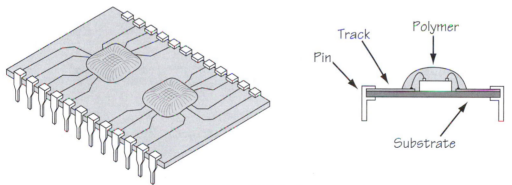

Figure 19-16. Bare die encapsulated in plastic

If the hybrid is targeted for a harsh environment, the entire substrate may be coated with polymer. If necessary, additional physical protection can be provided by enclosing the substrate with a ceramic cap filled with a dry nitrogen atmosphere (Figure 19-17).

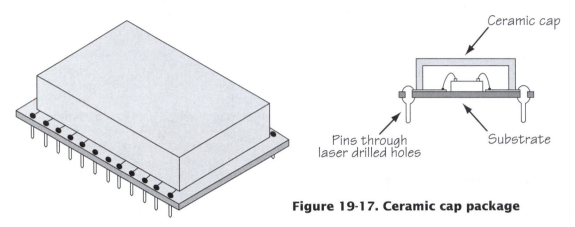

Figure 19-17. Ceramic cap package

In this case, any bare die may remain uncovered because the dry nitrogen environment protects them. If extreme protection is required, the hybrid may be hermetically sealed in alloy cans filled with dry nitrogen (Figure 19-18).

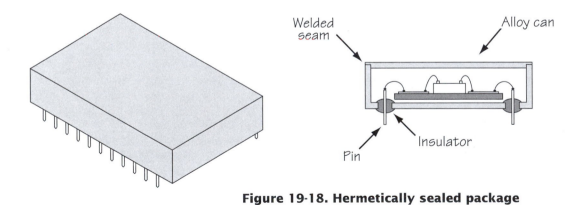

Figure 19-18. Hermetically sealed package

Finally, it should be noted that the majority of the assembly and packaging techniques introduced above are not restricted to hybrids, but may also be applied to individual integrated circuits (as discussed in Chapter 14) and multichip modules (as discussed in Chapter 20).

Multichip Modules (MCMs)

Traditionally, integrated circuits, hybrids, and printed circuit boards were differentiated on the basis of their substrate materials: semiconductors (mainly silicon), ceramics, and organics, respectively. Unfortunately, even this simple categorization is less than perfect. Some hybrids use laminate substrates, which are, to all intents and purposes, small printed circuit boards. So, at what point does a printed circuit board become a hybrid with a laminate substrate?

The problem is exacerbated in the case of multichip modules—a generic name for a group of advanced interconnection and packaging technologies featuring unpackaged integrated circuits[1] mounted directly onto a common substrate (Figure 20-1).

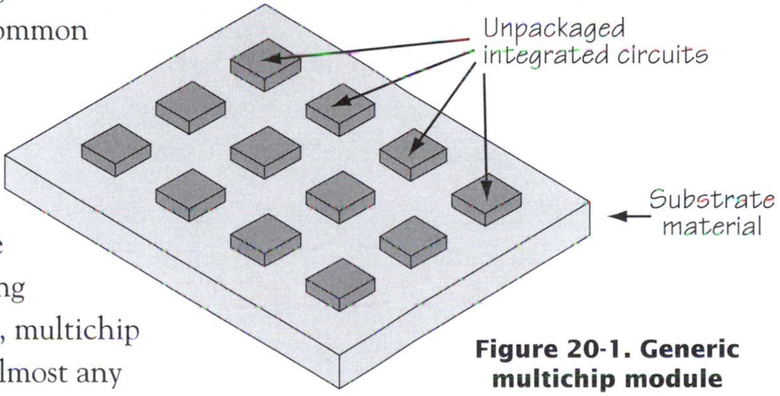

Multichip modules and hybrids can sometimes be differentiated by their choice of substrate materials and manufacturing processes. Having said this, multichip modules can be based on almost any substrate or even combinations of substrates, including laminates and ceramics, which are also used for hybrids. A slightly more reliable differentiation is that multichip modules usually have finer tracks and higher tracking and component densities[2] than hybrids. Even so, the classification of some devices is open for debate. In fact, many "hybrid designers"

Figure 20-1. Generic multichip module

Unpackaged integrated circuits

Substrate material

---

[1] Including, but not limited to, analog, digital, application-specific, simple logic, memory devices, processor cores, and communications functions.

[2] In this context, the phrase "component density" does not necessarily refer to the number of integrated circuits mounted on the multichip module, but rather to the number of logic gates they represent. For example, a multichip module may consist of only four integrated circuits, but each of these could contain tens of millions of logic gates. See also the discussions on *Equivalent Integrated Circuits* later in this chapter.

in the late 1980s were actually designing multichip modules without knowing it. This is because the term *multichip module* and the first definitions of what constituted these devices were only introduced in 1990.

Categorization by Substrate

The original multichip module definitions were based on categorization by their substrate material:

MCM-L   Laminates such as small, fine-line printed circuit boards with copper tracks and copper vias. These are usually made out of FR4 or polyimide and typically contain 5 to 25 tracking layers.

MCM-C   Ceramic substrates, some of which are similar to those used for hybrids: that is, formed from a single, seamless piece of ceramic and carrying tracks that are created using thick-film or thin-film processes (or a mixture of both). However, a large proportion of ceramic multichip modules are of the cofired variety (more below), in which case they are formed from a material such as aluminum nitride or beryllium oxide and can contain as many as 260 layers (or more).

MCM-D   Ceramic, glass, or metal substrates that are covered with a layer of dielectric material such as polyimide. The dielectric coat is used to modify the substrate's capacitive characteristics and tracks are created on the surface of the dielectric using thin-film processes. Modules of this type typically have around 5 tracking layers.

MCM-S   Semiconductor substrates, predominantly silicon, with very fine tracks formed using opto-lithographic processes similar to those used for integrated circuits. Semiconductor substrates are also known as *active substrates*, because components such as transistors and logic gates can be fabricated directly onto their surface.[3] One additional benefit of using silicon as a substrate is that its coefficient of thermal expansion exactly matches that of any silicon chips that are attached to it.

---

[3] These logic gates are particularly useful for constructing special circuits to test the multichip module after it has been integrated into a larger system.

Why Use Multichip Modules?

One obvious question that may arise at this point is why should anyone wish to use multichip modules at all? If the only requirement is to increase tracking and component densities, why not simply build bigger and better integrated circuits? After all, integrated circuit fabrication processes offer the smallest components and the finest tracks known to man. The first answer to this question is based on yield. The processes used to manufacture semiconductors are as near perfect as modern technology allows, but they still leave a small number of molecular-level defects called *inclusions* in the crystalline structure. The presence of an inclusion will affect the manufacturing process and prevent an integrated circuit from functioning as planned.

For example, consider a semiconductor wafer that is to be used to build 10,000 integrated circuits, each 1 mm × 1 mm in size. Assuming that this wafer has 10 inclusions spread evenly across its surface, 10 of the chips would fail, 9,990 would work,[4] and the resulting yield would be 9,990/10,000 = 99.9%. Now consider the effect if the same wafer were used to build 100 integrated circuits, each 10 mm × 10 mm. Assuming that each inclusion affects a separate circuit, 10 of the chips would fail, 90 would work, and the yield would be 90.0%. Continuing this progression, if the wafer were used to build 25 circuits, each 20 mm × 20 mm, the yield would fall to only 60.0%.[5] Further increasing the size of the integrated circuits would rapidly reduce the yield to zero.

In addition to inclusions, problems can also arise during the doping and metalization stages—the larger the area, the more failure-prone is the process. Thus, the semiconductor game is a case of the smaller the chips, the greater the chance for successful yields. The solution is to construct a number of smaller integrated circuits and package them as a single multichip module.

Another consideration is that it may be required to implement a portion of the design using a high-speed technology such as *emitter-coupled logic (ECL)*, which uses a lot of power. If all of the logic for a multimillion-gate design were built in this technology, enough heat could be released to melt the device into a puddle. The solution is to put the speed-critical parts of the design on high-

---

[4] This is obviously an ideal case. In reality, there are a great number of other manufacturing considerations that will also reduce the yield.

[5] This value is also based on the assumption that each inclusion would affect a separate circuit.

speed, high-power chips; put the less critical functions on slower, lower-power chips; and package all of the chips as a single multichip module.

However, there still remains one unanswered question—why not package each chip individually and mount them on a standard printed circuit board? The answer here is one of scale. On-chip tracks are incredibly fine compared to bonding wires, package leads, and circuit board tracks. Similarly, the lengths of on-chip tracks are measured in millimeters, unlike their circuit board equivalents, which are measured in tens of centimeters. These differences in scale have a number of consequences:

a) The bonding wires combined with the package leads present very significant parasitic effects in terms of capacitance and inductance, which means that the logic gates driving off-chip have to handle relatively large currents. Thus, the logic gates driving off-chip must be made larger, which, in turn, means that they eat into the available silicon real estate, consume more power, and generate more heat.

b) The longer a track is, the more susceptible it is to picking up noise. The main technique to overcome noise is to reduce its significance by increasing the voltage differential between the logic 0 and logic 1 values. Because on-chip tracks are so short, they are relatively immune to external noise, and the voltages used to represent the logic values may differ by as little as one volt. However, circuit board tracks are much longer and are correspondingly more susceptible. Thus, the voltages used to represent the logic values at the board level typically differ by as much as three or five volts. The greater the difference between the voltages used to represent the logic values, the longer it takes for a signal to transition between values and the slower the speed of the system.

c) Last but not least, signals propagate through tracks with a finite speed—the longer a track, the longer it takes for a signal to travel down it.

Multichip modules address all of these problems. Unpackaged integrated circuits can be mounted much closer together than their packaged counterparts, resulting in shorter tracks and faster signals. Also, the fabrication processes used for multichip modules result in much finer tracks than those found on printed circuit boards. The removal of the package leads and the shorter, finer, chip-to-chip tracks means that the logic gates driving off-chip can be smaller and less power-hungry, thereby saving silicon real estate and generating less heat. Additionally, the shorter chip-to-chip tracks are less susceptible to noise, allowing the voltage differential between logic values to be reduced, thereby minimizing the time it takes for a signal to transition between logic values.

Yet another consideration is that every solder joint and other form of connection in an assembly is a potential failure mechanism. Individually packaged integrated circuits already have a connection from each pad on the die to its corresponding package lead, and they require a second connection from this lead to the substrate. By comparison, in the case of a bare die, only a single connection is required between each pad on the die and its corresponding pad on the substrate. Thus, using bare die results in one less level of inter-connect, which can be extremely significant when using die with hundreds or thousands of pads.

Cofired Ceramics

Ceramic substrates have numerous benefits (as were discussed in Chapter 19). However, once the ceramic has been fired, it is very difficult to machine and cannot be satisfactorily milled, drilled, or cut using currently available technology.[6] One solution to this problem is *cofired ceramics*, in which a number of slices of ceramic are preformed and then fired together.

The process begins with unfired, or "green," ceramic tape, which is rolled out into paper-thin layers. After each layer has been cut to the required size, microscopic holes are punched through it to form vias.[7] One technique used to punch these holes involves a steel die[8] studded with round or square pins (Figure 20-2).

---

[6] Modern laser technology offers some success in drilling and cutting ceramics, but not as much as one might hope.

[7] Cofired substrates typically use a tremendous number of blind and buried vias.

[8] In this context, the term "die" refers to a piece of metal with a design engraved or embossed on it for stamping onto another material, upon which the design appears in relief.

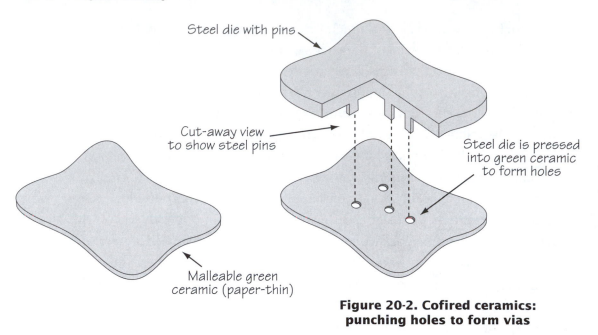

Steel die with pins

Cut-away view
to show steel pins

Steel die is pressed
into green ceramic
to form holes

Malleable green
ceramic (paper-thin)

**Figure 20-2. Cofired ceramics:
punching holes to form vias**

Unfortunately, although fast in operation, these dies take a long time to construct and the technique is incredibly expensive—each die may cost in the neighborhood of $10,000 and can typically be used for only a single layer of a single module. As an alternative, some manufacturers have experimented with programmable variants consisting of a matrix of pins, each of which may be retracted or extended under computer control. Programmable die are very expensive tools to build initially, but can be quickly reprogrammed for multiple layers and multiple modules. Their main drawback with this technique is that all of the pins are on a fixed grid. Another approach that is routinely used is that of a numerically (computer) controlled tool which punches the holes individually. This technique is relatively inexpensive, but requires much more time to complete the task.

After the holes have been punched, a screen-printing operation is used to backfill the holes with a conducting ink (a paste containing metallic powder). The mask used in this step is referred to as a *via-screen*. Next, tracks are screen-printed onto the surface of each layer using the same conducting ink. These processes are similar to the thick-film hybrid technique, but use much finer meshes for the screen print masks, thereby resulting in much finer lines and spaces (Figure 20-3).

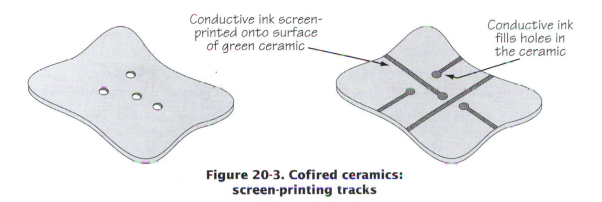

**Figure 20-3. Cofired ceramics:
screen-printing tracks**

After all of the layers have been prepared in this way, they are stacked up, pressed together, and pre-fired at a relatively low temperature to burn off the binders used in the conductive inks and any moisture in the ceramic. The substrate is then fully fired at around 1,600°C, which melts the tracks to form good conductors and vias, and transforms the individual layers into one homogeneous piece of ceramic (Figure 20-4).

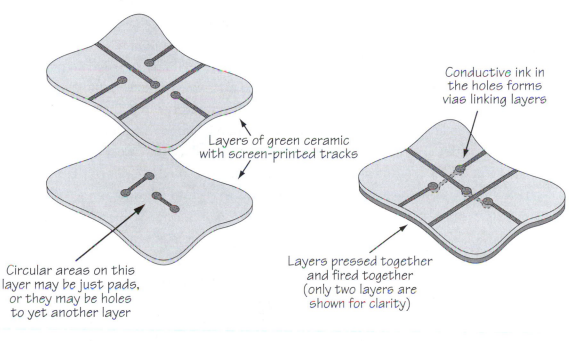

**Figure 20-4. Cofired ceramics:
assembling and firing the layers**

Cofired ceramic substrates constructed in this way can contain an amazing number of layers. Substrates with 100 layers are not uncommon, while leading-edge processes may support 260 layers or more. To provide a sense of scale, an 85-layer substrate may be approximately 4.5 mm thick, while a 120-layer substrate will be around 6.25 mm thick. Finally, after the substrate has been fired, any tracks and pads on the surface layers are plated with a *noble metal*[9] such as gold.

One not-so-obvious consideration with cofired ceramics is that the act of rolling out the ceramic tape lines up the particles of ceramic in the direction of the roll. This causes the layers to shrink asymmetrically when the ceramic is fired. To counter this effect, the layers are alternated such that adjacent layers are rolled in different directions. The net result is symmetrical shrinkage with respect to the major axes.

Low-fired Cofired Ceramics

One problem with traditional cofired processes is that the tracks must be created using *refractory metals*.[10] Although these are reasonable conductors, they are not ideal for high-reliability and high-performance applications. For high reliability, one would ideally prefer a noble metal like gold, while for high performance at a reasonable price one would ideally prefer copper. (Surprisingly, copper is one of the best conductors available for high-frequency applications, especially those operating in the microwave arena.) Unfortunately, metals like gold and copper simply vaporize at refractory temperatures.

Low-fired cofired processes use modern ceramic materials whose compositions are very different from those of traditional ceramics. To differentiate the two, the term *HTCC (high-temperature cofired ceramics)* is used to refer to high-temperature cofired ceramics requiring tracks to be formed using refractory metals, while the term *LTCC (low-temperature cofired ceramics)* refers to low-fired cofired ceramics. These new materials can be fired at temperatures as low as 850°C. This allows tracks to be screen-printed using copper-based or noble metal-based conductive inks (the firing must take place in a nitrogen-muffled furnace to prevent the metals from oxidizing).

---

[9] Noble metals are those such as gold and platinum, which are extremely inactive and are unaffected by air, heat, moisture, and most solvents.

[10] Refractory metals are those such as tungsten, titanium, and molybdenum, which are capable of withstanding the extremely high, or refractory, temperatures necessary to fire the ceramic.

Assembly and Packaging

To a large extent, multichip module packages are created on a design-specific basis. Due to the high lead counts involved, the package tends to be an integral part of the module itself. This means that packaging cannot be the afterthought it is in the case of some hybrids. For our purposes in the discussion below, we will assume a module with a cofired ceramic substrate carrying four bare die.

Because cofired substrates are constructed from multiple layers, designers have the ability to create quite sophisticated structures. For example, holes called *cavities*, or *wells*, can be cut into a number of the substrate's upper layers before they are combined together and fired (Figure 20-5).

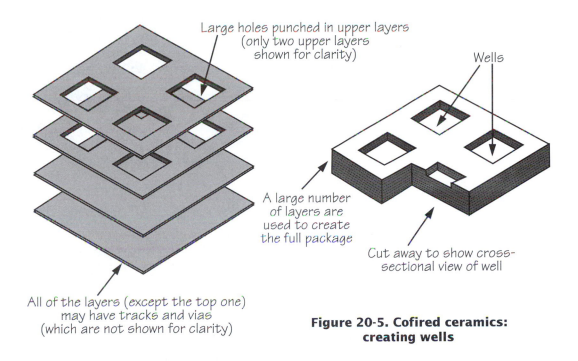

Large holes punched in upper layers
(only two upper layers shown for clarity)

Wells

A large number of layers are used to create the full package

Cut away to show cross-sectional view of well

All of the layers (except the top one) may have tracks and vias (which are not shown for clarity)

Figure 20-5. Cofired ceramics: creating wells

Remembering that the layers of ceramic are paper thin, the wells are actually formed from whatever number of layers is necessary to bring them to the required thickness. These wells will eventually accommodate the die, while the remaining ceramic will create a bed for the package's lid.[11] Before the die are mounted on the substrate, however, the external leads must be attached.

---

[11] Note that, with the exception of the upper layer, each of the layers forming the wells may also be used to carry tracks.

Pin Grid Arrays

In the case of a packaging style known as *pin grid arrays* (PGAs), the module's external connections are arranged as an array of conducting leads, or pins, in the base. In many PGAs, the leads are flat brazed directly onto the bottom layer of the device. Alternatively, in a similar manner to the way in which the wells were created in the upper layers, smaller wells may be cut into a number of the substrate's lower layers before they are combined together and fired. Each of these wells is connected to a track constructed on one of the inner layers. After firing, the wells are lined with a thin layer of metal and the leads are inserted (Figure 20-6).

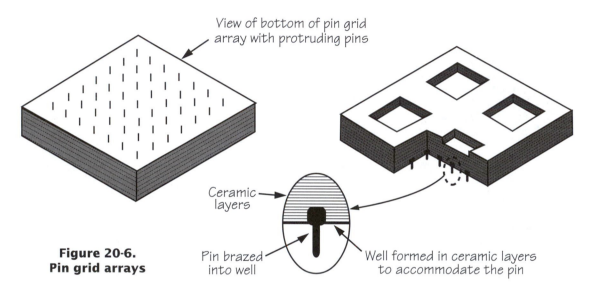

View of bottom of pin grid array with protruding pins

Ceramic layers

**Figure 20-6.
Pin grid arrays**

Pin brazed into well

Well formed in ceramic layers to accommodate the pin

The lead pitch (the spacing between adjacent leads) is typically in the order of 1.25 mm, and a module can contain more than a thousand such leads (although the average device contains rather less than this). After the leads have been inserted, the entire assembly is heated to approximately 1,000°C, which is sufficient to braze the leads to the layer of metal lining the wells. Finally, the leads are gold-plated to ensure that they will make good electrical connections when the module is eventually mounted on a circuit board.

Pad, Ball, and Column Grid Arrays

As an alternative to pin grid arrays, some modules are presented as *pad grid arrays* (PGAs), in which the external connections are arranged as an array of conducting pads on the base. In yet another alternative, which has become one of the premier packaging technologies currently in use, devices may be presented as *ball grid arrays (BGAs)*[12] or *column grid arrays (CGAs)*. These forms of package are similar to a pad grid array, but with the addition of small balls or columns of solder attached to the conducting pads (Figure 20-7).

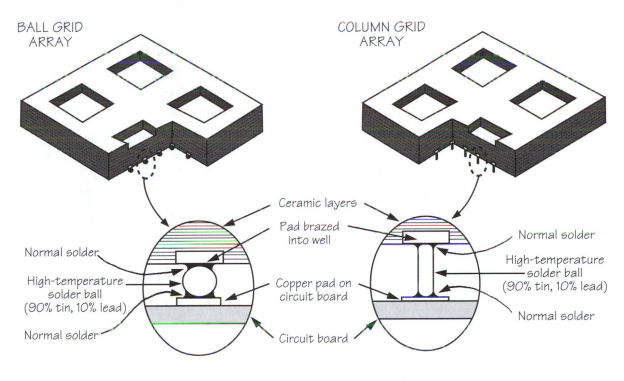

BALL GRID ARRAY

COLUMN GRID ARRAY

Ceramic layers

Pad brazed into well

Normal solder

Normal solder

High-temperature solder ball (90% tin, 10% lead)

High-temperature solder ball (90% tin, 10% lead)

Copper pad on circuit board

Normal solder

Normal solder

Circuit board

Figure 20-7. Ball grid arrays and column grid arrays

The balls (or columns) are made out of a high-temperature solder (90% tin and 10% lead), which does not melt during the soldering process. Instead, a standard low-temperature solder, referred to as *eutectic solder*, is used to connect the balls (or columns) to the pads. These packages are said to be self aligning because, when they are mounted on a circuit board and heated in a reflow oven, the surface tension of the eutectic solder causes the device to automatically align itself.

---

[12] Ball grid arrays may also be referred to as solder grid arrays, bump grid arrays, or land grid arrays.

Ball grid arrays are physically more robust than column grid arrays. However, column grid arrays are better able to accommodate any thermal strains caused by temperature mismatches between the multichip module and the circuit board.

Fuzz-Buttons

In yet another variant, which may be of use for specialist applications, small balls of fibrous gold known as *fuzz-buttons* are inserted between the pads on the base of the package and their corresponding pads on the board. When the package is forced against the board, the fuzz-buttons compress to form good electrical connections. One of the main advantages of the fuzz-button approach is that it allows broken devices to be quickly removed and replaced.

Populating the Die

However, we digress. Once the external leads have been connected, the module is ready to be populated with its die (Figure 20-8). In many ways the assembly processes used for multichip modules are similar to those used for hybrids. Bare die can be physically mounted on the substrate using adhesive, eutectic, or flipped-chip techniques, and electrically connected using wire bonding, tape automated bonding, or solder bumping methodologies. After the die have been mounted, the wells are back-filled with nitrogen, and protective lids are firmly attached.

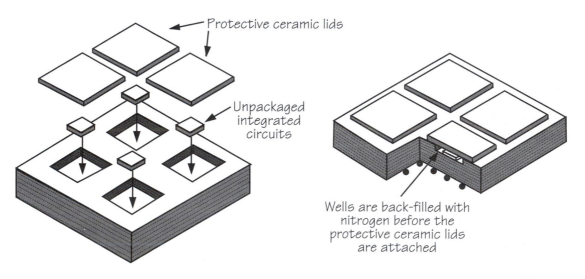

Figure 20-8. Populating the die

Equivalent Integrated Circuits

In previous discussions, it was noted that printed circuit boards may contain thousands of components, and hybrids may contain hundreds of components, but the multichip module described above contained only four integrated circuits. However, simply counting the components does not provide an appropriate basis for comparison. Although today's multichip modules typically contain relatively few devices, the actual logic represented by those devices can be huge. Thus, as opposed to *component density*, a more appropriate comparison is that of *functional density*.

There are a number of methods that may be used to compare these different interconnection strategies, including the average number of vias per square centimeter and the average number of tracks per layer per square centimeter. However, one of the most popular techniques is based on the concept of an *equivalent integrated circuit*.[13] The definition of an equivalent integrated circuit is one that contains a certain number of pins or leads;[14] a commonly used value is 20.

First, all of the leads associated with the real components attached to the substrate in question are added together and divided by the number of leads on the hypothetical equivalent integrated circuit. Thus, the result returned by this operation is in units of equivalent integrated circuits. For example, 300 leads divided by 20 equals 15 equivalent integrated circuits. This value is then divided by the area of the substrate to determine the number of equivalent integrated circuits per square centimeter.

To illustrate what this means in real terms, the author performed these calculations on a real printed circuit board, a real hybrid, and a real multichip module that he happened to find lying around his office. The printed circuit board was 27.5 cm × 20.0 cm and contained 627 components with a total of 1,736 leads; the hybrid was 2.5 cm × 5.0 cm and contained 36 components with a total of 352 leads; while the multichip module was 5.0 cm × 5.0 cm and contained four die each with 300 leads giving a total of 1,200 leads. By some

---

[13] This is similar in principle to the concept of an *equivalent gate*, which was introduced in Chapter 14.

[14] As usual, the actual definition of what constitutes an equivalent integrated circuit depends on whom one is talking to.

strange quirk of fate, this multichip module was the same as the one illustrated in the figures above. Performing the equivalent integrated circuit calculations returned the results shown in Table 20-1.

Table 20-1. Comparison based on equivalent integrated circuits

	Area of substrate (cm^2)	# of real component leads	# of equivalent integrated circuits	Equivalent integrated circuits per cm^2
Printed circuit board	550	1,736	86.8	0.16
Hybrid	12.5	352	17.6	1.41
Multichip module	25	1,200	60	2.40

Although this sample (one each of a circuit board, hybrid, and multichip module) is statistically meaningless, it does serve to illustrate the trend, which is that multichip modules offer the most sophisticated interconnection technology currently available.

The Mind Boggles

In addition to the substrates introduced above, some multichip modules use more esoteric materials such as quartz, garnet, and sapphire.[15] These modules are targeted towards specialized applications. For example, the ultra-high frequency capabilities of sapphire are of particular interest in the microwave arena. Additionally, some multichip modules use a mixture of substrates: for example, ceramic substrates with areas of embedded or deposited silicon or diamond film. Other devices may use a mixture of tracking techniques: for example, thin-film processes for surface signals combined with thick-film processes for inner layers, or vice versa.

In addition to complex materials selection and process problems, designers of multichip modules are faced with a number of other challenges. Parasitic and thermal management are key issues due to the high interconnection density and high circuit speeds. Thus, it is simply not possible to split a multichip module design into an interconnection problem, a thermal management problem, and a

---

[15] See also the discussions on diamond substrates in Chapter 21.

packaging problem. Each multichip module presents a single mammoth problem in concurrent design and typically requires a unique solution. The result is that working with multichip modules can be an expensive hobby in terms of tooling requirements and long design times.

The multichip module designer's task has migrated from one having primarily electronic concerns to one incorporating components of materials science, chemistry, and packaging. In addition to basic considerations—such as selecting the packaging materials and determining the optimal connection strategy (pin grid array versus pad grid array versus ball grid array)—the designer has to consider a variety of other packaging problems, including lead junction temperatures and thermal conduction and convection characteristics. The magnitude of the task is daunting, but if it was easy then everybody would be doing it, and you can slap me on the head with a kipper[16] if it doesn't make life exciting!

---

[16] A kipper is a smoked and salted fish (especially tasty when brushed with butter, toasted, and served as a breakfast dish with strong, hot English tea).

Alternative and Future Technologies

Electronics is one of the most exciting and innovative disciplines around, with evolutionary and revolutionary ideas appearing on almost a daily basis. Some of these ideas skulk around at the edges of the party, but never really look you in the eye or take the trouble to formally introduce themselves; some surface for a short time and then disappear forever into the twilight zone from whence they came; some tenaciously manifest themselves in mutated forms on a seasonal basis; and some leap out as if from nowhere with a fanfare of trumpets and join the mainstream so quickly that before you know it they seem like old friends.

In this, the penultimate chapter, we introduce a smorgasbord of technologies, many of which have only recently become commercially available or are on the cutting-edge of research and development. Even the most outrageous topics presented below have undergone experimental verification, but nature is a harsh mistress and natural selection will take its toll on all but the fittest. Although some of the following may seem to be a little esoteric at first, it is important to remember that a good engineer can easily believe three impossible things before breakfast. Also remember that the nay sayers proclaimed that it was impossible for bumble bees to fly (although they obviously could), that man would never reach the moon, and that I would never finish this book![1]

Reconfigurable Hardware and Interconnect

The term *hardware* is generally understood to refer to any of the physical portions constituting an electronic system, including nuts and bolds, connectors, components, circuit boards, power supplies, cabinets, and monitors (see sidebar). However, this discussion concentrates on the subset of hardware consisting of circuit boards, electronic components, and interconnect.

---

[1] Ha! I pluck my chest hairs threateningly in their general direction.

Unfortunately, the phrase *reconfigurable hardware* is akin to the phrase "stretch-resistant socks"—they both mean different things to different people. To the young and innocent, "stretch-resistant" would tend to imply a pair of socks that will not stretch. But, as those of us who are older, wiser, and a little sadder know, "stretch-resistant" actually refers to socks that will indeed stretch—they just do their best to resist it for a while! Similarly, the term *reconfigurable* is subject to myriad diverse interpretations depending on the observer's point of view.

As a starting point, *reconfigurable* hardware refers to an electronic product whose function can be customized to a specific system or application. There are obvious benefits to making one product (that can be customized) many times, as opposed to making many application-specific products once. The problem is, the perception of what is implied by "reconfigurable" is a moving target which evolves over time as new techniques and technologies become available. Throughout most of the 1980s, the most sophisticated level of customization was displayed by products based on *programmable logic devices (PLDs)* such as PROMs, PLAs, and PALs,[2] or variants of read-only memory such as PROMs,[3] EPROMs, and E²PROMs.[4] Products using these devices were usually targeted at a particular application and then focused towards a specific implementation. By comparison, the advent of *field-programmable gate arrays (FPGAs)* in the mid-1980s and early 1990s opened the door to products that could be almost totally customized for diverse applications.

As you have hopefully surmised by now, *Bebop to the Boolean Boogie* does not simply describe "where we are," but also attempts to show "how we got here." However, the pace of technological development since the birth of the

> The term *software* refers to programs, or sequences of instructions, that are executed by hardware. Additionally, *firmware* refers to programs that are hard-coded into non-volatile memory devices, while *vaporware* refers to either hardware or software that exists only in the minds of the people who are trying to sell them to you. Last but not least, *wetware* may refer either to human brains or programs that are very new and are not yet as robust as one might hope.

---

[2] Programmable logic devices were introduced in Chapter 16.

[3] PROMs may be regarded as being programmable logic devices or as memory devices depending on the applications for which they are being used.

[4] Memory devices were introduced in Chapters 15 and 16.

integrated circuit poses particular problems in the case of reconfigurable logic, because there is no consistent terminology that has survived the ravages of time. To permit the presentation of a consistent view, the solution offered by Bebop is to adopt the following terminology:[5]

Configurable Hardware:	A product whose function may be customized once or a very few times.
Reconfigurable Hardware:	A product whose function may be customized many times.
Remotely Reconfigurable Hardware:	A product whose function may be customized remotely, by telephone or radio, while remaining resident in the system.
Dynamically Reconfigurable Hardware:	A product whose function may be customized "on-the-fly" while remaining resident in the system.
Virtual Hardware:	An extension of dynamically reconfigurable hardware.

And so, armed with our trusty definitions, it only remains for us to gird up our loins and delve into the ramifications of hardware whose very reason for being can be transmogrified by an idle whim.

Configurable Hardware

A classic example of a product whose function may be customized only once is provided by a car radio, of which there may be several versions: a cheap, no-frills little number for the cost-conscious buyer, a mid-range model for the young man about town, and an ultra-flash, no-holds-barred version for the rising business executive (the sort who delights in thrusting his gold Rolex under your nose). But, little do they know that it is not unusual for all of these variations to be constructed on identical circuit boards, which can be configured by adding or removing certain components and modifying certain

---

[5] The terminology presented here differs to some extent from current industry standards, which only tend to reflect the technology of the day (or flavor of the month).

switches or jumpers. In fact, on some occasions, the only major difference between the different models is the quality of their cases and the number of buttons and dials that they support. From the viewpoint of the manufacturer, the circuit boards used in these radios are *configurable*; but from the perspective of the user, their function is cast in stone (Figure 21-1).

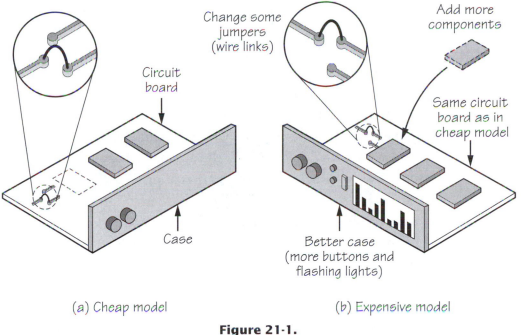

Figure 21-1.
Configurable hardware: car radios

A similar example that may be a little closer to home revolves around digital wristwatches. One down-side of the electronics era is the inevitability of being cornered by someone who insists on regaling you with the 1,001 details you didn't want to know about his new digital watch that can simultaneously display the current time in Tokyo, Paris, London, New York, and Moscow, play sixteen immediately annoying tunes, and has a calculator rivaling the control panel of the space shuttle thrown in for good measure. By some strange quirk of fate, this almost invariably occurs on those days when the timepiece you are sporting arrived in the form of a free gift at the bottom of a box of cornflakes. But there is no need to lower your head in shame, because it is not beyond the bounds of possibility that both of these instruments contain identical integrated circuits! In the case of the simpler model, a hard-wired voltage level applied to one of the device's pins instructs it to pretend to be "cheap and cheerful." Once again, the major difference between the two models is the quality of their cases and . . . the price tag.

A final example of configurable products that is too good not to share revolves around a number of well-known computer manufacturers who used to offer a choice between the fast, deluxe, and expensive version of a machine, or the slower, somewhat cheaper model. However, unbeknownst to the innocent purchaser, the only difference between the two systems was a simple switch on the main circuit board. Depending on the position of this switch, the system's clock either ran at full speed or was "slugged" to half its normal operating frequency. When the owners of one of the slower systems decided that they just had to upgrade, large amounts of money would change hands and, after appropriate sacrifices had been offered to the Gods, the computer engineer would (eventually) arrive. After making a ceremony of preparing to exchange the circuit boards, the engineer would suddenly leap to his feet, point excitedly out of the window, and cry "Good grief! What's that?" Then, while everybody's backs were turned, he would flick the switch and have the board halfway back into the system before anybody knew what was happening![6]

Reconfigurable Hardware

Hardware that is simply configurable is obviously limited, because everything that the product can do has to be designed into its base configuration, which has to encompass all possible variants. One technique for producing a product whose function may be extended beyond its original design objectives is to base that product on devices that can be reprogrammed. For example, a PROM could be employed in the role of a hardware truth table (Figure 21-2).

Similarly, a PLA could be used to implement a state machine that performs a certain sequence of operations, or, based on the inputs presented to it, a PAL device could be used to implement Boolean equations and to generate appropriate outputs. In all of these cases, the functions of the truth table, state machine, or Boolean equations could be modified by simply exchanging the programmable device with an upgraded version.

---

[6] And there are those who would say that electronics is dull and boring!

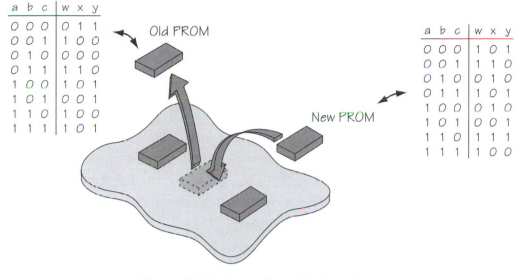

a	b	c	w	x	y
0	0	0	0	1	1
0	0	1	1	0	0
0	1	0	0	0	0
0	1	1	1	1	0
1	0	0	1	0	1
1	0	1	0	0	1
1	1	0	1	0	0
1	1	1	1	0	1

Old PROM

New PROM

a	b	c	w	x	y
0	0	0	1	0	1
0	0	1	1	1	0
0	1	0	0	1	0
0	1	1	1	0	1
1	0	0	0	1	0
1	0	1	0	0	1
1	1	0	1	1	1
1	1	1	1	0	0

**Figure 21-2. Reconfigurable hardware:
PROM as a hardware truth table**

Another technique would be to use non-volatile memory devices to store firmware programs for use by a microprocessor or a microcontroller. An example could be a set of instructions used by a microprocessor to play a tune such as the National Anthem on a musical door chime. Different versions of the PROM could be used to allow the product to be marketed in different countries (hopefully with the correct anthem). Additionally, in the case of those countries that count revolution as a national sport, the product could be easily reconfigured to reflect the *"tune of the day."*

The non-volatile memory examples introduced above could use PROM, EPROM, E^{2}PROM, or FLASH devices; similarly, the programmable logic device examples could employ PLD, EPLD, E^{2}PLD, or FLASH-PLD components. In all of these cases, from a board-level perspective, the board itself would be classified as *reconfigurable*. However, from a device-level viewpoint, PROMs (which can only be programmed a single time) would fall into the category of *configurable*, while their more sophisticated cousins—EPROM, E^{2}PROM, and FLASH—would be categorized as *reconfigurable* (similarly in the case of PLDs versus EPLDs, E^{2}PLDs, and FLASH-PLDs). Additionally, E^2-based and FLASH-based components may be referred to as *in-system programmable (ISP)*, because they can be reprogrammed while remaining resident on the circuit board.

Dynamically Reconfigurable Hardware

The advent of SRAM-based FPGAs presented a new capability to the electronics fraternity: *dynamically reconfigurable hardware*, which refers to designs that can be reconfigured "on-the-fly."

FPGAs contain a large number of diverse logic gates and registers, which can be connected together in widely different ways to achieve a desired function. SRAM-based variants allow new configuration data to be downloaded into the device by the main system in a fraction of a second. In the case of these devices, a few of the external pins are dedicated to the task of loading the data, including an *enable* control and *clock* and *data* inputs. When the *enable* control is placed in its active state, edges on the *clock* are used to load the device's SRAM with a stream of 0s and 1s, which is presented to the serial *data* input. Although all of the logic gates and SRAM cells are created on the surface of a single piece of silicon substrate, it may be useful to visualize the device as comprising two distinct strata: the logic gates and the programmable SRAM "switches" (Figure 21-3).

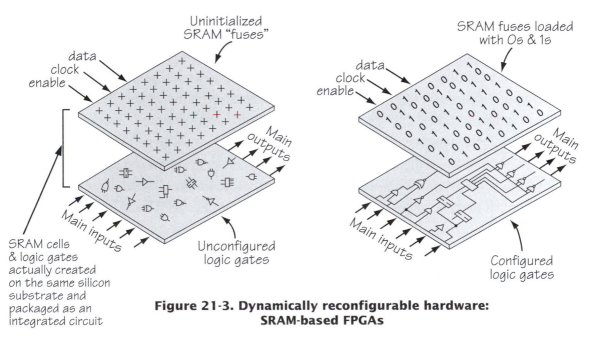

Figure 21-3. Dynamically reconfigurable hardware: SRAM-based FPGAs

The versatility of these devices opens the floodgates to a wealth of possibilities. For example, the creation of circuit boards whose interfaces can be configured to meet a variety of communications protocols, or devices acting as *digital signal processors (DSPs)*, whose core algorithms can be modified to process data in a variety of ways. As alternative protocols become available or improved algorithms are invented, the patterns that are used to configure the FPGAs can be modified to take full advantage of these new developments.

However, the true power of these devices, which are referred to as *in-circuit reconfigurable (ICR)*, resides in their ability to be reconfigured on-the-fly. For example, when a system is first turned on, it might configure all of the FPGAs to perform diagnostic functions, both on themselves and on the circuit board. Once the diagnostic checks have been completed, the system can dynamically reconfigure the FPGAs to fulfill the main function of the design.

Another example is illustrated by the Tomahawk cruise missile, which uses one navigational technique to control itself while flying over water and another while soaring over land. When the Tomahawk crosses the boundary from water to land or vice versa, it causes its on-board FPGAs to be dynamically reconfigured, changing from water-navigation mode to land-navigation mode in a fraction of a second.[7]

Dynamically Reconfigurable Interconnect

Wonderful as all of the above is, even these techniques only scratch the surface of the possibilities offered by today's emerging technologies. Designers would ideally like to create board-level products that can be reconfigured to perform radically improved, or completely different functions from the ones that they were originally designed for. The solution is to be able to dynamically configure the board-level connections between devices.

A breed of devices offer just this capability: *field-programmable interconnect devices (FPIDs)*, which may also be known as *field-programmable interconnect chips (FPICs)*.[8] These devices, which are used to connect logic devices together, can be dynamically reconfigured in the same way as standard SRAM-based FPGAs. Due to the fact that each FPID may have around 1,000 pins, only a few such devices are typically required on a circuit board (Figure 21-4).

---

[7] Of course, some of us might take the view that it is inherently unwise to have an armed cruise missile rocketing around the sky in a mindless state while it reprograms its own brain!

[8] FPIC is a trademark of Aptix Corporation.

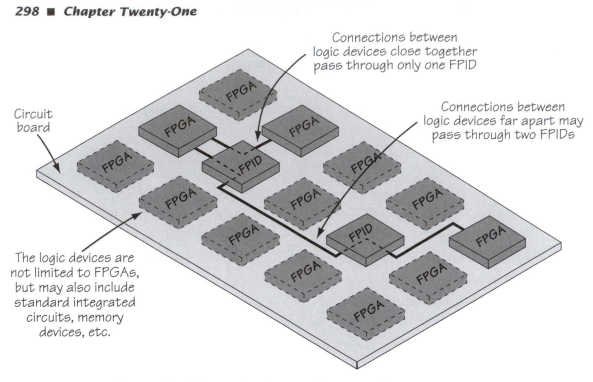

Connections between
logic devices close together
pass through only one FPID

Connections between
logic devices far apart may
pass through two FPIDs

Circuit
board

The logic devices are
not limited to FPGAs,
but may also include
standard integrated
circuits, memory
devices, etc.

**Figure 21-4. Dynamically reconfigurable interconnect:
SRAM-based FPIDs**

In fact, the concepts discussed here are not limited to board-level implementations. Any of the technologies discussed thus far may also be implemented in hybrids and multichip modules. Additionally, ultra-large-scale ASIC and system-on-chip (SoC) devices are becoming available; these little rapscallions can combine microprocessor cores, blocks of memory, and communication functions with embedded FPGA-style and FPID-style functions.

Virtual Hardware

The main limitation with the majority of SRAM-based FPGAs is that it is necessary to load the entire device. Apart from anything else, it is usually necessary to halt the operation of the entire circuit board while these devices are being reconfigured. Additionally, the contents of any registers in the FPGAs are irretrievably lost during the process.

To address these issues, a new generation of FPGAs was introduced around the beginning of 1994. In addition to supporting the dynamic reconfiguration of selected portions of the internal logic, these devices also feature:

a) No disruption to the device's inputs and outputs.

b) No disruption to the system-level clocking.

c) The continued operation of any portions of the device that are not undergoing reconfiguration.

d) No disruption to the contents of internal registers during reconfiguration, even in the area being reconfigured.

The latter point is of particular interest, because it allows one instantiation of a function to hand over data to the next function. For example, a group of registers may initially be configured to act as a binary counter. Then, at some time determined by the main system, the same registers may be reconfigured to operate as a linear feedback shift register (LFSR),[9] whose seed value is determined by the final contents of the counter before it is reconfigured.

Although these devices are evolutionary in terms of technology, they are revolutionary in terms of the potential they offer. To reflect their new capabilities, appellations such as *virtual hardware*, *adaptive hardware*, and *Cache Logic*[10] have emerged.

The phrase *virtual* hardware is derived from its software equivalent, *virtual memory*, and both are used to imply something that is not really there. In the case of virtual memory, the computer's operating system pretends that it has access to more memory than is actually available. For example, a program running on the computer may require 500 megabytes to store its data, but the computer may have only 128 megabytes of memory available. To get around this problem, whenever the program attempts to access a memory location that does not physically exist, the operating system performs a sleight-of-hand and exchanges some of the contents in the memory with data on the hard disk. Although this practice, known as *swapping*, tends to slow things down, it does allow the program to perform its task without having to wait while someone runs down to the store to buy some more memory chips.

Similarly, the phrase *Cache Logic* is derived from its similarity to the concept of *Cache Memory*, in which high-speed, expensive SRAM is used to store active

---

[9] Linear feedback shift registers (LFSRs) are introduced in detail in Appendix F.
[10] Cache Logic is a trademark of Atmel Corporation, San Jose, CA, USA.

data, while the bulk of the data resides in slower, lower-cost memory devices such as DRAM. (In this context, "active data" refers to data or instructions that a program is currently using, or which the operating system believes the program will want to use in the immediate future.)

In fact, the concepts behind virtual hardware are actually quite easy to understand. Each large macro-function in a device is usually formed by the combination of a number of smaller micro-functions such as counters, shift registers, and multiplexers. Two things become apparent when a group of macro-functions are divided into their respective micro-functions. First, functionality overlaps, and an element such as a counter may be used several times in different places. Second, there is a substantial amount of *functional latency*, which means that, at any given time, only a portion of the micro-functions are active. To put this another way, relatively few micro-functions are in use during any given clock cycle. Thus, the ability to dynamically reconfigure individual portions of a virtual hardware device means that a relatively small amount of logic can be used to implement a number of different macro-functions.

By tracking the occurrence and usage of each micro-function, then consolidating functionality and eliminating redundancy, virtual hardware devices can perform far more complex tasks than they would appear to have logic gates available. For example, in a complex function requiring 100,000 equivalent gates, only 10,000 gates may be active at any one time. Thus, by storing, or caching, the functions implemented by the extra 90,000 gates, a small, inexpensive 10,000-gate device can be used to replace a larger, more expensive 100,000-gate component (Figure 21-5).

In fact, it is even possible to "compile" new design variations in real-time, which may be thought of as dynamically creating subroutines in hardware! Hence the phrase *adaptive hardware* referred to above.

Adaptive Computing Machines (ACMs)

Ah, how I love the smell of freshly minted silicon chips in the morning. So I really wish I'd been present when the folks at QuickSilver Technology (www.qstech.com) unveiled their first *adaptive computing machine (ACM)* test chip in April 2002. This little rapscallion demonstrated a breakthrough performance three times that of a dedicated ASIC, which is no small beans, let me tell you! (Also check out picoChip at www.picochip.com.)

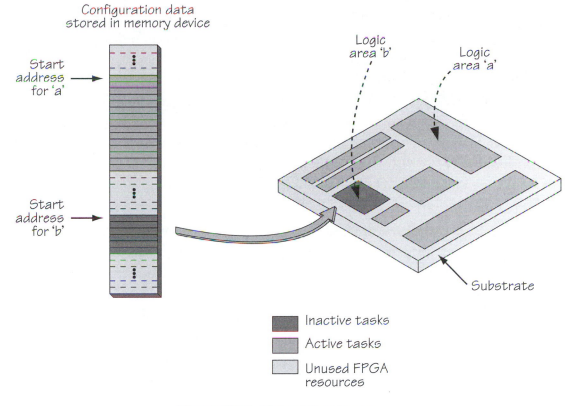

Figure 21-5. Virtual Hardware

Let's start with the fact that a product like a *software defined radio (SDR)* wireless handset typically contains a mixture of ASIC, DSP, and RISC micro-processor material. In order to reduce size and power, these will probably be implemented as a number of cores on a system-on-chip (SoC) device.

The RISC material appears in the form of a general-purpose microprocessor providing high-level, non-compute-intensive control functions. The DSP material is used because it's way faster than the general-purpose micro, and its software can be developed way faster than an ASIC can be designed. The ASIC material is reserved for the most compute-intensive, mission-critical functions only, because ASICs have two big problems: they take a lot of time and money to develop, and any algorithms they contains are effectively "frozen in silicon." This latter point is a bit of a downer in the wireless world where standards can change faster than you can say "*I love this standard . . . Good grief, it just changed again!*"

Spatial and temporal segmentation (SATS)

The ACM is a new type of digital IC whose architecture is designed to allow dynamic software algorithms to be directly mapped into dynamic hardware resources on-the-fly, resulting in the most efficient use of hardware in terms of cost, size (silicon real estate), speed, and power consumption. One of the most important aspects of the ACM is its ability to have its architecture change on demand tens or hundreds of thousands of times per second, while consuming very little power. This distinguishes the ACM from any other IC implementation technology, because it can perform *spatial and temporal segmentation (SATS)*.

SATS is the process of rapidly adapting the ACM's dynamic hardware resources to perform the various portions of an algorithm in different "segments" of time and in different locations on the ACM. As a simple example, consider that some operations on a wireless phone are modal, which means they only need to be performed some of the time. The four main modes are *acquisition* (search), *idle*, *traffic* (receive), and *traffic* (transmit) (Figure 21-6).

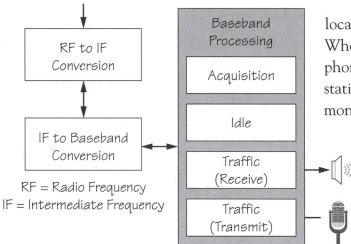

RF = Radio Frequency
IF = Intermediate Frequency

Figure 21-6. Some of the operations on a wireless phone are modal

The *acquisition* mode refers to locating the nearest base station. When in *idle* mode, the wireless phone keeps track of the base station it's hooked up to and monitors the paging channel, looking for a signal that says *"Wake up because a call is being initiated."* The *traffic* mode has two variations: *receiving* and *transmitting* (although you may think the widgets inside a phone are performing both tasks simultaneously, the reality is that, inside the digital chip, the processors are often sequencing back and forth between the transmit and receive functions).

In the case of a wireless phone based on conventional IC technologies, each of these baseband processing functions requires its own DSP or ASIC (or some area on a common ASIC). This means that even when a function isn't being used, it still occupies silicon real estate and consumes power. In turn, this affects decisions made by the design engineers. For example, when in *acquisition* mode, you can increase the number of "searcher" functions to speed things up, but each function requires ASIC material real estate. This causes engineers to limit the number of search functions they use (hence the fact that it can take your cell phone a seemingly endless time to locate the nearest base station when you first power it on).

By comparison, a next-generation ACM-based wireless phone requires only a single ACM that can be adapted on-the-fly to perform each baseband function as required. For example, when in *acquisition* mode, the ACM would be adapted to perform only acquisition; during *transmit* mode the ACM would be adapted to the transmit functions . . . and so forth. This means that at any point in time, only the function that is required would be resident in the ACM and consuming resources. More importantly, it also means that when the phone is in *acquisition* mode, for example, all of the ACM's resources can be configured as searchers, which means that you can use many more searchers than you could afford to support in ASIC material, which in turn means that the ACM can out-perform the ASIC for this type of real-world function.

Three-Dimensional Molded Interconnect

For many years, designers have wished for the ability to create robust, three-dimensional (3D) circuit boards for use with products such as handheld cellular telephones, radios, and calculators. In addition to providing the interconnect, these 3D circuit boards would also act as the product's package. (Alternatively, depending on your point of view, we could say that designers want to be able to create a 3D product package that also acts as the interconnect). In addition to reducing the product's size and weight, there can also be significant benefits in terms of cost and manufacturability.

A number of processes to create 3D circuit boards have emerged over the years, but none have gained any significant level of commercial acceptance. One hopeful contender involved creating a standard two-dimensional (2D) circuit board and then molding it into a 3D shape. This process achieved some acceptance for simple product packages, such as those for power supplies, with limited aesthetic requirements. However, the process is not suitable for products that require aesthetic appeal and ergonomic shapes involving complex surfaces.

Another technique involved the prefabrication of tracks on the inside surface of a mold. The process of injecting plastic into the mold caused the tracks to become an integral part of the packaging. But this process had its own problems—not the least of which was creating the interconnects on the 3D surface of the mold in the first place.

One development in 3D photo-imaging technology in the early 1990s reawakened interest in the injection molding technique. The process commences with the injection molding of a plastic material capable of withstanding high enough temperatures to undergo reflow soldering or vapor-phase soldering processes.[11] In addition to the physical shape of the product, it is also possible to mold in features such as holes, ribs, recesses, standoffs, and chamfered edges. These features offer substantial savings compared to their equivalent drilling, routing, and grinding operations, and also reduce part counts and labor.

A 3D photo-tool, or *mask*, is formed by molding a PVC-based material to conform exactly to the contours of the plastic part. Next, a computer-controlled laser is used to draw an image of the desired circuit onto the photo-tool (Figure 21-7).

The remainder of the process is very similar to that for a standard circuit board, and can therefore leverage off existing technology. The surface of the plastic is covered by a special resist, the photo-tool is inserted, the whole assembly is exposed to ultraviolet light, the degraded resist is removed, and copper tracks are grown using an additive process.

One of the major problems with the process at this time is the lack of appropriate computer-aided design tools. Existing tools are geared to standard 2-D circuit boards and do not have the ability to transform 2D designs into a 3D space. Sadly, there is not yet a large enough market potential to cause this position to change. However, this technology is still in its infant stage and, with sufficient market interest, the requisite electromechanical toolsets could certainly become available.

---

[11] These soldering processes were introduced in Chapter 18.

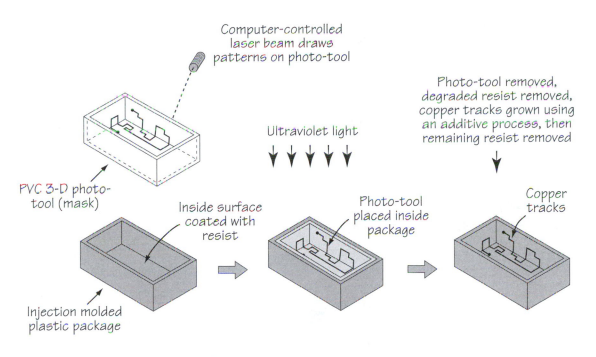

Figure 21-7. Three-dimensional molded interconnect

Optical Interconnect

Electronics systems exhibit ever-increasing requirements to process ever-increasing quantities of data at ever-increasing speeds. Interconnection technologies based on conducting wires are fast becoming the bottleneck that limits the performance of electronic systems.

To relieve this communications bottleneck, a wide variety of opto-electronic interconnection techniques are undergoing evaluation. In addition to the extremely fast propagation of data,[12] optical interconnects offer greater signal isolation, reduced sensitivity to electromagnetic interference, and a far higher bandwidth than do conducting wires.

---

[12] Light travels at 299,792,458 meters per second in a vacuum. Thus, a beam of light would take approximately 2.6 seconds to make a round trip from the earth to the moon and back again!

Fiber-Optic Interconnect

The fibers used in fiber-optic systems are constructed from two different forms of glass (or other materials) with different refractive indices. These fibers, which are finer than a human hair, can be bent into weird and wonderful shapes without breaking. When light is injected into one end of the fiber, it repeatedly bounces off the interface between the two glasses, undergoing almost total internal reflection with minimal loss, until it reemerges at the other end (Figure 21-8).

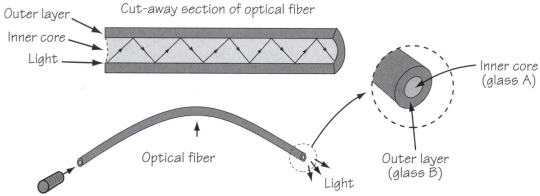

Figure 21-8. Light propagating through an optical fiber

Experimental systems using fiber-optic interconnect have been evaluated at all levels of a system—for example, to link bare die in a multichip module (Figure 21-9).

The transmitting device employs a surface-emitting laser-diode, which is constructed along with the transistors and other components on the integrated circuit's substrate. The receiving device uses a photo-transistor to convert the incoming light back into an electrical signal.

Each die can support numerous transmitters and receivers, which can be located anywhere on the surface of their substrates. However, there are several problems with this implementation, including the difficulty of attaching multiple optical fibers, the difficulties associated with repair and rework (replacing a defective die), and the physical space occupied by the fibers. Although the individual fibers are extremely thin, multichip modules may require many thousands of connections. Additionally, in this form, each optical fiber can be used only to connect an individual transmitter to an individual receiver.

At the board level, a variation of discrete wired technology[13] has been developed, in which optical fibers are ultrasonically bonded into the surface of a board. When combined with *chip-on-board* (COB) techniques, this process offers significant potential for the future. In fact, due to the way in which the optical fibers are connected to the die, it is possible for an individual transmitter to be connected to multiple receivers. Unfortunately, the techniques used to achieve this may not be disclosed here, because the author has been sworn to secrecy using arcane oaths and strange handshakes, and an Englishman's word is his bond!

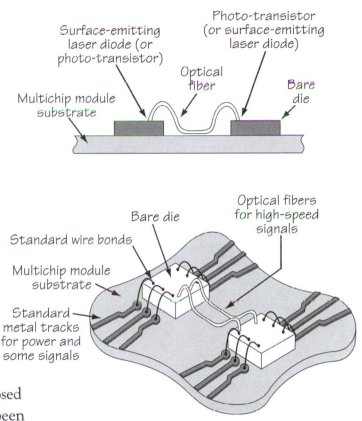

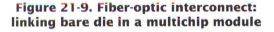

Figure 21-9. Fiber-optic interconnect: linking bare die in a multichip module

Last but not least, optical fibers may be used to provide intra-board connections, which may be referred to as *optical backplanes*. Once again, discrete wire technology, modified to accommodate optical fibers, may be employed to create the backplanes. Alternatively, the daughter boards may be mounted in a rack without an actual backplane, and groups of optical fibers may be connected into special couplers (Figure 21-10).

Each optical fiber from a transmitter is connected into a coupler, which amplifies the optical signal and can re-transmit it to multiple receivers. This form of backplane offers great latitude in regard to the proximity of the boards. In fact, boards connected in this way can be separated by as much as tens of meters.

---

[13] Discrete wired technology was introduced in Chapter 18.

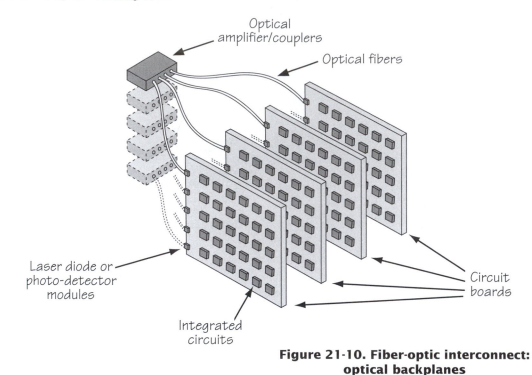

Optical amplifier/couplers

Optical fibers

Laser diode or photo-detector modules

Integrated circuits

Circuit boards

Figure 21-10. Fiber-optic interconnect: optical backplanes

In 1834, the Scottish scientist John Scott Russell was observing a barge being pulled along a canal by a pair of horses. When the barge stopped, he noticed that the bow wave continued forward without appearing to deteriorate in any way. Most of us wouldn't have paid this any attention, but Russell jumped on a horse and followed the wave for miles.

What Russell had observed was a special form of wave called a *soliton* that, due to its contour, can retain its shape and speed. One example of a soliton is the *Severn bore*: a wave caused by unusual tidal conditions (where the Severn is a river that separates England and Wales). Thus a bore (wave) can travel miles up the river without significant diminishment or deformation.

The reason this is of interest is that light waves become distorted and attenuated as they travel through optical fibers—similar to the way in which electrical pulses deteriorate as they propagate through conducting wires. Scientists and engineers are now experimenting with light solitons—pulses that can travel long distances through optical fibers without distortion.

Using light solitons, data can retain its integrity while being transmitted over 20,000 kilometers through optical fibers at a rate of 10 gigabits per second, which is four times faster than is possible using conventional techniques.

Free-Space Interconnect

In the case of the *free-space* technique, a laser-diode transmitter communicates directly with a photo-transistor receiver without employing an optical fiber. Consider a free-space technique used to link bare die mounted on the substrate of a multichip module (Figure 21-11).

In this case, the transmitters are constructed as side-emitting laser diodes along the upper edges of the die; similarly with the photo-transistors on the receivers. Each die may contain multiple transmitters and receivers. The free-space technique removes some of the problems associated with its fiber-optic equivalent: there are no fibers to attach and replacing a defective die is easier.

However, as for optical fibers, each transmitter can still be used only to connect to an individual receiver. Additionally, the free-space technique has its own unique problems: the alignment of the devices is critical, and there are thermal tracking issues. When a laser-diode is turned on, it rapidly cycles from ambient temperature to several hundred degrees Celsius. The heat generated by an individual laser-diode does not greatly affect the die because each diode is so small. However, the cumulative effect of hundreds of such diodes does affect the die, causing it to expand, thereby disturbing the alignment of the transmitter-receiver pairs.

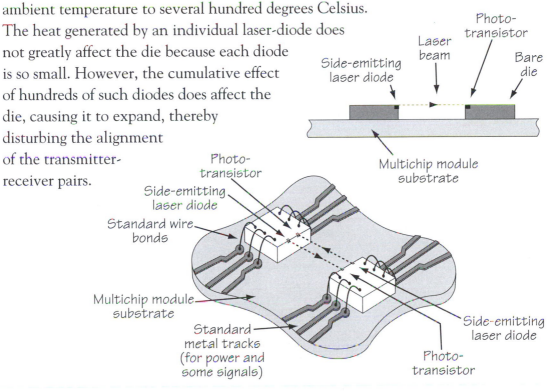

Figure 21-11. Free-space interconnect: linking bare die in a multichip module

Guided-Wave Interconnect

Another form of optical interconnect that is receiving significant interest is that of *guided-wave*, whereby optical waveguides are fabricated directly on the substrate of a multichip module. These waveguides can be created using variations on standard opto-lithographic thin-film processes. One such process involves the creation of silica waveguides (Figure 21-12).[14]

Using a *flipped-chip* mounting technique, the surface-emitting laser-diodes and photo-transistors on the component side of the die are pointed down towards the substrate. One major advantage of this technique is that the waveguides can be constructed with splitters, thereby allowing a number of transmitters to drive a number of receivers. On the down side, it is very difficult

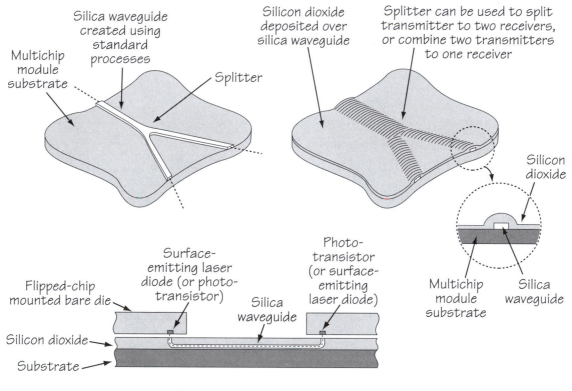

**Figure 21-12. Guided wave interconnect:
silica waveguides**

---

[14] Thanks for the information on silica waveguides go to Dr. Terry Young of the GEC-Marconi Research Center, Chelmsford, Essex, England, with apologies for all the extremely technical details that were omitted here.

to route one waveguide over another, because the crossover point tends to act like a splitter and allows light from the waveguides to "leak" into each other.

An alternative form of waveguide technology, *photo-imagable polyimide interconnect*, which was announced towards the tail-end of 1993, is of particular interest in the case of multichip modules of type MCM-D. As you may recall from Chapter 20, MCM-D devices are ceramic, glass, or metal substrates that are covered with a layer of dielectric material such as polyimide. The dielectric coat is used to modify the substrate's capacitive characteristics, and tracks are created on the surface of the dielectric using thin-film processes.

When exposed to light passed through an appropriate mask, a layer of photo-imagable polyimide can be imprinted with patterns in a similar way to exposing a photograph. After being developed, the polyimide contains low-loss optical waveguides bounded by relatively opaque reflective surfaces (Figure 21-13).

Apart from its inherent simplicity, one of the beauties of this technique is that both the exposed and unexposed areas of polyimide have almost identical dielectric constants. Thus, in addition to leveraging off existing technology, the polyimide waveguides have relatively little impact on any thin-film metalization tracking layers that may be laid over them.

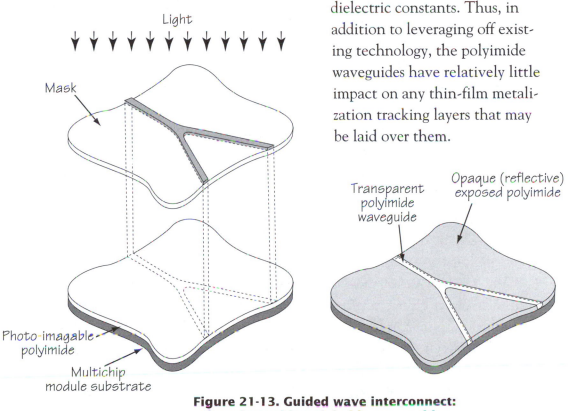

Figure 21-13. Guided wave interconnect: photo-imagable polyimide waveguides

This technique is currently finding its major audience in designers of multichip modules, but it is also being investigated as a technique for multilayer circuit boards. In the future, circuit boards could be fabricated as a mixture of traditional copper interconnect and very high speed optical interconnect.

Holographic Interconnect

Unfortunately, all of the optical interconnection technologies introduced above have their own unique problems and limitations. The technique that shows most potential is that of guided waves, but even this has the problem of routing waveguides over each other. Additionally, both waveguides and optical fibers share a common shortcoming, the fact that the light tends to bounce around an awful lot. In fact, the light bounces around so much that it actually travels approximately four to six times the straight-line distance between the transmitter and the receiver.

Quite apart from these concerns, it is doubtful if any of the above techniques will be capable of dealing with the sheer number of interconnection paths that will be required. A new contender that is rising to the challenge is *holographic interconnect.* Using the term "holographic" in this context may at first seem a little strange, because holography is traditionally considered to be a method of obtaining three-dimensional images known as holograms (from the Greek: *holos,* meaning "whole" and *gram,* meaning "message"). However, the term holographic is appropriate, because this form of interconnect is actually based on a three-dimensional image.

In the case of holographic interconnect for a multichip module, the process commences with an extremely thin slice of quartz, into which sophisticated patterns are cut using a laser. The quartz slice is then mounted approximately 1 mm above the die, and a *face-surface mirror*[15] is mounted approximately 2 cm above the quartz (Figure 21-14).

When a surface-emitting laser diode on one of the die is turned on, it transmits a laser beam straight up into one of the patterns in the quartz. This pattern causes the laser beam to be deflected so that it bounces backwards and forwards between the face-surface mirror and the quartz. When the laser hits

---

[15] The "face-surface" appellation refers to the fact that the reflective coating is located on the face of the mirror, and not behind it as would be the case in a mirror you look into when you shave or apply your makeup (or both, depending on how liberated you are).

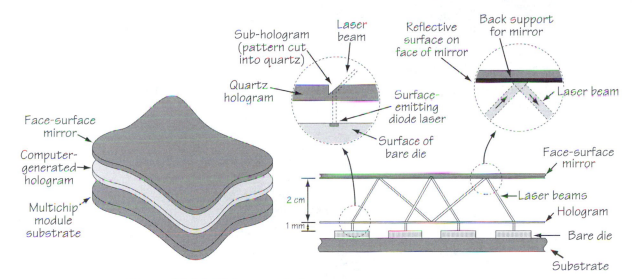

Figure 21-14. Holographic interconnect: multichip module

the point on the quartz directly above a photo-transistor on the receiving die, another pattern cut into the quartz causes it to be deflected back through the quartz and down into the receiver.

Note that the patterns cut into the quartz in the illustration above are gross simplifications. Additionally, there are typically two patterns cut into the quartz over the receiver, one on each side of the slice. The pattern on the upper side of the slice captures the laser beam and deflects it into the slice, and the pattern on the lower side then captures it again and redirects it out of the slice down onto the receiver. These patterns also perform the same function as a lens, focusing the beam precisely onto the receiver.

The angles of the patterns cut into the quartz are precisely calculated, as are the number of reflections between the face-surface mirror and the quartz, thereby ensuring that each laser beam exactly hits the target for which it is intended and no other. All of these calculations are performed by a computer, and the laser used to cut the three-dimensional patterns into the quartz is also controlled by a computer. Thus, the slice of quartz is referred to a *computer-generated hologram (CGH)* and the individual patterns above each transmitter and receiver are known as *sub-holograms*.

In fact, the laser beam actually undergoes relatively few reflections between the face-surface mirror and the quartz: far fewer than an equivalent beam transmitted through a waveguide or an optical fiber. Thus, the time taken for a signal to propagate from a transmitter to a receiver—known as the *time-of-flight* —is actually close to that for free-space interconnect. But the clever part is still to come. The sub-hologram above a laser diode can be created in such a way that it splits the laser beam into sub-beams, each of which can target a different receiver (Figure 21-15).

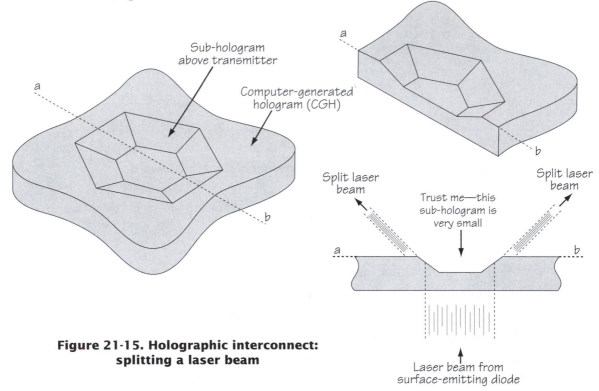

Figure 21-15. Holographic interconnect: splitting a laser beam

For the sake of clarity, Figure 21-15 shows a hexagonal sub-hologram with equal angles between the vertices, and equal angles of incidence for each face. However, the angles between the vertices can vary, and each face can have a different angle of incidence. Furthermore, the sub-holograms above receivers can be correspondingly complex, and capable of receiving signals from multiple transmitters.

Finally, in addition to multichip module applications, holographic interconnect is also undergoing evaluation as a possible backplane technology for the interconnection of circuit boards. These backplanes—known as *holo-backplanes*—offer an additional, amazingly powerful capability. It is technically feasible to create multiple holograms in the backplane's equivalent of the quartz slice. Thus, by moving the holographic slice a fraction of a millimeter in a direction parallel to the face-surface mirror, the interconnection pattern could be reconfigured and a completely new set of board-to-board interconnects could be established. The mind boggles!

Optical Memories

It has been estimated that the total sum of human knowledge is doubling approximately every ten years. Coupled with this, the amount of information that is being generated, stored, and accessed is increasing at an exponential rate. This is driving the demand for fast, cheap memories that can store gigabits (a billion bits), terabits (a thousand billion bits), or even petabits (a million billion bits) of data.

One medium with the potential to cope with this level of data density is *optical storage*. Among many other techniques, evaluations are being performed on extremely thin layers of glass-based materials,[16] which are doped with organic dyes or rare-earth elements. Using a technique known as *photochemical hole-burning (PHB)*, a laser in the visible waveband is directed at a microscopic point on the surface of the glass (Figure 21-16).

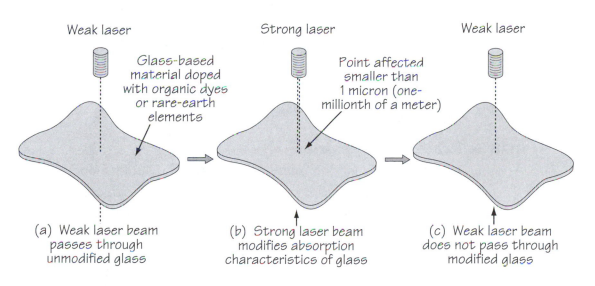

Figure 21-16. Optical memories: photochemical hole-burning (PHB)

---

[16] One such material, boric-acid glass, is also widely used in heat-resistant kitchenware!

If the laser is weak, its light will pass through the glass without affecting it and reappear at the other side. If the laser is stronger (but not intense enough to physically damage the glass), electrons in the glass will be excited by the light. The electrons can be excited such that they change the absorption characteristics of that area of the glass and leave a band, or hole, in the absorption spectrum. To put this another way, if the weak laser beam is redirected at the same point on the glass surface, its light would now be absorbed and would not reappear at the other side of the glass.

Thus, depending on whether or not the light from the weaker beam passes through the glass, it can be determined whether or not that point has been exposed to the strong laser. This means that each point can be used to represent a binary 0 or 1. Because the point affected by the laser is so small, this process can be replicated millions upon millions of times across the surface of the glass.

If the points occur at one micron (one millionth of a meter) intervals, then it is possible to store 100 megabits per square centimeter, but this still does not come close to the terabit storage that will be required. However, it turns out that each point can be "multiplexed" and used to store many bits of information. A small change in the wavelength of the laser can be used to create another hole in a different part of the spectrum. In fact, 100x multiplexing has been achieved, where each point on the glass was used to store 100 bits of data at different wavelengths. Using 100x multiplexing offers a data density of 10 gigabits per square centimeter, and even higher levels of multiplexing may be achieved in the future!

Protein Switches and Memories

Another area receiving a lot of interest is that of switches and memories based on proteins. We should perhaps commence by pointing out that this concept doesn't imply anything gross like liquidizing hamsters to extract their proteins! Out of all the elements nature has to play with, only carbon, hydrogen, oxygen, nitrogen, and sulfur are used to any great extent in living tissues, along with the occasional smattering of phosphorous, minuscule portions of a few choice metals, and elements like calcium for bones. The basic building blocks of living tissues are twenty or so relatively simple molecules called amino acids. For example, consider three of the amino acids called threonine, alanine, and serine (Figure 21-17).

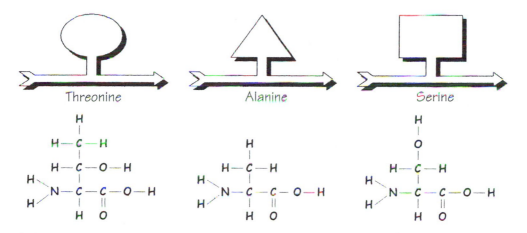

Figure 21-17. Threonine, alanine, and serine are three of the twenty or so biological building blocks called amino acids

These blocks can join together to form chains, where the links between the blocks are referred to as *peptide bonds*, which are formed by discarding a water molecule (H_2O) from adjacent COOH and NH_2 groups (Figure 21-18).

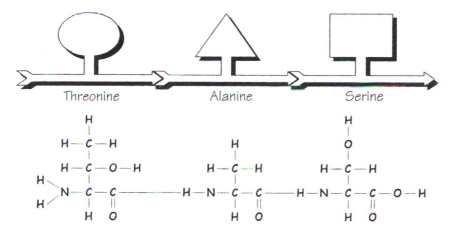

Figure 21-18: Amino acids can link together using peptide bonds to form long polypeptide chains

Proteins consist of hundreds or thousands of such chains of amino acids. Note that the distribution of electrons in each amino acid varies depending on the size of that acid's constituent atoms, leaving areas that are slightly more positively or negatively charged (similar to a water molecule, as is discussed later in this chapter). The linear chain shown in Figure 21-18 is known as the *primary structure* of the protein, but this chain subsequently coils up into a spring-like helix, whose shape is maintained by the attractions between the positively and negatively charged areas in the chain. This helix is referred to as the protein's *secondary structure*, but there's more, because the entire helix subsequently "folds" up into an extremely complex three-dimensional structure, whose shape is once again determined by the interactions between the positively and negatively charged areas on the helix. Although this may seem to be arbitrarily random, this resulting *tertiary structure* represents the lowest possible energy level for the protein, so proteins of the same type always fold up into identical (and stable) configurations.

Organic molecules have a number of useful properties, not the least of which is that their structures are intrinsically "self healing" and reject contamination. Also, in addition to being extremely small, many organic molecules have excellent electronic properties.

In the case of certain proteins, it's possible to coerce an electron to move to one end of the protein or the other, where it will remain until it's coerced back again (note that the term "end" is somewhat nebulous in this context). Thus, a protein of this type can essentially be used to store and represent a *logic 0* or a *logic 1* value based on the location of this electron.[17] Similarly, it's possible for some protein structures to be persuaded to act in the role of switches.

In the case of traditional semiconductor-based transistors, even when one considers structures measured in fractions of a millionth of a meter, each transistor consists of millions upon millions of atoms. By comparison, protein-based switches and registers can be constructed using a few thousand atoms, which means that they are thousands of times smaller, thousands of times faster, and consume a fraction of the power of their semiconductor counterparts.

Unlike metallic conductors, some proteins transfer energy by moving electron excitations from place to place rather than relocating entire electrons. This can result in switching speeds that are orders of magnitude faster than their semiconductor equivalents.

Some proteins react to electric fields, while others respond to light. For example, there is a lot of interest in the protein *Rhodopsin*, which is used

---

[17] In the case of some proteins, rather than physically moving an electron from one "end" to the other, it's possible to simply transfer an *excitation* from one electron to another. This requires far less power and occurs much faster that moving the electron itself, but it's a little too esoteric a concept to explore in detail here.

by certain photosynthetic bacteria to convert light into energy. The bacteria that contain Rhodopsin are the ones that cause ponds to turn red, and their saltwater cousins are responsible for the purple tint that is sometimes seen in San Francisco Bay.

In certain cases, light from a laser can be used to cause such optically responsive proteins to switch from one state to another (which they do by changing color) and back again. Additionally, some varieties of proteins are only responsive to the influence of two discrete frequencies. This feature is extremely attractive, because it offers the possibility of three-dimensional optical protein memories.

Experiments have been performed in which 3D cubes have been formed as ordered arrays of such bi-frequency proteins suspended in transparent polymers. If the protein were affected by a single laser, then firing a beam into the cube would result in a line of proteins changing state. But in the case of bi-frequency proteins, two lasers mounted at 90° to each other can be used to address individual points in the 3-D space (Figure 21-19).

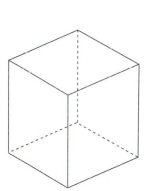

3-D cube of optically-responsive proteins suspended in transparent polymer

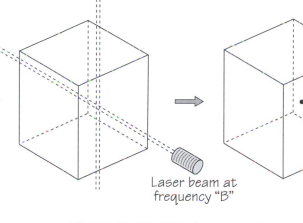

Laser beam at frequency "A"

Proteins change color at intersection point of two laser beams

Laser beam at frequency "B"

Figure 21-19: Protein memories: cubic arrays of light-sensitive proteins

By only slightly enhancing the technology available today, it may be possible to store as much as 30 gigabits in a 1 cm × 1 cm × 1 cm cube of such material, where even one gigabit would be equivalent to more than 100 of today's 256 megabit RAM devices!

Electromagnetic Transistor Fabrication

For some time it has been known that the application of strong electromagnetic fields to special compound semiconductors can create structures that behave like transistors. The original technique was to coat the surface of a semiconductor substrate with a layer of dopant material, and then to bring an extremely strong, concentrated electromagnetic field in close proximity.

The theory behind this original technique was that the intense field caused the electromigration of the dopant into the substrate. However, much to everyone's surprise, it was later found that this process remained effective even without the presence of the dopant!

Strange as it may seem, nobody actually understands the mechanism that causes this phenomenon. Physicists currently suspect that the strong electromagnetic fields cause microscopic native defects in the crystals to migrate through the crystal lattice and cluster together.

Heterojunction Transistors

If there is one truism in electronics, it is that "faster is better," and a large proportion of research and development funds are invested in increasing the speed of electronic devices.

Ultimately there are only two ways to increase the speed of semiconductor devices. The first is to reduce the size of the structures on the semiconductor, thereby obtaining smaller transistors that are closer together. The second is to use alternative materials that inherently switch faster. However, although there are a variety of semiconductors, such as *gallium arsenide (GaAs)*, that offer advantages over silicon for one reason or another, silicon is cheap, readily available, and relatively easy to work with. Additionally, the electronics industry has billions of dollars invested in silicon-based processes.

For these reasons, speed improvements have traditionally been achieved by making transistors smaller. However, some pundits believe that we are reaching the end of this route using conventional technologies. At one time, the limiting factors appeared to be simple process limitations: the quality of the resist, the ability to manufacture accurate masks, and the features that could be achieved with the wavelength of ultraviolet light. Around 1990, when structures with dimensions of 1.0 microns first became available, it was believed that structures of 0.5 microns would be the effective limit that could be achieved with opto-lithographic processes, and that the next stage would be a move to X-ray lithography. However, there have been constant improvements in the techniques associated with mask fabrication, optical systems and lenses, servo motors and positioning systems, and advances in chemical engineering such as chemically-amplified resists.[18] The combination of all these factors means that it is now possible to achieve structures smaller than 0.1 microns by continuing to refine existing processes.

However, there are other considerations. The speed of a transistor is strongly related to its size, which affects the distance electrons have to travel. Thus, to enable transistors to switch faster, technologists have concentrated on reducing size, a strategy which is commonly referred to as *scaling*. However, while scaling reduces the size of structures, it is necessary to maintain certain levels of dopants to achieve the desired effect. This means that, as the size of the structures is reduced, it is necessary to increase the concentration of dopant atoms. Increasing the concentration beyond a certain level causes leakage, resulting in the transistor being permanently ON and therefore useless. Thus, technologists are increasingly considering alternative materials and structures.

An interface between two regions of semiconductor having the same basic composition but opposing types of doping is called a *homojunction*. By comparison, the interface between two regions of dissimilar semiconductor materials is called a *heterojunction*. Homojunctions dominate current processes because they are easier to fabricate. However, the interface of a heterojunction has naturally occurring electric fields, which can be used to accelerate electrons, and transistors created using heterojunctions can switch much faster than their homojunction counterparts of the same size.

One form of heterojunction that is attracting a lot of interest is found at the interface between silicon and germanium. Silicon and germanium are in the same family of elements and have similar crystalline structures. In theory this should make it easy to combine them, but it's a little more difficult in practice. A process currently being evaluated is to create a standard silicon wafer with

---

[18] In the case of a chemically amplified resist, the application of a relatively small quantity of ultraviolet light stimulates the formation of chemicals in the resist, which accelerates the degrading process. This reduces the amount of ultraviolet light required to degrade the resist and allows the creation of finer features with improved accuracy.

doped regions, and then to grow extremely thin layers of a silicon-germanium alloy where required.

The two most popular methods of depositing these layers are *chemical vapor deposition (CVD)* and *molecular beam epitaxy (MBE)*. In the case of chemical vapor deposition, a gas containing the required molecules is converted into a plasma[19] by heating it to extremely high temperatures using microwaves. The plasma carries atoms to the surface of the wafer where they are attracted to the crystalline structure of the substrate. This underlying structure acts as a template. The new atoms continue to develop the structure to build up a layer on the substrate's surface (Figure 21-20).

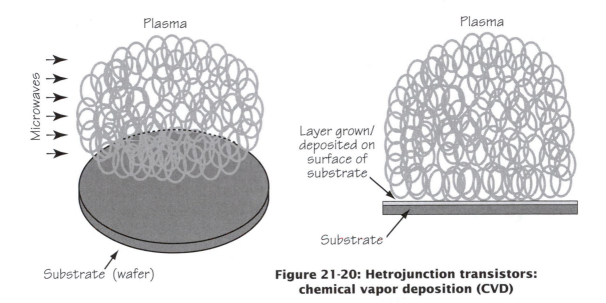

**Figure 21-20: Hetrojunction transistors:
chemical vapor deposition (CVD)**

By comparison, in the case of molecular beam epitaxy, the wafer is placed in a high vacuum, and a guided beam of ionized molecules is fired at it, effectively allowing molecular-thin layers to be "painted" onto the substrate where required.[20]

---

[19] A gaseous state in which the atoms or molecules are dissociated to form ions.

[20] Molecular beam epitaxy is similar to *electron beam epitaxy (EBE)*, in which the wafer is first coated with a layer of dopant material before being placed in a high vacuum. A guided beam of electrons is then fired at the wafer causing the dopant to be driven into it.

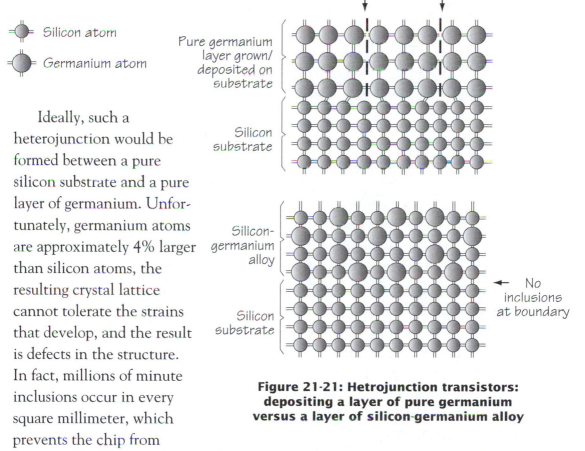

- Silicon atom
- Germanium atom

Figure 21-21: Hetrojunction transistors: depositing a layer of pure germanium versus a layer of silicon-germanium alloy

Ideally, such a heterojunction would be formed between a pure silicon substrate and a pure layer of germanium. Unfortunately, germanium atoms are approximately 4% larger than silicon atoms, the resulting crystal lattice cannot tolerate the strains that develop, and the result is defects in the structure. In fact, millions of minute inclusions occur in every square millimeter, which prevents the chip from working. Hence, the solution of growing a layer of silicon-germanium alloy, which relieves the stresses in the crystalline structure, thereby preventing the formation of inclusions (Figure 21-21).

Heterojunctions offer the potential to create transistors that switch as fast, or faster, than those formed using gallium arsenide, but use significantly less power. Additionally, they have the advantage of being able to be produced on existing fabrication lines, thereby preserving the investment and leveraging current expertise in silicon-based manufacturing processes.

Buckyballs and Nanotubes

Prior to the mid-1980s, the only major forms of pure carbon known to us were graphite and diamond. In 1985, however, a third form consisting of spheres formed from 60 carbon atoms (C_{60}) was discovered. Officially known as Buckministerfullerine[21]—named after the American architect R. Buckminister Fuller who designed geodesic domes with the same fundamental symmetry— these spheres are more commonly known as "buckyballs." In 2000, scientists with the U.S. Department of Energy's Lawrence Berkeley National Laboratory (Berkeley Lab) and the University of California at Berkeley reported that they had managed to fashion a transistor from a single buckyball.

Sometime later, scientists discovered another structure called the *nanotube*, which is like taking a thin sheet of carbon and rolling it into a tube. Nanotubes can be formed with walls that are only one atom thick. The resulting tube has a diameter of 1 nano (one thousandth of one millionth of a meter).

Nanotubes are an almost ideal material. They are stronger than steel, have excellent thermal stability, and they are also tremendous conductors of heat and electricity. For example, a conductor formed from millions of nanotubes arranged to form a 1cm cross-section could conduct more than 1 billion amps.

In addition to acting as wires, nanotubes can be persuaded to act as transistors—this means that we now have the potential to replace silicon transistors with molecular-sized equivalents at a level where standard semiconductors cease to function. We're only just starting to experiment with nanotube-based transistors, but they can theoretically run at clock speeds of one terahertz or more (one thousand times faster than today's processors).

In the longer-term future, it may be that some integrated circuits will use nanotubes to form both the transistors and the wires linking them together. In the shorter-term, we can expect to see interesting mixtures of nanotubes with conventional technologies. For example, in the summer of 2002, IBM announced an experimental transistor called a *carbon nanotube FET (CNFET)*. This device is based on a field-effect transistor featuring a carbon nanotube measuring only 1.4 nanometers in diameter acting as the channel (the rest of the transistor is formed using conventional silicon and metal processing technologies). It will be a number of years before these devices become commercially available, but they are anticipated to significantly outperform their silicon-only counterparts.[22]

---

[21] Science magazine voted Buckministerfullerine the *"Molecule of the Year"* in 1991.

[22] As an aside, constructing the masks used to create integrated circuits with 0.06 micron (60 nano) geometries and smaller (especially those featuring esoteric technologies like CNFETs) looks set to be to be a major problem. One company that may have the answer with their nanotechnology-based innovations is Nano Innovation Systems (www.nanoinnovationsys.com).

Diamond Substrates

As was noted in the previous section, there is a constant drive towards smaller, more densely packed transistors switching at higher speeds. Unfortunately, packing the little devils closer together and cracking the whip to make them work faster substantially increases the amount of heat that they generate. Similarly, the increasing utilization of optical interconnect relies on the use of laser diodes, but today's most efficient laser diodes only convert 30% to 40% of the incoming electrical power into an optical output, while the rest emerges in the form of heat. Although each laser diode is relatively small (perhaps as small as only 500 atoms in diameter), their heating effect becomes highly significant when tens of thousands of them are performing their version of Star Wars.

And so we come to *diamond*, which derives its name from the Greek *adamas*, meaning "invincible." Diamond is famous as the hardest substance known, but it also has a number of other interesting characteristics: it is a better conductor of heat at room temperatures than any other material,[23] in its pure form it is a good electrical insulator, it is one of the most transparent materials available, and it is extremely strong and non-corrosive. For all of these reasons, diamond would form an ideal substrate material for multichip modules.[24]

In addition to multichip modules, diamond has potential for a variety of other electronics applications. Because diamond is in the same family of elements as silicon and germanium, it can function as a semiconductor and could be used as a substrate for integrated circuits. In fact, in many ways, diamond would be far superior to silicon: it is stronger, it is capable of withstanding high temperatures, and it is relatively immune to the effects of radiation (the bane of components intended for nuclear and space applications). Additionally, due to diamond's high thermal conductivity, each die would act as its own heat sink and would rapidly conduct heat away. It is believed that diamond-based devices could switch up to 50 times faster than silicon and operate at temperatures over 500°C.

---

[23] Diamond can conduct five times as much heat as copper, which is the second most thermally conductive material known.

[24] Actually, other exotic substrates are also of interest to electronic engineers, including sapphire, which is of particular use in microwave applications.

Chemical Vapor Deposition

Unfortunately, today's integrated circuit manufacturing processes are geared around fabricating large numbers of chips on wafers that can be up to 300 mm in diameter. By comparison, a natural diamond 10 mm in diameter would be considered to be really, REALLY large. It simply wouldn't be cost-effective to take one of these beauties, slice it up, and make diamond-based chips one at a time. Furthermore, if you were to be the proud owner of a large natural diamond, the last thing that would come to mind would be to chop it up into thin slices for electronics applications!

However, there are a number of methods for depositing or growing diamond crystals, one of the most successful being chemical vapor deposition (CVD), which was introduced in the earlier discussions on heterojunction transistors. With this CVD process, microwaves are used to heat mixtures of hydrogen and hydrocarbons into a plasma, out of which diamond films nucleate and form on suitable substrates. Although the plasma chemistry underlying this phenomena is not fully understood, polycrystalline diamond films can be nucleated on a wide variety of materials, including metals such as titanium, molybdenum, and tungsten, ceramics, and other hard materials such as quartz, silicon, and sapphire.

Chemical Vapor Infiltration

CVD processes work by growing layers of diamond directly onto a substrate. A similar, more recent technique—known as *chemical vapor infiltration (CVI)*[25] —commences by placing diamond powder in a mold. Additionally, thin posts, or columns, can be preformed in the mold, and the diamond powder can be deposited around them. When exposed to the same plasma as used in the CVD technique, the diamond powder coalesces into a polycrystalline mass. After the CVI process has been performed, the posts can be dissolved, leaving holes through the diamond for use in creating vias. CVI processes can produce diamond layers twice the thickness of those obtained using CVD techniques at a fraction of the cost.

---

[25] Thanks go to Crystallume, Menlo Park, CA, USA, for the information on their CVD and CVI processes.

Ubiquitous Laser Beams

An alternative, relatively new technique for creating diamond films involves heating carbon with laser beams in a vacuum. Focusing the lasers on a very small area generates extremely high temperatures, which rip atoms away from the carbon and also strip away some of their electrons. The resulting ions fly off and stick to a substrate placed in close proximity. Because the lasers are tightly focused, the high temperatures they generate are localized on the carbon, permitting the substrate to remain close to room temperature. Thus, this process can be used to create diamond films on almost any substrate, including semiconductors, metals, and plastics.

The number of electrons stripped from the carbon atoms varies, allowing their ions to reform in *nanophase* diamond structures that have never been seen before. Nanophase materials are a new form of matter that was only discovered relatively recently, in which small clusters of atoms form the building blocks of a larger structure. These structures differ from those of naturally occurring crystals, in which individual atoms arrange themselves into a lattice. In fact, it is believed that it may be possible to create more than thirty previously unknown forms of diamond using these techniques.

The Maverick Inventor

Last but not least, in the late 1980s, a maverick inventor called Ernest Nagy[26] invented a simple, cheap, and elegant technique for creating thin diamond films. Nagy's process involves treating a soft pad with diamond powder, spinning the pad at approximately 30,000 revolutions per minute, and maintaining the pad in close contact with a substrate. Although the physics underlying the process is not fully understood, diamond is transferred from the pad to form a smooth and continuous film on the substrate. The diamond appears to undergo some kind of phase transformation, changing from a cubic arrangement into a hexagonal form with an unusual structure. Interestingly enough, Nagy's technique appears to work with almost any material on almost any substrate!

---

[26] Nagy, whose full name is Ernest Nagy de Nagybaczon, was born in 1942 in Hungary. He left as a refugee in the 1956 uprising and now lives in England.

The requirement for Single-Crystal Diamond

All of the techniques described above result in films that come respectfully close, if not equal, to the properties of natural diamond in such terms as heat conduction. Thus, these films are highly attractive for use as substrates in multichip modules. However, the unusual diamond structures that are created fall short of the perfection required for them to be used as a substrate suitable for the fabrication of transistors.

Substrates for integrated circuits require the single, large crystalline structures found only in natural diamond. Unfortunately, there are currently no known materials onto which a single-crystal diamond layer will grow, with the exception of single crystal diamond itself (which sort of defeats the point of doing it in the first place). The only answer appears to be to modify the surface of the substrate onto which the diamond layer is grown, and many observers believe that this technology may be developed in the near future. If it does prove possible to create consistent, single-crystal diamond films, then, in addition to being *"a girl's best friend,"* diamonds would also become *"an electronic engineer's biggest buddy."*

Chip-On-Chip (COC)

The intra-chip connections linking bare die on a multichip module are a source of fairly significant delays. One obvious solution is to mount the die as closely together as possible, thereby reducing the lengths of the tracks and the delays associated with them. However, each die can have only eight others mounted in close proximity on a 2D substrate. The solution is to proceed into three dimensions. Each die is very thin and, if they are mounted on top of each other, it is possible to have over a hundred die forming a 3D cube (Figure 21-22).

One problem with this *chip-on-chip* (COC) technique is the amount of heat that is generated, which drastically affects the inner layers forming the cube. This problem could be eased by constructing the die out of diamond as discussed above. Another problem with traditional techniques is that any tracks linking the die must be connected down the outer surfaces of the cube. The result is that the chip-on-chip technique has typically been restricted to applications utilizing identical die with regular structures. For example, the most common application to date has been large memory devices constructed by stacking SRAM or DRAM die on top of each other.

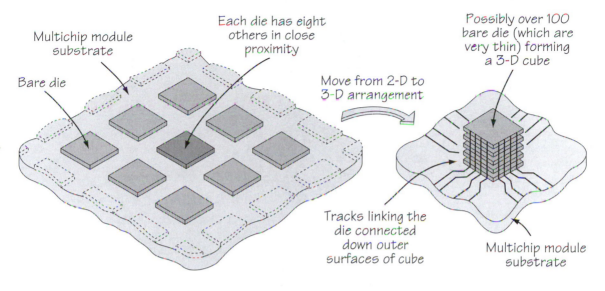

Figure 21-22: Chip-on-chip: die organized as 3D cubes

A relatively new technique which may serve to alleviate the problem of chip-on-chip interconnect is a process for creating vias through silicon substrates. Experiments have been performed in which aluminum "blobs" are placed on the surface of a silicon substrate and, by means of a gradient furnace, the aluminum migrates through the silicon providing vias from one side to the other (Figure 21-23).

Another technique more in keeping with the times is to create the vias by punching the aluminum through the silicon by means of a laser. These developments pave the way for double-sided silicon substrates with chips and interconnect on both sides. Additionally, they offer strong potential for interconnecting the die used in chip-on-chip structures.

Conductive Adhesives

Many electronics fabrication processes are exhibiting a trend towards mechanical simplicity with underlying sophistication in materials technology. A good example of this trend is illustrated by conductive, or *anisotropic*, adhesives, which contain minute particles of conductive material.

These adhesives find particular application with the flipped-chip techniques used to mount bare die on the substrates of hybrids, multichip modules, or circuit boards. The adhesive is screen printed onto the substrate at the site

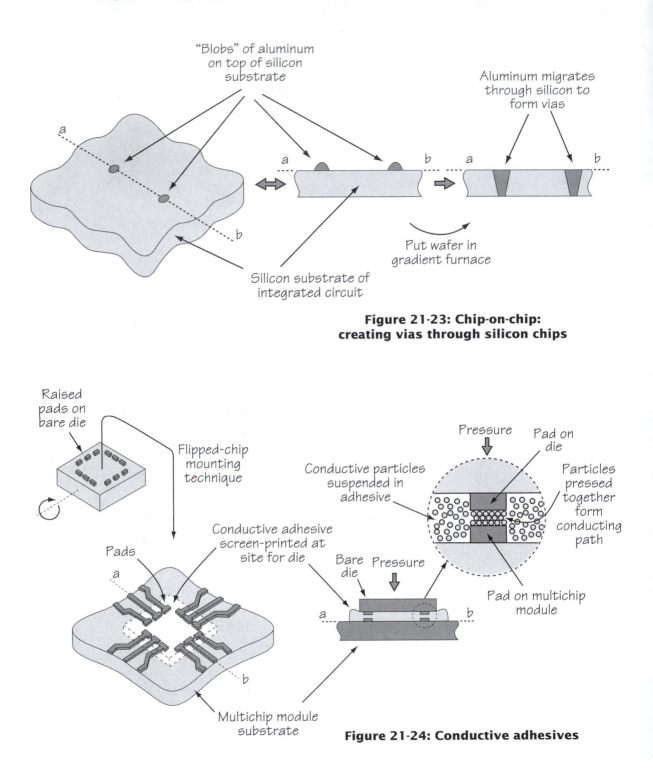

"Blobs" of aluminum on top of silicon substrate

Aluminum migrates through silicon to form vias

a

b

a

b

a

b

Put wafer in gradient furnace

Silicon substrate of integrated circuit

Figure 21-23: Chip-on-chip: creating vias through silicon chips

Raised pads on bare die

Flipped-chip mounting technique

Pressure

Pad on die

Conductive particles suspended in adhesive

Particles pressed together form conducting path

Conductive adhesive screen-printed at site for die

Pads

a

b

Bare die

Pressure

a

b

Pad on multichip module

Multichip module substrate

Figure 21-24: Conductive adhesives

where the die is to be located, the die is pressed into the adhesive, and the adhesive is cured using a combination of temperature and pressure (Figure 21-24).

The beauty of this scheme is that the masks used to screen print the adhesive do not need to be too complex, and the application of the adhesive does not need to be excessively precise, because it can be spread across all of the component pads. The conducting particles are only brought in contact with each other at the sites where the raised pads on the die meet their corresponding pads on the substrate, thereby forming good electrical connections.

The original conductive adhesives were based on particles such as silver. But, in addition to being expensive, metals like silver can cause electron migration problems at the points where they meet the silicon substrates. Modern equivalents are based on organic metallic particles, thereby reducing these problems.

In addition to being simpler and requiring fewer process steps than traditional methods, the conductive adhesive technique removes the need for solder, whose lead content is beginning to raise environmental concerns.

Superconductors

One of the "Holy Grails" of the electronics industry is to have access to conductors with zero resistance to the flow of electrons, and for such conductors, known as *superconductors*, to operate at room temperatures. As a concept, superconductivity is relatively easy to understand: consider two sloping ramps into which a number of pegs are driven. In the case of the first ramp, the pegs are arranged randomly across the surface, while in the second the pegs are arranged in orderly lines. Now consider what happens when balls are released at the top of each surface (Figure 21-25).

In the case of the randomly arranged pegs, the ball's progress is repeatedly interrupted, while in the case of the pegs arranged in orderly lines, the ball slips through *"like water off a duck's back."* Although analogies are always suspect (and this one doubly so), the ramps may be considered to represent conducting materials, the gravity accelerating the balls takes on the role of voltage differentials applied across the ends of the conductors, the balls play the part of electrons, and the pegs portray atoms.

The atoms in materials vibrate due to the thermal energy contained in the material: the higher the temperature, the more the atoms vibrate. An ordinary

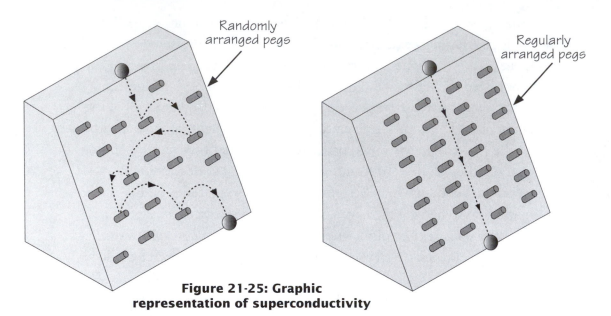

**Figure 21-25: Graphic
representation of superconductivity**

conductor's electrical resistance is caused by these atomic vibrations, which obstruct the movement of the electrons forming the current. Using the Kelvin,[27] or absolute, scale of temperature, 0 K (corresponding to –273ºC) is the coldest possible temperature and is known as absolute zero. If an ordinary conductor were cooled to a temperature of absolute zero, atomic vibrations would cease, electrons could flow without obstruction, and electrical resistance would fall to zero. A temperature of absolute zero cannot be achieved in practice, but some materials exhibit superconducting characteristics at higher temperatures.[28]

In 1911, the Dutch physicist Heike Kamerlingh Onnes discovered super-conductivity in mercury at a temperature of approximately 4K (–269ºC). Many other superconducting metals and alloys were subsequently discovered but, until 1986, the highest temperature at which superconducting properties were achieved was around 23K (-250ºC) with the niobium-germanium alloy (Nb_3Ge).

---

[27] Invented by the British mathematician and physicist William Thomas, first Baron of Kelvin.

[28] If the author were an expert in superconductivity, this is the point where he might be tempted to start muttering about *"Correlated electron movements in conducting planes separated by insulating layers of mesoscopic thickness, under which conditions the wave properties of electrons assert themselves and electrons behave like waves rather than particles."* But he's not, so he won't.

In 1986, Georg Bednorz and Alex Müller discovered a metal oxide that exhibited superconductivity at the relatively high temperature of 30K (−243°C). This led to the discovery of ceramic oxides that superconduct at even higher temperatures. In 1988, an oxide of thallium, calcium, barium and copper ($Tl^2Ca^2Ba_2Cu_3O_{10}$) displayed superconductivity at 125K (−148°C), and, in 1993, a family based on copper oxide and mercury attained superconductivity at 160K (−113°C). These "high-temperature" superconductors are all the more noteworthy because ceramics are usually extremely good insulators.

Like ceramics, most organic compounds are strong insulators; however, some organic materials known as *organic synthetic metals* do display both conductivity and superconductivity. In the early 1990s, one such compound was shown to superconduct at approximately 33K (−240°C). Although this is well below the temperatures achieved for ceramic oxides, organic superconductors are considered to have great potential for the future.

New superconducting materials are being discovered on a regular basis, and the search is on for room temperature superconductors, which, if discovered, are expected to revolutionize electronics as we know it. In fact, in the summer of 2002 while penning the final words in this chapter, the author was made aware of a new class of polymer materials called *ultraconductors*, which exhibit superconducting properties at ambient (room) temperature. These materials are claimed to conduct electricity at least 100,000 times better than gold, silver, or copper! Unfortunately, the author has been bound to secrecy and may not reveal any more at this time.

Nanotechnology

Nanotechnology is an elusive term that is used by different research and development teams to refer to whatever it is that they're working on at the time. However, regardless of their particular area of interest, nanotechnology always refers to something extremely small; for example, motors and pumps the size of a pinhead, which are created using similar processes to those used to fabricate integrated circuits. In fact, around the beginning of 1994, one such team unveiled a miniature model car which was smaller than a grain of short-grain rice. This model contained a micro-miniature electric motor, battery, and gear train, and was capable of traversing a fair-sized room (though presumably not on a shag-pile carpet).

In 1959, the legendary American physicist Richard Feynman gave a vision-ary talk, in which he described the possibility by which sub-microscopic com-puters could perhaps be constructed. Feynman's ideas have subsequently been extended to become one of the more extreme branches of nanotechnology featuring micro-miniature products that assemble themselves! The theory is based on the way in which biological systems operate. Specifically, the way in which *enzymes*[29] act as *biological catalysts*[30] to assemble large, complex mol-ecules from smaller molecular building blocks.

Back to the Water Molecule

Before commencing this discussion, it is necessary to return to the humble water molecule.[31] As you may recall, water molecules are formed from two hydrogen atoms and one oxygen atom, all of which share electrons between themselves. However, the elec-trons are not distributed equally, because the oxygen atom is a bigger, more robust fellow which grabs more than its fair share (Figure 21-26).

The angle formed between the two hydrogen atoms is 105°. This is because, of the six electrons that the oxygen atom owns, two are shared with the hydrogen atoms and four remain the exclusive property of the oxygen. These four huddle together on one side of the oxygen atom and put

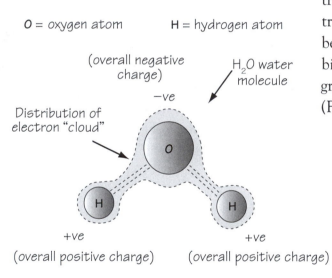

O = oxygen atom   H = hydrogen atom

(overall negative charge)

H_2O water molecule

−ve

Distribution of electron "cloud"

O

H   H

+ve   +ve

(overall positive charge)   (overall positive charge)

Figure 21-26: Distribution of electrons in a water molecule

---

[29] Enzymes are complex proteins that are produced by living cells and catalyze biochemical reactions at body temperatures.

[30] A catalyst is a substance that initiates a chemical reaction under different conditions (such as lower temperatures) than would otherwise be possible. The catalyst itself remains unchanged at the end of the reaction.

[31] Water molecules were introduced in Chapter 2.

"pressure" on the bond angle. The bond angle settles on 105° because this is the point where the pressure from the four electrons is balanced by the natural repulsion of the two positively charged hydrogen atoms (similar charges repel each other).

The end result is that the oxygen atom has an overall negative charge, while the two hydrogen atoms are left feeling somewhat on the positive side. This unequal distribution of charge means that the hydrogen atoms are attracted to anything with a negative bias-for example, the oxygen atom of another water molecule. Although the strength of the resulting bond, known as a *hydrogen bond*, is weaker than the bond between the hydrogen atom and its "parent" oxygen atom, it is still quite respectable.

When water is cooled until it freezes, its resulting crystalline structure is based on these hydrogen bonds. Even in its liquid state, the promiscuous, randomly wandering water molecules are constantly forming hydrogen bonds with each other. These bonds persist for a short time until another water molecule clumsily barges into them and knocks them apart. From this perspective, a glass of water actually contains billions of tiny ice crystals that are constantly forming and being broken apart again.

Imagine a Soup

However, we digress. Larger molecules can form similar electrostatic bonds with each other. Imagine a "soup" consisting of large quantities of many different types of molecules, two of which, M_a and M_b, may be combined to form larger molecules of type M_{ab} (Figure 21-27).

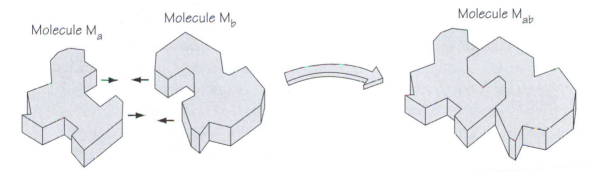

Molecule M_a

Molecule M_b

Molecule M_{ab}

Figure 21-27: Combining molecules M_a and M_b to form M_{ab}

This is similar in concept to two pieces of a jigsaw, which will only fit together if they are in the correct orientation to each other. Similarly, M_a and M_b will only bond to form M_{ab} if they are formally presented to each other in precisely the right orientation. However, the surfaces of the molecules are extremely complex three-dimensional shapes, and achieving the correct orientation is a tricky affair. Once the molecules have been brought together their resulting bonds are surprisingly strong, but the chances of the two molecules randomly achieving exactly the correct orientation to form the bonds are extremely small.

It is at this point of the story that enzymes reenter the plot. There are numerous enzymes, each dedicated to the task of "matchmaking" for two of their favorite molecules. The surface of an enzyme is also an extremely complex three-dimensional shape, but it is much larger than its target molecules and has a better chance of gathering them up. The enzyme floats around until it bumps into a molecule of type M_a to which it bonds. The enzyme then continues on its trek until it locates a molecule of type M_b. When the enzyme bonds to molecule M_b, it orientates it in exactly the right way to complete the puzzle with molecule M_a (Figure 21-28).

The bonds between M_a and M_b are far stronger than their bonds to the enzyme. In fact, as soon as these bonds are formed, the enzyme is actually repelled by the two little lovebirds and promptly thrusts M_{ab} away. However, the enzyme immediately forgets its pique, and commences to search for two more molecules (some enzymes can catalyze their reactions at the rate of half a million molecules per minute).

The saga continues, because another, larger enzyme may see its task in life as bringing M_{ab} together with yet another molecule M_{cd}. And so it continues, onwards and upwards, until the final result, whatever that may be, is achieved.

As our ability to create "designer molecules" increases, it becomes increasingly probable that we will one day be able to create "designer enzymes." This would enable us to mass-produce "designer proteins" that could act as alternatives to semiconductors (see also the *Protein Switches and Memories* topic earlier in this chapter). As one of the first steps along this path, a process could be developed to manufacture various proteins that could then be bonded to a substrate or formed into three-dimensional blocks for optical memory

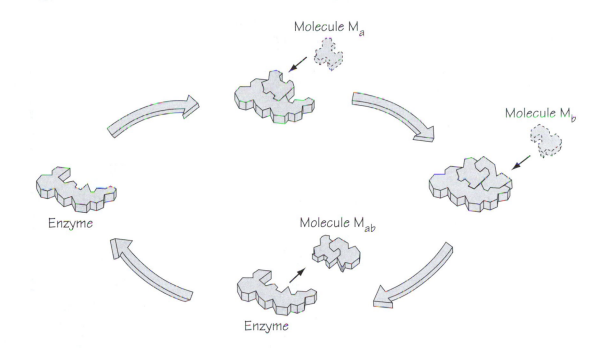

Figure 21-28: Using an enzyme to form molecule M_{ab}

applications. At a more sophisticated level, it may be possible for such a process to directly create the requisite combinations of proteins as self-replicating structures across the face of a substrate.

However, the possibilities extend far beyond the mass-production of proteins. It is conceivable that similar techniques could be used to assemble non-organic structures such as microscopic electromechanical artifacts. All that would be required (he said casually) would be for the individual components to be shaped in such a way that naturally occurring electrostatic fields would cause them to form bonds when they were brought together with their soul mates. In fact, this is one step along the path towards molecular-sized robots known as *nanobots*. Taken to extremes, the discipline of electronics in the future may not involve the extreme temperatures, pressures, and noxious chemicals that are in vogue today. Instead, electronics may simply involve "cookbook" style recipes; for example, the notes accompanying an electronics course in 2050 AD may well read as follows:

338 ■ Chapter Twenty-One

Intermediate Electronics (Ages 12 to 14)
Supercomputers 101

***Instructions for creating a micro-miniature massively parallel supercomputer.**

a) Obtain a large barrel.

b) In your barrel, mix two parts water and one part each of chemicals A, B, C . . .

c) Add a pinch of nanobot-mix (which you previously created in Nanobots 101).

d) Stir briskly for one hour with a large wooden spoon.

Congratulations, you will find your new supercomputers in the sediment at the bottom of the barrel. Please keep one teaspoon of these supercomputers for your next lesson.

*These instructions were reproduced from *Bebop to the Boolean Boogie*, 50th edition, 2050, the most popular electronics book in the history of the universe!

Of course, some of this is a little far-fetched (with the hopeful exception of the references to *Bebop to the Boolean Boogie*). However, for what it's worth, the author would bet his ex-wife's life savings that this type of technology will occur one day, and also that it will be here sooner than you think!

Again, the Mind Boggles

In reality, we've only touched on a very few of the myriad ideas that are out there running wild and free. For example, one area that is currently seen as a growth industry is that of *MicroElectroMechanical Systems (MEMS)* and their optical counterparts OMEMS. These refer to devices that contain both electrical, mechanical, and—in the case of OMEMS—optical elements and are physically very small (sometimes measuring only a few millionths of a meter in size). MEMS can be used to create microscopic sensors to monitor the surrounding environment and actuators that can modify the environment with great precision, while OMEMS find use in communications systems.

As an example, in early 2001, biophysicists at the Hungarian Academy of Sciences devised a way to create microscopic machines that are constructed and operated by light. In one case they created gears that are each less than 1/5,000 of an inch (5 μm) in diameter. The rotors spin when illuminated by a low-power laser as photons hit their flanges. Such devices could pump materials across miniature chemical arrays.

There are several theories as to how life developed on Earth, including "seeding" by meteors from outer space[32] or random combinations of atoms forming simple molecules, which in turn formed rudimentary proteins, then . . . and so it goes.

Maybe a combination of both of these theories is true. However, to test the *"random combination"* theory, scientists decided to replicate the conditions on Earth billions of years ago before life developed, with the intention of duplicating the happy event. So, they took a large glass jar, added some water, filled the rest with the gases that were around at the time, and applied electrical discharges to the mixture to simulate the effects of lightning strikes. Much to their surprise . . . nothing happened.

Obviously this was something of a downer. While they were all scratching their heads, a tentative hand was raised in the air, and a high-pitched voice belonging to the tea boy said "Excuuussse me, but what about adding rocks?" After they had slapped him round the back of the head for his insolence, they thought "What the heck, let's give it a whirl." So, they added a mixture of rocks, zapped everything with electric discharges again, and, lo and behold, they found amino acids and simple proteins.

It turned out that when certain shales are split, their rough surfaces form exactly the right templates to bond atoms and molecules together to create the rudimentary building blocks of life!

Now, this does not either prove or disprove the existence of God. However, there is almost certainly something somewhere, because everything fits together so perfectly. At the very least, if there is a God, then it certainly serves to confirm the fact that *"He/She works in mysterious ways, His/Her wonders to perform."*

---

[32] Scientists have found amino acids (the basic building blocks of proteins, and therefore life itself) in a number of meteors.

Small as they are, MEMS and OMEMS are still huge on the molecular scale. In early 2001, the University of Illinois Beckman Institute for Advanced Science and Technology reported research with organic molecules like tiny mechanical switches. These molecular switches are only attached to the substrate by a single atom and they can spin as fast as 100 trillion times a second, which provides the potential for switching arrays running at 100 terahertz!

And just to illustrate the range of things that are being investigated, biomedical engineers at the Georgia Institute of Technology and Emory University are working on linking electronics to living neurons to create incredibly sophisticated neural networks.

Summary

The potpourri of technologies introduced here have been offered for your delectation and delight. Some of these concepts may appear to be a little on the wild side, and you certainly should not believe everything that you read or hear. On the other hand, you should also be careful not to close your mind, even to seemingly wild and wacky ideas, in case something sneaks up behind you and bites you on the ####.[33]

As the *Prize* said in *Where is Earth* by Robert Sheckley: *"Be admiring but avoid the fulsome, take exception to what you don't like, but don't be stubbornly critical; in short, exercise moderation except where a more extreme attitude is clearly called for."*

And so, with our lower lips quivering and little tears rolling down our cheeks, we come to the close of this, the penultimate chapter. But turn that frown upside down into a smile, because Chapter 22 (on the CD accompanying this book) is going to have you sitting on the edge of your seat, and there are still oodles of yummy appendices containing enough mouth-watering information to form a book in their own right. As that great British Prime Minister Winston Spencer Churchill (1874–1965) would have said: *"Now this is not the end. It is not even the beginning of the end. But it is, perhaps, the end of the beginning."*[34]

---

[33] Arsek no questions!

[34] Speech at the Lord Mayor's Day Luncheon, London, England (November 10, 1942).

Assertion-Level Logic

The purpose of a circuit diagram is to convey the maximum amount of information in the most efficient fashion. One aspect of this is assigning meaningful names to wires; for example, naming a wire system_reset conveys substantially more information than calling it big_boy. Another consideration is that the signals carried by wires may be *active-high* or *active-low* (see sidebar).

Consider a portion of a circuit containing a tri-state buffer with an active-low ~enable input (Figure A-1).

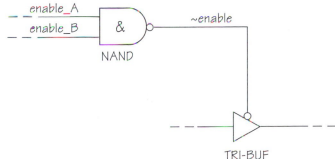

Figure A-1. Naming active-low signals

Beware — here be dragons!

The definitions of *active-high* and *active-low* are subject to confusion. Some academics (and even text books) define an active-low signal as one whose asserted (TRUE or logic 1) state is at a lower voltage level than its unasserted (FALSE or logic 0) state. However, this definition serves only to cause confusion when combined with negative logic implementations (see also Appendix B).

Engineers in the trenches generally take the position that an active-high state is one whose active state is considered to be TRUE or logic 1, while an active-low state is one whose active

state is considered to be FALSE or logic 0. These definitions allow all forms of logic—including the assertion-level logic introduced in this chapter—to be represented without any confusion, regardless of whether positive or negative logic implementations are employed. These manly-man definitions are the ones used throughout this book. However, when using the terms active-high and active-low in discussions with other folks, you are strongly advised to make sure that you all understand them to mean the same thing before you find yourself up to your ears in alligators.

Instead of a tilde, some designers prefer to use an exclamation mark '!' nicknamed a *shriek* to indicate active-low signals—for example, !enable. Alternatively, designers may prefix or postfix active-low signal names with a special letter; for example, Nenable or enableB. Yet another technique is to draw a horizontal line, or bar, over the name, but this is not recommended for a number of reasons. First, signal names with bars over them are difficult to replicate in textual form on a computer. Second, bars are often used to indicate negations in Boolean equations. In this latter case it would be impossible to determine whether a bar formed an integral part of a signal's name or if it was intended to indicate that the signal should be negated. Tilde characters may also be used to indicate complementary outputs; for example, the q and ~q outputs from a latch or a flip-flop.

To convey as much information as possible, it is preferable to indicate the nature of an active-low signal in its name. One method of achieving this is to prefix the name with a tilde character '~'; hence, ~enable (see sidebar). When both the enable_A and enable_B signals are set to their active-high states, the ~enable signal is driven to its active-low state and the tri-state buffer is enabled.

It is important to note that the active-low nature of ~enable is not decided by the NAND. The only thing that determines whether a signal is active-high or active-low is the way in which it is used by any target gates, such as the tri-state buffer in this example. Thus, active-low signals can be generated by ANDs and ORs as easily as NANDs and NORs.

The problem with the standard symbols for BUF, NOT, AND, NAND, OR, and NOR is that they are tailored to reflect operations based on active-high signals being presented to their inputs. To address this problem, special *assertion-level* logic symbols can be used to more precisely indicate the function of gates with active-low inputs. Consider how we might represent a NOT gate used to invert an ~enable signal into its *enable* counterpart (Figure A-2).

Figure A-2. Standard versus assertion-level NOT symbols

Both symbols indicate an inversion, but the assertion-level symbol is more intuitive because it reflects the fact that an active-low input is being transformed into an active-high output. In both cases the bobbles on the symbols indicate the act of inversion. One way to visualize this is that the symbol for a NOT has been pushed into a symbol for a BUF until only its bobble remains visible (Figure A-3).

In the real world, both standard and assertion-level symbols are implemented using identical logic gates. Assertion-level logic does not affect the final implementation, but simply offers an alternative way of viewing things. Visualizing bobbles as representing inverters is a useful technique for handling more complex functions. Consider a variation on our original circuit in which the ~enable_A and ~enable_B signals are active-low and the tri-state buffer has an active-high enable input (Figure A-4).

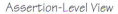
Standard View

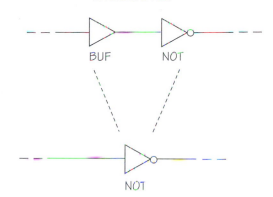

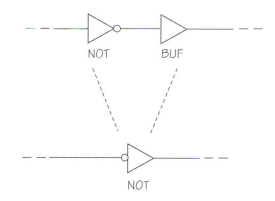

Figure A-3. One way to visualize standard versus assertion-level NOT symbols

The purpose of this circuit is to set the *enable* signal to its active-high state if either of the ~enable_A or ~enable_B signals are in their active-low states.

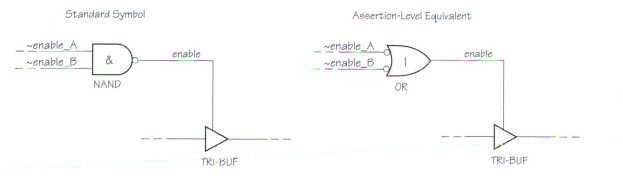

Figure A-4. Standard NAND versus assertion-level OR symbols

Remember that both of these circuit representations are functionally identical (you can easily prove this by drawing out their truth tables). Once again, however, the assertion-level representation is the more intuitive, especially for someone who is unfamiliar with the function of the circuit. This is because the assertion-level symbol unambiguously indicates that *enable* will be set to logic 1 if either of the *~enable_A* or *~enable_B* signals are presented with logic 0s.

Any standard primitive gate symbol can be transformed into its assertion-level equivalent by inverting all of its inputs and outputs, and exchanging any & (AND) operators for | (OR) operators (and vice versa). In fact, a quick review of Chapter 9 reveals that these steps are identical to those used in a DeMorgan Transformation. Thus, assertion-level symbols may also be referred to as DeMorgan equivalent symbols. The most commonly used assertion-level symbols are those for BUF, NOT, AND, NAND, OR, and NOR (Figure A-5).

Note that the assertion-level symbols for XOR and XNOR are identical to their standard counterparts. The reason for this is that if you take an XOR and invert its inputs and outputs, you end up with an XOR again; similarly for an XNOR. A little experimentation with their truth tables will quickly reveal the reason why.

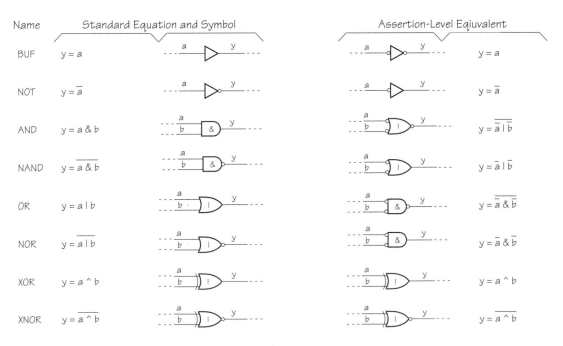

Figure A-5. Commonly used assertion-level symbols

Positive Logic versus Negative Logic

The terms *positive logic* and *negative logic* refer to two conventions that dictate the relationship between logical values and the physical voltages used to represent them. Unfortunately, although the core concepts are relatively simple, fully comprehending all of the implications associated with these conventions requires an exercise in lateral thinking sufficient to make even a strong man break down and weep!

Before plunging into the fray, it is important to understand that logic 0 and logic 1 are always equivalent to FALSE and TRUE, respectively.[1] The reason these terms are used interchangeably is that digital functions can be considered to represent either *logical* or *arithmetic* operations (Figure B-1).

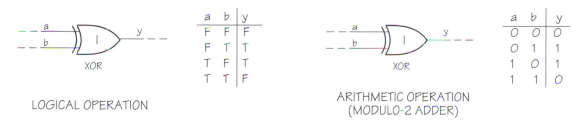

Figure B-1. Logical versus arithmetic views of a digital function

Having said this, it is generally preferable to employ a single consistent format to cover both cases, and it is easier to view logical operations in terms of 0s and 1s than it is to view arithmetic operations in terms of Fs and Ts. The key point to remember as we go forward is that, by definition, logic 0 and logic 1 are logical concepts that have no direct relationship to any physical values.

---

[1] Unless you're really taking a walk on the wild side, in which case all bets are off.

Physical to Abstract Mapping (NMOS Logic)

Let's gird up our loins and meander our way through the morass one step at a time . . . The process of relating logical values to physical voltages begins by defining the frames of reference to be used. One absolute frame of reference is provided by truth tables, which are always associated with specific functions (Figure B-2).

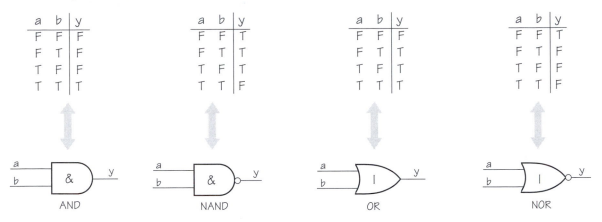

Figure B-2. Absolute relationships between truth tables and functions

Another absolute frame of reference is found in the physical world, where specific voltage levels applied to the inputs of a digital function cause corresponding voltage responses on the outputs. These relationships can also be represented in truth table form. Consider a logic gate constructed using only NMOS transistors (Figure B-3).

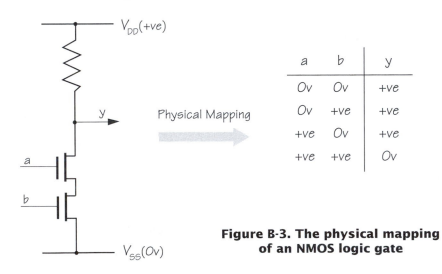

Figure B-3. The physical mapping of an NMOS logic gate

With NMOS transistors connected as shown in Figure B-3, an input connected to the more negative V_{SS} turns that transistor OFF, and an input connected to the more positive V_{DD} turns that transistor ON. The final step is to define the mapping between the physical and abstract worlds; either 0v is mapped to FALSE and +ve is mapped to TRUE, or vice versa (Figure B-4).

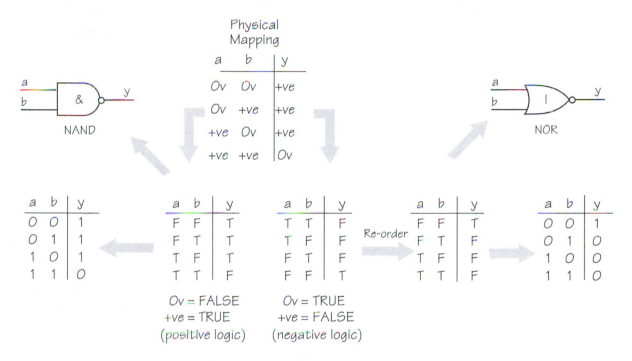

Figure B-4. The physical to abstract mapping of an NMOS logic gate

Using the positive logic convention, the more positive potential is considered to represent TRUE[2] and the more negative potential is considered to represent FALSE. By comparison, using the negative logic convention, the more negative potential is considered to represent TRUE[3] and the more positive potential is considered to represent FALSE. Thus, this circuit may be considered to be performing either a NAND function in positive logic or a NOR function in negative logic. (Are we having fun yet?)

---

[2] Hence, positive logic is also known as *positive-true*.

[3] Hence, negative logic is also known as *negative-true*.

Physical to Abstract Mapping (PMOS Logic)

From the previous example it would appear that positive logic is the more intuitive as it is easy to relate logic *0* to 0v (no volts) and logic 1 to +ve (presence of volts). On this basis one may wonder why negative logic was ever invented. The answer to this is, as are so many things, rooted in history. When the MOSFET technology was originally developed, PMOS transistors were easier to manufacture and were more reliable than their NMOS counterparts. Thus, the majority of early MOSFET-based logic gates were constructed from combinations of PMOS transistors and resistors. Consider a logic gate constructed using only PMOS transistors (Figure B-5).

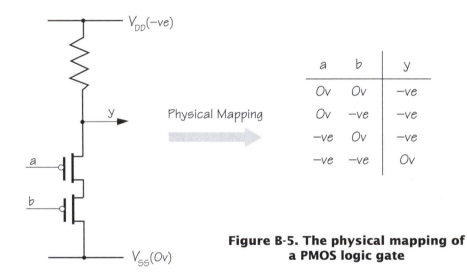

a	b	y
0v	0v	−ve
0v	−ve	−ve
−ve	0v	−ve
−ve	−ve	0v

Physical Mapping

Figure B-5. The physical mapping of a PMOS logic gate

Circuits constructed using the original PMOS transistors typically made use of a negative power supply (that is, a power supply with a 0V terminal and a negative (−ve) terminal). With PMOS transistors connected as shown above, an input connected to the more positive V_{SS} turns that transistor OFF, and an input connected to the more negative potential V_{DD} turns that transistor ON. Once again, the final step is to define the mapping between the physical and abstract worlds; either 0v is mapped to FALSE and −ve is mapped to TRUE, or vice versa (Figure B-6).

Thus, this circuit may be considered to be performing either a NOR function in positive logic or a NAND function in negative logic. In this case, negative logic is the more intuitive as it is easy to relate logic 0 to 0v (no volts)

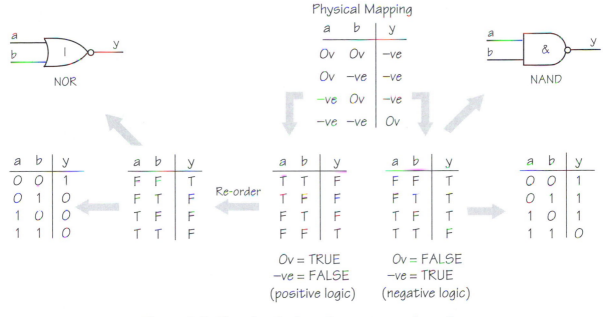

Figure B-6. The physical to abstract mapping of a PMOS logic gate

and logic 1 to −ve (presence of volts). Additionally, the physical structure of the PMOS gate is identical to that of the NMOS gate; if the NMOS gate is represented in positive logic and the PMOS gate is represented in negative logic, then both representations equate to NAND functions, which is, if nothing else, aesthetically pleasing.

Physical to Intermediate to Abstract Mapping

The discussions above offer the most consistent way to view the positive and negative logic conventions, and the casual reader is advised to stop at this point and proceed to something less demanding on the brain.

You're still here? Well, don't say I didn't warn you! Occasionally, an additional level of indirection is included in the progression from the physical to the abstract worlds. One reason for this is that supply and switching voltage values are relative rather than absolute quantities. In the examples above, the NMOS gate is shown as being powered by $V_{SS} = 0v$ and $V_{DD} = +ve$, while the PMOS gate is shown as being powered by $V_{SS} = 0v$ and $V_{DD} = -ve$.

However, within certain physical constraints, the actual values of V_{DD} and V_{SS} are largely immaterial. The only real requirements are for V_{DD} to be more positive than V_{SS} for NMOS gates, and for V_{DD} to be more negative than V_{SS} for PMOS gates. Assuming that both technologies require a 5-volt differential between V_{DD} and V_{SS}, the following shows a small selection of the many possible voltage combinations that would suffice (Figure B-7).

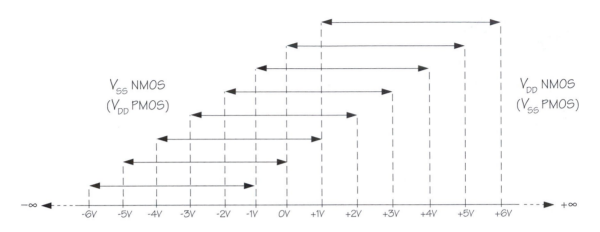

Figure B-7. Alternative values for V$_{DD}$ and V$_{SS}$

Now consider an NMOS gate using V_{DD} = +2 volts and V_{SS} = –3 volts. Although it is preferable to map from the physical world directly to FALSE and TRUE values, it is also possible to map to intermediate HIGH and LOW values. The problem arises in the definitions of HIGH and LOW; some would regard HIGH as being the more positive potential and LOW as being the more negative, while others would adopt absolute values as measured from 0 volts as being the criteria. In this latter case, –3 volts would be considered to be HIGH based on the fact that it has a greater absolute value with respect to 0 volts than +2 volts, which would therefore be considered to be LOW.

This means that, for an NMOS gate supplied by V_{DD} = +2v and V_{SS} = –3v, the progression of mapping from *physical* through *intermediate* to *abstract* representations could be as shown in Figure B-8.

It is easy to see how this progression can become extremely confusing. To further illustrate the possible complexities involved, consider a design containing a combination of NMOS and PMOS logic gates.[4] Employing three voltage

---

[4] Note that this is not the same as connecting NMOS and PMOS transistors together in a complementary manner so as to form CMOS gates (this was discussed in Chapter 6).

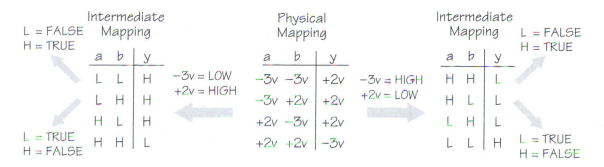

L = FALSE H = TRUE	Intermediate Mapping				Physical Mapping				Intermediate Mapping			L = FALSE H = TRUE
	a	b	y		a	b	y		a	b	y	
	L	L	H	−3v = LOW	−3v	−3v	+2v	−3v = HIGH	H	H	L	
	L	H	H	+2v = HIGH	−3v	+2v	+2v	+2v = LOW	H	L	L	
	H	L	H		+2v	−3v	+2v		L	H	L	
L = TRUE H = FALSE	H	H	L		+2v	+2v	−3v		L	L	H	L = TRUE H = FALSE

**Figure B-8. Physical to intermediate to abstract
mapping for an NMOS logic gate**

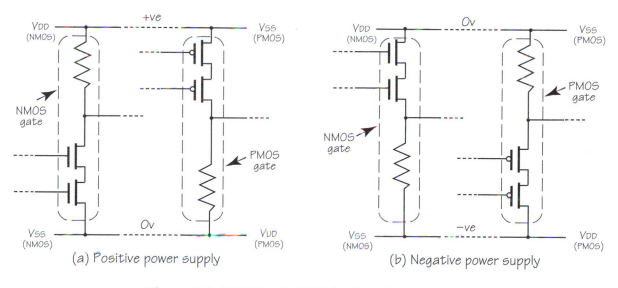

(a) Positive power supply   (b) Negative power supply

**Figure B-9. NMOS and PMOS logic gates using
common voltage levels**

levels, 0v, +ve and −ve, would be less than satisfactory, because the gates would not be able to communicate with each other. However, there are solutions requiring only two voltage levels (Figure B-9).

As was noted above, the only real requirements are for V_{DD} to be more positive than V_{SS} for NMOS gates, and for V_{DD} to be more negative than V_{SS} for PMOS gates. Both these requirements can be satisfied using only two voltage levels as long as the labels associated with those levels are redefined as necessary. The two examples shown in Figure B-9 are actually different views of the same

352 ■ Appendix B

thing; any two voltage levels with the appropriate relationship to each other would suffice. The application of the positive or negative logic systems in a mixed scenario of this type would prove to be an interesting exercise.

Fortunately, the majority of modern technologies favor positive logic representations and, unless otherwise stated, the positive logic convention is typically assumed. However, it is ultimately the designers' responsibility to define and/or ascertain which convention is being employed in any particular design.

Reed-Müller Logic

Some digital functions can be difficult to optimize if they are represented in the conventional sum-of-products or product-of-sums forms, which are based on ANDs, ORs, NANDs, NORs, and NOTs. In certain cases it may be more appropriate to implement a function in a form known as Reed-Müller logic, which is based on XORs and XNORs. One indication as to whether a function is suitable for the Reed-Müller form of implementation is if that function's Karnaugh Map displays a checkerboard pattern of 0s and 1s. Consider a familiar two-input function (Figure C-1).

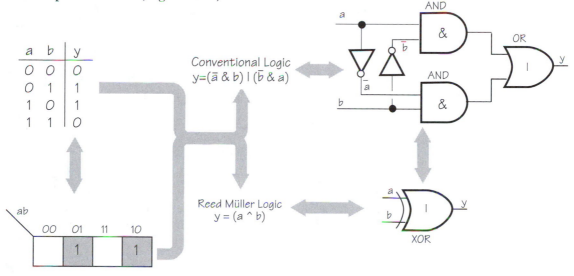

**Figure C-1. Two-input function suitable
for a Reed-Müller implementation**

Since the above truth table is easily recognizable as being that for an XOR function, it comes as no great surprise to find that implementing it as a single XOR gate is preferable to an implementation based on multiple AND, OR and NOT gates. A similar checkerboard pattern may apply to a three-input function (Figure C-2).

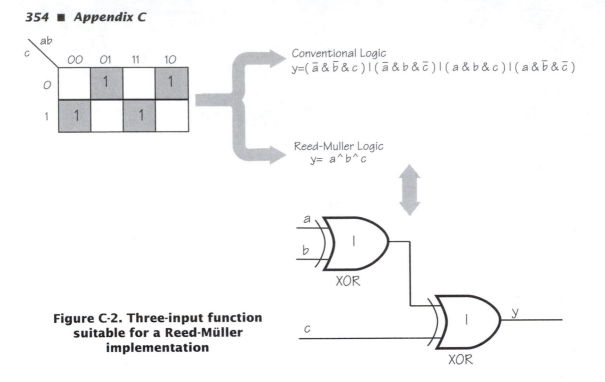

Conventional Logic
$$y = (\overline{a} \& \overline{b} \& c) \,|\, (\overline{a} \& b \& \overline{c}) \,|\, (a \& b \& c) \,|\, (a \& \overline{b} \& \overline{c})$$

Reed-Muller Logic
$$y = a \wedge b \wedge c$$

Figure C-2. Three-input function suitable for a Reed-Müller implementation

As XORs are both commutative[1] and associative,[2] it doesn't matter which combinations of inputs are applied to the individual gates. The checkerboard pattern for a four-input function continues the theme (Figure C-3).

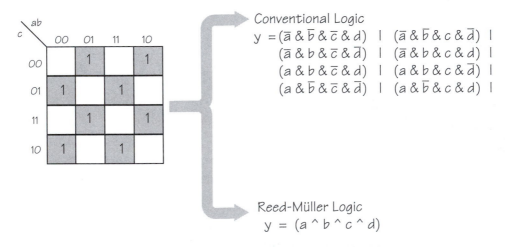

Conventional Logic
$$\begin{aligned}
y = &(\overline{a} \& \overline{b} \& \overline{c} \& d) \,|\, (\overline{a} \& \overline{b} \& c \& \overline{d}) \,| \\
&(\overline{a} \& b \& \overline{c} \& \overline{d}) \,|\, (\overline{a} \& b \& c \& d) \,| \\
&(a \& b \& \overline{c} \& d) \,|\, (a \& b \& c \& \overline{d}) \,| \\
&(a \& \overline{b} \& \overline{c} \& \overline{d}) \,|\, (a \& \overline{b} \& c \& d) \,|
\end{aligned}$$

Reed-Müller Logic
$$y = (a \wedge b \wedge c \wedge d)$$

Figure C-3. Four-input function suitable for a Reed-Müller implementation

---

[1] For example, $(a \wedge b) \equiv (b \wedge a)$. (Remember that the "$\equiv$" symbol means "is equivalent to.")
[2] For example, $(a \wedge b) \wedge c \equiv a \wedge (b \wedge c)$.

Larger checkerboard patterns involving groups of 0s and 1s also indicate functions suitable for a Reed-Müller implementation (Figure C-4).

Once you have recognized a checkerboard pattern, there is a quick "rule of thumb" for determining the variables to be used in the Reed-Müller implementation. Select any group of 0s or 1s and identify the significant and redundant variables,[3] and then simply XOR the significant variables together.

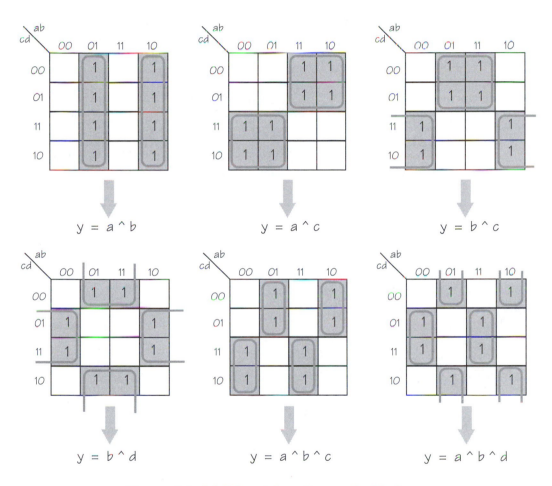

Figure C-4. Additional functions suitable for Reed-Müller implementations

---

[3] The significant variables are those whose values are the same for all of the boxes forming the group, while the redundant variables are those whose values vary between boxes.

As all of the checkerboard patterns above include a logic 0 in the box in the upper left corner,[4] the resulting Reed-Müller implementations can be realized using only XORs. However, any pair of XORs may be replaced with XNORs, the only requirement being that there is an even number of XNORs.

Alternatively, if the checkerboard pattern includes a logic 1 in the box in the upper left corner, the Reed-Müller implementation must contain an odd number of XNORs. Once again, it does not matter which combinations of inputs are applied to the individual XORs and XNORs (Figure C-5).

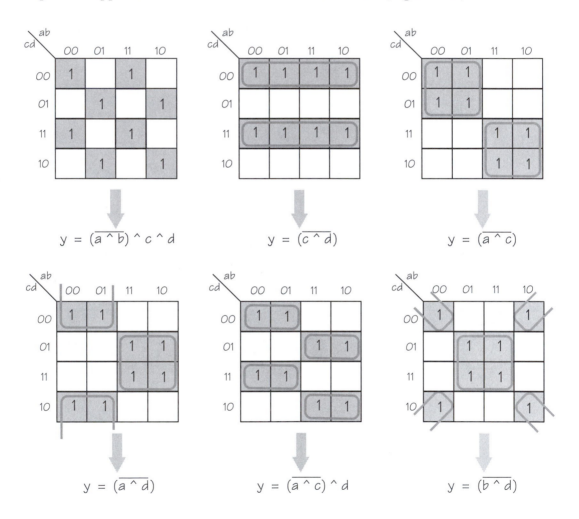

Figure C-5. Reed-Müller implementations requiring XNOR gates

---

[4] Corresponding to all of the inputs being logic 0.

Although these examples provide a very limited introduction to the concept of Reed-Müller logic, checkerboard Karnaugh Map patterns are easy to recognize and appear surprisingly often. Reed-Müller implementations are often appropriate for circuits performing arithmetic or encoding functions (see also Appendix D).

Gray Codes

When moving between states in a standard binary sequence, multiple bits may change from 0 to 1 or vice versa; for example, three bits change value when moving from 0011_2 to 0100_2. In the physical world there is no way to ensure that all the bits will transition at exactly the same time, so a system may actually pass through a sequence of intermediate states. Thus, an intended state change of 0011_2 to 0100_2 could actually result in the sequence 0011_2 to 0111_2 to 0100_2.

One way to avoid this problem is to use a *gray code*, in which only a single bit changes when moving between states (Figure D-1).

Gray codes are of use for a variety of applications, such as the ordering of the input variables on Karnaugh Maps.[1] Another application is encoding the angle of a mechanical shaft, where a disc attached to the shaft is patterned with areas of conducting material (Figure D-2).

b [3:0]	g [3:0]
0 0 0 0	0 0 0 0
0 0 0 1	0 0 0 1
0 0 1 0	0 0 1 1
0 0 1 1	0 0 1 0
0 1 0 0	0 1 1 0
0 1 0 1	0 1 1 1
0 1 1 0	0 1 0 1
0 1 1 1	0 1 0 0
1 0 0 0	1 1 0 0
1 0 0 1	1 1 0 1
1 0 1 0	1 1 1 1
1 0 1 1	1 1 1 0
1 1 0 0	1 0 1 0
1 1 0 1	1 0 1 1
1 1 1 0	1 0 0 1
1 1 1 1	1 0 0 0

Binary ➡ (b [3:0] column)   (g [3:0] column) ⬅ Gray Code

Figure D-1. Binary versus gray code

The conducting areas are arranged in concentric circles, where each circle represents a binary digit. A set of electrical contacts, one for each of the circles, is used to detect the logic values represented by the presence or absence of the conducting areas. A digital controller can use this information to determine the

---

[1] This was discussed in detail in Chapter 10.

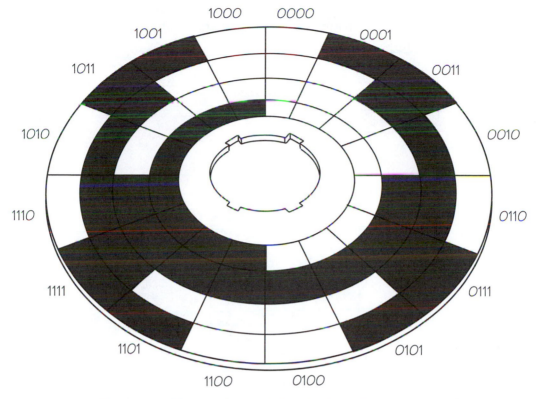

Figure D-2. Gray code used for shaft-angle encoding

angle of the shaft. The precision of the measurement depends on the number of bits (circles). A 4-bit value representing 16 unique states allows the shaft's angle to be resolved to 22.5 degrees, while a 10-bit value representing 1,024 unique states allows the shaft's angle to be resolved to 0.35 degrees. Using a gray code sequence to define the conducting and non-conducting areas ensures that no intermediate values are generated as the shaft rotates.

Commencing with a state of all zeros, a gray code can be generated by always changing the least significant bit that results in a new state. An alternative method which may be easier to remember and use is as follows:

a) Commence with the simplest gray code possible; that is, for a single bit.

b) Create a mirror image of the existing gray code below the original values.

c) Prefix the original values with 0s and the mirrored values with 1s.

d) Repeat steps b) and c) until the desired width is achieved.

An example of this mirroring process used to generate a 4-bit gray code is shown in Figure D-3.

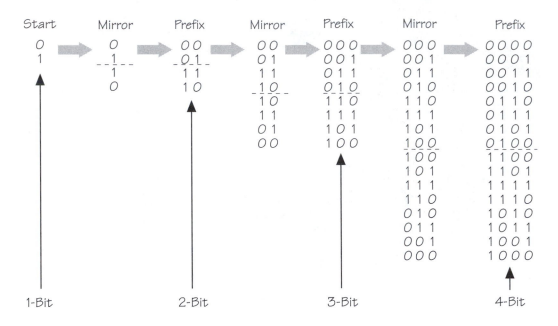

Figure D-3. Generating a 4-bit gray code by means of the mirroring process

It is often required to convert a binary sequence into a gray code or vice versa. Such converters are easy to create and are primarily of interest here due to their affinity to the Reed-Müller implementations that were introduced in Appendix C. Consider a binary-to-gray converter (Figure D-4). The checkerboard patterns of 0s and 1s in the Karnaugh Maps immediately indicate the potential for Reed-Müller implementations.

Similar checkerboard patterns are also seen in the case of a gray-to-binary converter (Figure D-5).

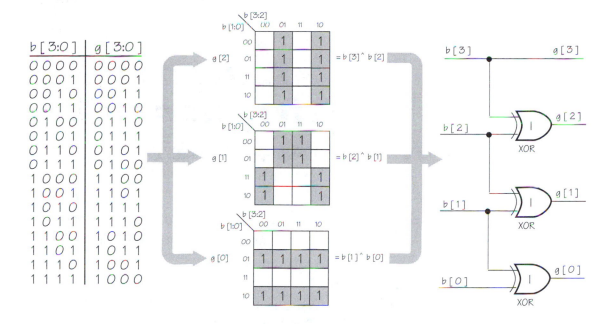

Figure D-4. Binary-to-gray converter

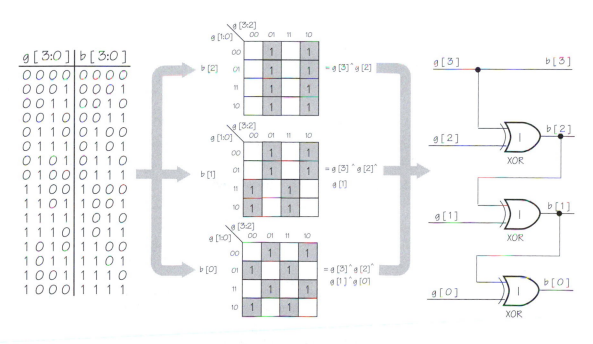

Figure D-5. Gray-to-binary converter

A Reed-Müller Extraction Utility

In this appendix we discover a thought-provoking little mind game to delight the secret computer programmer that lurks within all of us.[1,2,3] After reading the previous two appendices, you should be a *"master of the mystic arts"* when it comes to Reed-Müller logic and gray codes. This is rather fortunate because, without this knowledge, what you are poised on the brink of reading would be likely to make your brains leak out of your ears. The simple program discussed below contains the never-before-seen-in-public, soon-to-be-famous, mind-bogglingly-convoluted, *Max's Algorithm* for automatically extracting Reed-Müller expressions.

Before examining the program which, by some strange quirk of fate, is called *max_prog*, it would be useful to know what it actually does. First, you create a text file representation of one or more truth tables that you would like to examine and, second, you run the program using this file as its input. In return, the program determines which truth tables are suitable for Reed-Müller implementations and returns the appropriate expressions. For example, consider the truth tables shown in Figure E-1.

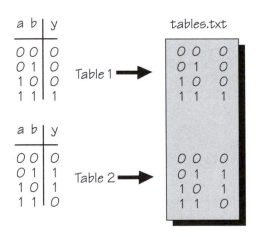

Figure E-1: Example truth tables for Reed-Müller evaluation

---

[1] The program is written in the C programming language and is intended to be compatible with any ANSI-standard C compiler.

[2] In certain cases, good programming practices have been thrown out of the window for the sake of brevity. For example, functions are generally not declared before they are called, and no error checking is performed on the input data.

[3] If you'd like to know more about writing programs in C, you can't do much better than reading "*The C Programming Language*" (Second Edition) by Brian W. Kernighan and Dennis M. Ritchie. For those of you who wish to voyage a little further, "*Reusable Data Structures For C*" by Roger Sessions is well worth perusing.

As you will see, the part of the program that reads the input file is rather simple, hence the file's somewhat minimalist look. When you've created the input file, you run the program from the operating system's command line as follows:

```
max_prog < tables.txt                                {Calls the program 'max_prog' and      }
                                                     {feeds it with the text file 'tables.txt'. }

Truth table 1: No Reed-Muller solution              {These are the results returned by the }
Truth table 2: Reed-Muller solution is y = (a ^ b)  {program for the tables shown above.   }
```

As truth tables 1 and 2 in Figure E-1 are obviously those for an AND and an XOR, respectively, the results returned by the program should come as no surprise. Note that this version of the program assumes that the inputs to the truth tables are named a, b, c, ... etc., and that the output is named y. In fact, this version of the program can handle truth tables with up to ten inputs (but only one output), and the number of inputs could be easily extended if necessary. Now, onto the program itself, the first part of which simply declares anything that the program is going to use.[4]

```
/* Max's algorithm to detect and evaluate functions suitable for Reed-Muller evaluation.  */
/* Accepts truth tables with up to ten inputs and returns any Reed-Muller equations.       */
/* Beware! Contains minimal parsing and no error detection — you have been warned!         */

#include <stdio.h>             /* Include the standard files supplied with C.              */
#include <ctype.h>
#include <string.h>

#define XSIZE 1024             /* Set the maximum number of elements necessary             */
                              /* to handle truth tables with two to ten inputs.           */

int bin_to_gray[XSIZE];        /* This will be the binary to gray-code lookup table.       */
int gray_to_bin[XSIZE];        /* This will be the gray-code to binary lookup table.       */

char workspace[XSIZE];         /* This will be the working area. Each element              */
                              /* corresponds to a box in a Karnaugh Map.                  */
```

---

[4] If you do decide to play with this program, it is necessary to enter all the program segments in the same order that they are presented here.

```
struct xy {
  int x, y;
} xy_table[11] = {                     /* When presented with the number of inputs       */
    { 0, 0},  /*  0 inputs */          /* associated with a truth table, this structure will  */
    { 2, 1},  /*  1 inputs */          /* return the most appropriate dimensions for an    */
    { 2, 2},  /*  2 inputs */          /* equivalent Karnaugh Map (as nearly rectangular   */
    { 4, 2},  /*  3 inputs */          /* as possible).                                    */
    { 4, 4},  /*  4 inputs */          /*                                                  */
    { 8, 4},  /*  5 inputs */          /* The 'x' and 'y' values returned by the structure */
    { 8, 8},  /*  6 inputs */          /* represent the number of columns (across) the     */
    {16, 8},  /*  7 inputs */          /* map and the number of rows (down) the map        */
    {16,16},  /*  8 inputs */          /* respectively.                                    */
    {32,16},  /*  9 inputs */
    {32,32}   /* 10 inputs */
};
```

We will return to discuss the lookup tables and the working area in a little while, but first let us move briskly on to the function that controls everything:

```
int main()
{
  int   num_tt = 0,     /* The number of the truth table currently being processed.     */
        num_in,         /* The number of inputs in the current truth table.             */
        mask;           /* A binary mask which is used to track the significant inputs.  */

  load_lookup_tables();                         /* Loads the lookup tables.             */

  while (num_in = get_truth_tables()) {         /* Load one truth table at a time.      */
    num_tt++;                                   /* Keep track of the No. of the table.  */
    if (test_reed_muller(num_in)) {             /* If this table has a R-M solution ... */
      mask = extract_reed_muller(num_in);       /* Extract the expression.              */
      output_reed_muller(mask,num_tt,num_in);   /* Output the equation.                 */
    }
    else                                        /* If no R-M solution then say so.      */
      printf("Truth table %i: No Reed-Muller solution\n",num_tt);
  }                                             /* Repeat for all the tables until none  */
  return 0;                                     /* remain, then return 0 and exit.       */
}
```

The variable num_tt, which keeps track of the number of the truth table being processed, is initialized with zero, and then the function load_lookup_tables()

is called to initialize the lookup tables (what a surprise). The program then enters a loop as follows . . .

The function get_truth_tables() loads the outputs from the first truth table into the working area. This function also returns the number of inputs for the truth table and assigns the value to the variable num_in. When get_truth_tables() returns a value of zero for the number of inputs, then all of the truth tables in the input file have been processed and the program terminates. Otherwise, the first action in the loop is to increment num_tt, which keeps track of the number of the truth table that is being analyzed.

Once a truth table has been loaded, the function test_reed_muller() examines it to determine if it has a Reed-Müller solution. If there is such a solution, then the function extract_reed_muller() is used to determine which of the truth table's inputs are significant. The information on the significant inputs is presented in the form of a bit-mask which is assigned to the variable mask. Finally, the function output_reed_muller() displays the Reed-Müller equation for all the world to see. Good grief, if only everything in life could be so simple. Now, before plunging into the inner workings of the program, there are two utility functions that are called from time to time:

```
/* power_two: Generates a power of two.                                           */
/*          : Accepts an integer 'exponent'.                                      */
/*          : Returns an integer equal to 2 raised to the power of 'exponent'.    */
int power_two(int exponent)
{
    return 1 << exponent;      /* Returns 1 shifted to the left by the number of bits defined  */
}                              /* by 'exponent' (which equals 2 to the power of 'exponent').   */
/* get_binary_value : Extracts a binary value presented as a string of 0 and 1 characters.  */
/*                  : Accepts a pointer to the string of characters, '*p'.                   */
/*                  : Returns the integer represented by the 0 and 1 characters.             */
int get_binary_value(char *p)
{
    int value=0;                         /* Declare and initialize a temporary variable.   */

    for (; *p ; p++)                     /* Keep on processing the character pointed to by  */
        value = (value<<1) + (*p – '0'); /* '*p' and incrementing '*p' until it points to a */
    return value;                        /* null character, then return the contents of 'value'.  */
}
```

The get_binary_value() function accepts a pointer to a string of characters such as "1001", which have been read from the input file and which represent a set of binary input values associated with a line from one of the truth tables. The function converts this string into the numerical value that it represents; for example, "1001" = 9. The function is also used to process a single "0" or "1" character representing an output value associated with a line in a truth table, and to convert it into its numerical equivalent (0 or 1).

The first interesting action performed by the main controlling routine is to call the function load_lookup_tables(), and it doesn't take a rocket scientist to guess what this one does:

```
/* load_lookup_tables : Loads the 'bin_to_gray' lookup table with gray codes, and the     */
/*                    : 'gray_to_bin' lookup table with binary codes.                      */
/*                    : Returns a dummy parameter */
int load_lookup_tables()
{
   int bin, gray;                  /* Declare two temporary variables.                     */

   for (bin = 0; bin < XSIZE; bin++) {  /* Increment 'bin' through a binary sequence . . . */
      gray = bin ^ (bin >> 1);          /* . . . 'gray' equals the gray-code equivalent of 'bin' */
      bin_to_gray[bin] = gray;          /* . . . load gray code into 'bin_to_gray' lookup table. */
      gray_to_bin[gray] = bin;          /* . . . load binary value into 'gray_to_bin' lookup. */
   }

   return 0;
}
```

The load_lookup_tables() function revolves around a loop, in which the variable bin is incremented from zero to the maximum number of elements in the tables. For each value of bin, its equivalent gray code is generated and assigned to the variable gray. Finally, the element in the bin_to_gray lookup table pointed to by bin is loaded with the value in gray, and the element in the gray_to_bin lookup table pointed to by gray is loaded with the value in bin. Note that the statement "gray = bin ^ (bin >>1)" is the software equivalent of the binary-to-gray converter illustrated in Figure D-4 in Appendix D.

Throughout the rest of this program, binary-to-gray and gray-to-binary conversions can be easily performed by accessing the appropriate value in the bin_to_gray and gray_to_bin lookup tables respectively. For example, assuming that an integer variable called index has already been assigned the equivalent of the binary value 1100_2:

```
temp = bin_to_gray[index];   {Would result in 'temp' containing the binary value 1010₂   }
temp = gray_to_bin[index];   {Would result in 'temp' containing the binary value 1000₂   }
```

You can check these results using Figure D-1 in Appendix D. After everything has been declared and the lookup tables have been initialized, the program is ready to rock-and-roll. The first function of any significance is get_truth_tables(), which loads the truth tables, one at a time, from the input file:

```c
/* get_truth_tables : Loads one truth table at a time from the input file into the working   */
/*                  : area, 'workspace'. Be very careful—this function contains minimal      */
/*                  : parsing and no error detection. As a wise man once said—"You can        */
/*                  : make it fool proof . . . but you can't make it damn-fool proof."        */
/*                  : Returns the number of inputs in the table being processed.              */
int get_truth_tables()
{
    #define XLINE 132               /* Set the max No. of characters allowed on a line.       */
    char *p;                        /* Declare a character pointer variable.                  */
    char *strtok();                 /* Declare a function (from the standard C library).      */
    char line[XLINE];               /* Reserve space to load a line from the input file.      */
    int num_in;                     /* Keeps track of the number of truth table inputs.       */
    int inputs, output;             /* Binary equivalents of the inputs and output            */
    int in_table = 0;               /* Keeps track of whether we're in a truth table          */

    while(gets(line)) {             /* Use the standard function 'gets()' to read a line.     */
        if (!(p=strtok(line," \t"))) {  /* If std function 'strtok()' says this line empty . . . */
            if (in_table)           /* . . . if we were already processing a table then       */
                return num_in;      /* we've finished, so return number of inputs.            */
            else                    /* . . . but if we weren't already processing a table,    */
                continue;           /* then skip the rest of the "while" loop, go back        */
        }                           /* to the start of the loop, and get the next line.       */

        in_table = 1;               /* Have got a non-blank line, so must be in a table.      */
        num_in = strlen(p);         /* Use std function 'strlen()' to find No. of inputs.     */
        inputs = get_binary_value(p);  /* Get the integer value represented by the inputs.    */
        p = strtok((char *)0," \t");   /* Tell 'strtok()' to get next token (the output).     */
        output = get_binary_value(p);  /* Get the integer value represented by the output.    */
        workspace[gray_to_bin[inputs]] = output; /* Load the output data into the workspace.  */
    }                               /* Go back to the beginning of the loop.                  */

    return in_table ? num_in : 0;   /* Exit loop at end-of-file. If in a table then return    */

}                                   /* the number of inputs, otherwise return zero.           */
```

Functions that read-in and write-out data are generally rather boring, and this one is no exception. On the other hand, even though it is "mean and lean," get_truth_tables() permits a reasonable amount of latitude, and it's worth taking the time to note its main features and limitations:

FEATURES	LIMITATIONS
Allows any number of blank lines, including lines containing spaces and tabs, at the beginning and end of the file, and also between truth tables.	No blank lines are allowed in the middle of a truth table, but there must be at least one blank line separating one truth table from another.
Allows multiple spaces and tabs at the beginning and end of each truth table line, and also between the inputs and outputs.	All the input values must be grouped together without any spaces or tabs, but there must be at least one space or tab separating the inputs from the output.
The lines of a truth table can be specified in any order	All the lines in a truth table must be specified.
This version of the program supports truth tables with two to ten inputs.	This version of the program only supports truth tables with one output.

Actually, considering the size of this function, or its lack thereof, you couldn't really ask for too much more (what do you want for nothing, your money back?). The function processes each line of a truth table and loads the output value associated with that line into the working area workspace[]. When the function detects a blank line indicating the end of the truth table, it terminates and returns the number of inputs associated with this table to the main controlling loop.

The statement of most interest is "workspace[gray_to_bin[address]] = data;", which loads the output value into the working area. This statement uses the gray-code to binary lookup table to transpose the address representing the truth table's input values into an address used by the computer to point to the working area, and to load the output value from the truth table into the element in the working area pointed to by this transposed address. The organization of the working area is the key to the way in which the rest of the program works.

The problem is that it is not easy to check for Reed-Müller implementations if the data is represented in the form of a truth table. The most efficient method (as far as the author knows) is to visualize and process the data in the form of a Karnaugh Map. However, for a variety of reasons, it is easier for the program to handle the data as a one-dimensional array. Additionally, any solution must easily accommodate truth tables (or Karnaugh Maps) of different sizes; that is, with different numbers of inputs.

The working area used by this program is a one-dimensional array of characters called workspace[] (Figure E-2). An array of characters is used because they are only 8-bits wide, and therefore occupy less of the computer's memory than would integers, which may each require 16-bits or 32-bits depending on the compiler being used:

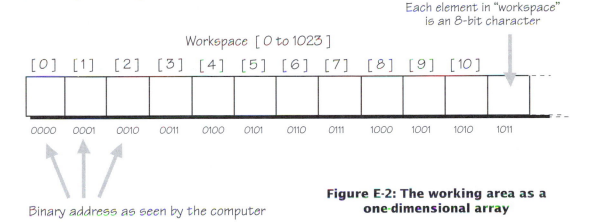

Figure E-2: The working area as a one-dimensional array

In fact, this program uses only the least-significant bit of each character's 8-bit field to store the contents (output values) of the truth table, and the other seven bits are therefore available for possible future enhancements. The next thing is to understand how the contents of the truth table are mapped into the working area. The easiest way to visualize this is to first view the truth table as a Karnaugh Map, and then to imagine grabbing the upper left-hand and upper right-hand corner boxes of map and pulling them to the left and right respectively, thereby expanding the map like a concertina (Figure E-3).

The next point to note is that there are number of alternative ways of viewing the Karnaugh Map associated with a particular truth table. For example,

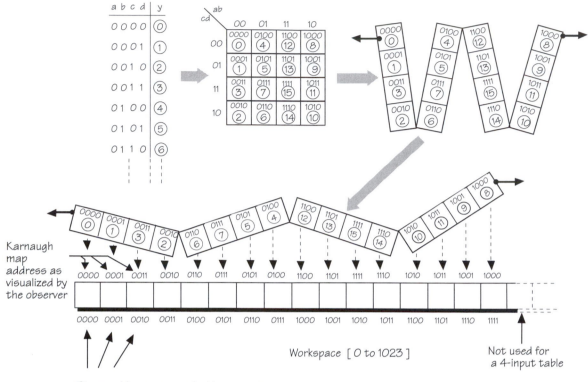

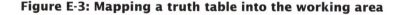

Figure E-3: Mapping a truth table into the working area

a table with four inputs may be represented as a Karnaugh Map organized with four columns and four rows, two columns and eight rows, eight columns and two rows, etc. Consider a 3-input truth table represented as 4×2 and 2×4 Karnaugh Maps (Figure E-4).

The beauty of the technique used by this program to map the data from a truth table into the working area is that, irrespective of the size of the truth table, and irrespective of how the truth table is visualized as a Karnaugh Map, the data is always mapped into the same locations in the working area (Figure E-5).

Thus, the statement "workspace[gray_to_bin[address]] = data;" in the get_truth_tables() function transposes the truth table (or Karnaugh Map) address, which is considered to be in the form of a gray code, into a binary

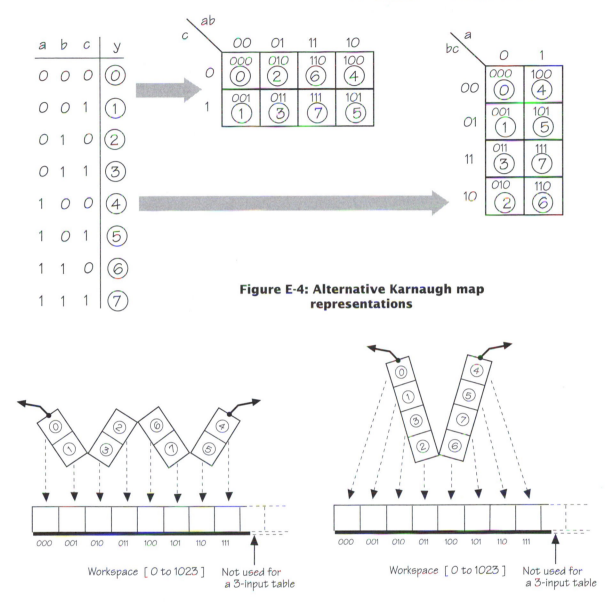

Figure E-4: Alternative Karnaugh map representations

Figure E-5: Consistent mapping to the working area

equivalent that is used to point to the working area. The output value from the truth table is loaded into the working area at this transposed address.

After all of the data from a truth table has been loaded into the working area, the function test_reed_muller() is called to determine whether this particular truth table has a Reed-Müller solution:

```
/* test_reed_muller : Checks the contents of the truth table to see if this table has a          */
/*                  : Reed-Müller solution.                                                       */
/*                  : Accepts one parameter, 'num_in,' the number of inputs in this table.         */
/*                  : Returns 1 if this table has a Reed-Müller solution, and 0 if it doesn't.    */
int test_reed_muller(int num_in)
{
   char *bottom, *top;                        /* Declare temporary character pointers.           */
   int start, end, look_for;                  /* Declare local variables.                        */

   for (start = 4, end = power_two(num_in); start <= end; start <<=1) {   /* Main loop.          */
      bottom = &workspace[0];                  /* Point 'bottom' at the LHS of 'workspace'.       */
      top = &workspace[start – 1];             /* Point 'top' at the RHS for this pass.           */
      look_for = *bottom++ ^ *top — ;          /* Find out what we're looking for.                */
      do {                                     /* Enter a sub-loop.                               */
         if ((*bottom++ ^ *top — ) != look_for) /* If the rest aren't what we want . . .          */
            return 0;                          /* . . . Return 0 to indicate no solution.         */
      } while (bottom < top);                  /* . . . Otherwise continue with the sub-loop.     */
   }

                                               /* If we get this far, then there is a solution,   */
   return 1;                                   /* so return 1 to tell the world.                  */
}
```

The `test_reed_muller()` function performs a series of actions which are more easily described pictorially. Consider the case of a 2-input truth table (Figure E-6).

The main loop first initializes *bottom* and *top* to point at locations 0 and 3 in the working area respectively. The first action is to perform an exclusive-OR of the contents of the working area pointed to by *bottom* and *top*, and to store the result in the variable look_for. It doesn't actually matter whether the result stored in look_for is a 0 or a 1, because whatever it is, that's what we're looking for. The values of *bottom* and *top* are then incremented and decremented respectively, an exclusive-OR is performed on the contents of the working area that they are now pointing to, and the result is compared to the value stored in look_for.

If the result of this exclusive-OR is not the same as the value stored in look_for, then there is no Reed-Müller

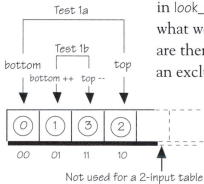

Figure E-6: Testing a 2-input truth table

solution, and `test_reed_muller()` returns a 0 and terminates. But if the result is the same as the value stored in look_for, then the function continues onward.

In fact, in the case of a 2-input truth table, all of the tests have been performed, so the function would return a 1 indicating that there is a Reed-Müller solution. Now consider the case of a 3-input truth table (Figure E-7).

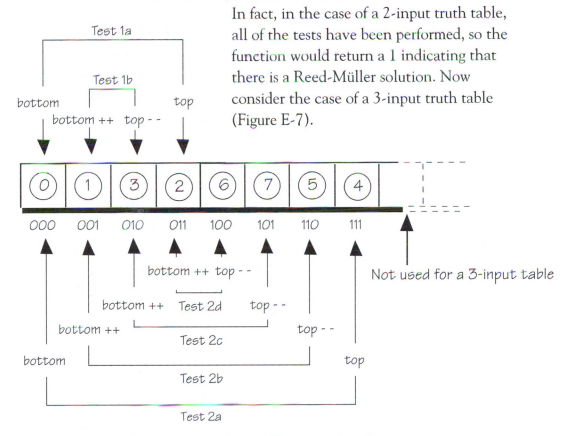

Figure E-7: Testing a 3-input truth table

The first series of tests are exactly the same for a 2-input truth table. Assuming that the table passes these tests, the main loop reinitializes bottom and top to point at locations 0 and 7 in the working area respectively. Once again, the first action in this iteration of the loop is to perform an exclusive-OR of the contents of the working area pointed to by bottom and top, and to store the result in look_for. And once again, it doesn't actually matter whether the result stored in look_for is a 0 or a 1, because whatever it is, that's what we're looking for this time around.

The values of bottom and top are then incremented and decremented respectively, an exclusive-OR is performed on the contents of the working area that they are now pointing to, and the result is compared to the value stored in look_for. This continues until it is determined that there is no Reed-Müller solution, or until *bottom* and *top* meet in the middle, in which case there is a solution.

Similarly, in the case of a 4-input truth table, the first series of tests would be the same as for a 2-input table, and the second series of tests would be the same as for a 3-input table. Assuming that the table passes these tests, the main loop would commence a third series of tests by reinitializing *bottom* and *top* to point at locations 0 and 15 in the working area respectively. As usual, the first action in this third iteration of the loop would be to perform an exclusive-OR of the contents of the working area pointed to by *bottom* and *top*, and to store the result in look_for. The process would then continue as before: incrementing *bottom*, decrementing *top*, and testing their contents against the value in look_for until it is determined that there is no Reed-Müller solution, or until *bottom* and *top* meet in the middle, in which case there is a solution.

The problem is that, although these tests are simple, there can be a lot of them. On the positive side, if there is no Reed-Müller solution then this function usually finds out fairly quickly. But if there is a solution, then it can take a lot of tests to prove it. In fact, an n-input truth table will require $(2^n - 2)$ tests if it has a solution. This is not ideal, but there does not appear to be a better way . . . unless you can find one (more below). Assuming that there is a Reed-Müller solution, then the extract_reed_muller() function is used to determine the significant inputs:

```
/* extract_reed_muller   : Extracts the significant inputs for the Reed-Müller solution.   */
/*                       : Accepts one parameter, 'num_in', the No. of inputs in the table.  */
/*                       : Returns a bitmask, 'mask,' representing the significant inputs.    */
int extract_reed_muller(int num_in)
{
   int i, x, y, mask, index, look_for;   /* Declare all the local variables.         */
   mask = power_two(num_in) - 1;         /* Set 'mask' to contain 1s for each input.  */
   look_for = workspace[0];              /* Set 'look_for' to the value in [0,0] of the map.  */
   x = xy_table[num_in].x;               /* Get the most appropriate number of columns.  */
   y = xy_table[num_in].y;               /* Get the most appropriate number of rows.     */

/* The following routines view the working area as a Karnaugh Map. When I wrote   */
/* them, only God and myself knew how they worked . . . now God only knows!5       */
```

[5] This is an old programming joke, but it's also true far more times than you'd care to believe.

```
for (i = 1; i < x; i++) {            /* This loop commences at box [0,0] in the map    */
  if (!(i & 1))                      /* (the upper-left corner) and searches "East."    */
    index = i * y;                   /* For every box on its journey it must generate   */
  else                               /* a value for 'index', which acts as a pointer    */
    index = ((i + 1) * y) − 1;       /* into the working area.                          */

  if (workspace[index] == look_for)  /* If the box contains the same as 'look_for' ...  */
    mask &= ~(bin_to_gray[index]);   /* ... Remove non-significant input from 'mask.'   */
  else                               /* ... Otherwise break out of this loop and move   */
    break;                           /* onto the next one.                              */
}

if (i < (x >> 1)) {                  /* Before beginning this loop, a test is performed */
  for (i = (x − 1); i > 0; i — ) {   /* to see if it's actually necessary. If the last  */
    if (!(i & 1))                    /* loop searched half way or all the way across    */
      index = i * y;                 /* the map, then there's no point in performing    */
    else                             /* the tests in this loop.                         */
      index = ((i + 1) * y) − 1;

                                     /* Otherwise, this loop searches "West"            */
    if (workspace[index] == look_for)/* from box [0,0] in the map. Because of           */
      mask &= ~(bin_to_gray[index]); /* the wrap-around effect associated with          */
    else                             /* Karnaugh Maps, this means that the              */
      break;                         /* search actually commences in the                */

  }                                  /* upper-right corner of the map, and              */
}                                    /* then heads back towards box [0,0].              */

for (i = 1; i < y; i++) {            /* This loop commences at box [0,0], and           */
  if (workspace[i] == look_for)      /* and searches "South," that is, towards the      */
    mask &= ~(bin_to_gray[i]);       /* bottom-left corner.                             */
  else
    break;
}

if (i < (y >> 1)) {                  /* Tests to see if this loop is actually necessary.*/
  for (i = (y − 1); i > 0; i — ) {   /* If it is, then search "North" from box [0,0].   */
    if (workspace[i] == look_for)    /* Because of the Karnaugh Map wrap-around         */
      mask &= ~(bin_to_gray[i]);     /* effect, the search actually commences in the    */
    else                             /* lower-left corner of the map, and then heads    */
      break;                         /* back towards box [0,0].                         */
  }
}

return mask;                         /* Finally, return the significant inputs in the   */
}                                    /* form of the bitmask, 'mask.'                    */
```

The extract_reed_muller() function is the hardest to follow, but it is certainly a good exercise in reverse-polish-logic-lateral-thinking which will keep your brain in peak mental fitness. The easiest way to visualize the East-West-South-North searches is shown below. Consider the Karnaugh Map associated with a 4-input truth table (Figure E-8).

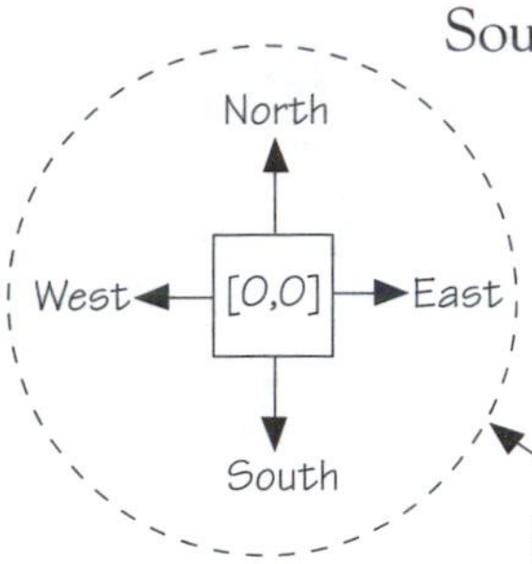

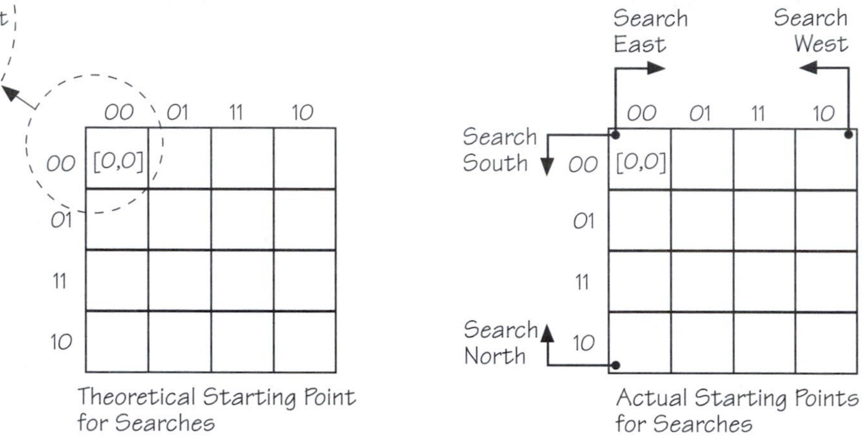

Figure E-8: Visualizing the searches to locate the significant inputs

In fact, the extract_reed_muller() function is reasonably efficient. One of the rules for a Reed-Müller solution is that there must be the same number of 1s stored in the Karnaugh Map boxes as there are 0s. This means that the largest possible block of 1s or 0s can occupy only half of the boxes, which in turn means that this function actually performs relatively few tests. In fact, for a 4-input truth table, this function will perform a maximum of only four tests (excluding the tests to see whether further tests are necessary). Now it only remains to output the equation for the Reed-Müller implementation:

```
/* output_reed_muller : Output the equation for the Reed-Muller implementation.        */
/*                    : Accepts three parameters: 'mask,' which contains the significant */
/*                    : inputs, 'num_tt,' which contains the No. of the truth table in the */
/*                    : input file, and 'num_in,' which contains the No. of table inputs.  */
/*                    : Returns a dummy value.                                            */
int output_reed_muller(int mask, int num_tt, int num_in)
{
```

```
int i;                               /* Declare a temporary integer variable.     */
char c;                              /* Declare a temporary character variable.   */

if (!workspace[0])                                    /* If box [0,0] holds        */
    printf("Truth table %i: Reed-Muller solution is y = (",num_tt);  /* a zero, then only   */
else                                                  /* need XORs, else           */
    printf("Truth table %i: Reed-Muller solution is y = !(",num_tt); /* add an inversion.   */

for (i = (power_two(num_in) >> 1), c = 'a'; !(mask & i); i >>= 1, c++);
printf("%c ",c);
for (i >>= 1, c++; i > 0; i >>= 1, c++)   /* Explaining this rather cunning routine   */
    if (mask & i)                         /* is left as an exercise for the reader ... */
        printf("^ %c ",c);                /* ... Ha!                                   */
printf(")\n");

return 0;                            /* Return a dummy value.                     */
}
```

How to Become Famous

There are a number of ways in which the program described above could be modified or extended. For example, the input function, get_truth_tables(), could be easily modified to accept truth tables with multiple outputs. Alternatively, it could be augmented to accept the textual equivalent of Karnaugh Maps. A more interesting exercise would be to modify the program to accept *don't care* values (represented using question marks, "?") on the inputs (Figure E-9).[6]

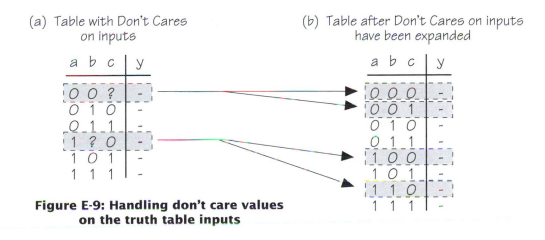

Figure E-9: Handling don't care values
on the truth table inputs

---

[6] The concepts of *don't cares* on inputs and outputs were introduced in more detail in Chapter 10.

This modification would be relatively simple, because it would only affect the input function. Note that handling a single don't care value on a line of inputs is trivial, but handling multiple don't cares is a little trickier. For those who enjoy a real challenge, modifying the program to cope with don't cares on the outputs should offer one! This is where the spare bits that are available in the working area may start to earn their keep.

Another interesting area to explore would be the test_reed_muller() function. As was noted during the discussions on this function, an n-input truth table will require $(2^n - 2)$ tests (assuming that the table has a Reed-Müller solution). It may be that you can discover a more efficient technique to determine whether the table has a solution.[7]

During the course of this discussion, the question may have arisen as to why the maximum number of truth table inputs is set as high as ten. After all, this would equate to a truth table with 1,024 lines, and there is little chance that anyone would take the time to type these in (and even less chance of their doing so without errors). However, one way in which the program could be modified would be for it to read in simple Boolean equations. For example, the input function could be modified to process the following text file:

```
INPUTS          = a, b, c, d;
OUTPUTS         = x, y;
INTERNAL_NODES  = j, k;

j =    (a & b);
k = a | (c ^ d);
x = j & (c | d);
y =    !(k & b);

END
```

The format shown here is fairly generic, but you could invent your own if you wish. The point is, if you start processing equations in this form, you can easily reach the ten-input limit.

---

[7] Beware when you alter this function, there are many traps for the unwary. A useful test is to apply your routine to the 4-input truth table equivalent of a 4 × 4 Karnaugh Map which contains only four 1s, one in each corner. If your routine decides that this table has a Reed-Müller solution, then it's wrong!

In fact, extending the program to accept Boolean equations is a large stride into the arena of *logic synthesis*, in which programs are used to optimize the logic used to implement a design. Many commercially available logic synthesis utilities use a method of representing the data internally known as *binary decision diagrams* (BDDs), in which equations are stored in decision trees. For example, consider the binary decision diagram for the simple expression "y = (a & b) | c" (Figure E-10).

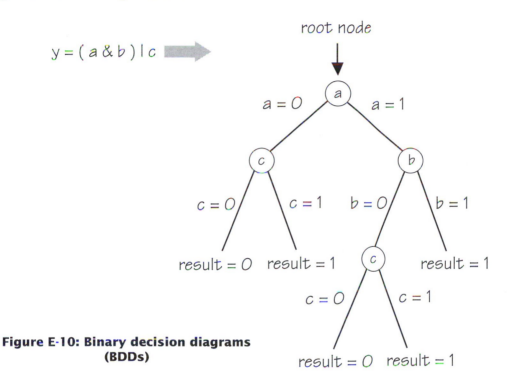

Figure E-10: Binary decision diagrams (BDDs)

The circles representing the inputs are known as *nodes*, and the 0 and 1 values associated with the lines linking two nodes indicate the possible state of the upper node. The 0 and 1 values at the bottom of the diagram indicate the results of the equation for each combination of inputs. The initial node is known as the *root*, the links between nodes may be referred to as *branches*, and the lowest links and their values are referred to as *leafs* or *leaf nodes*.[8]

---

[8] This terminology makes sense if the diagram is visualized as an inverted tree.

The binary decision diagram form of representation finds many uses. For example, as an alternative to actually evaluating a complex expression, it is possible to quickly determine the output from the expression by marching through the diagram using the values on the inputs as signposts. The reason why this is of interest, is that it may be possible to use these diagrams as inputs to *Max's Algorithm*. As the size of an expression increases, it becomes increasingly unlikely that the entire expression will have a Reed-Müller solution. However, it is more than likely that some portions of the expression may have such solutions. Under the direction of a controlling function, it would be possible to examine the decision diagram for Reed-Müller solutions working from the leaf nodes upwards.

You may feel free to use this program in your own projects. However, it would be nice if you could reference Max's Algorithm and its source, *Bebop to the Boolean Boogie*. Additionally, if you do adopt any of the suggestions above, or make modifications of your own, or use the algorithm for a real application, please contact the author (care of the publisher) and spread a little joy in the world. The most interesting derivations may well find themselves featured in the next edition of *Bebop*, thereby making you famous beyond your wildest dreams (and making your mother very proud).

Linear Feedback Shift Registers (LFSRs)

The *Ouroboros*—a symbol of a serpent or dragon devouring its own tail and thereby forming a circle—has been employed by a variety of ancient cultures around the world to depict eternity or renewal.[1] The equivalent to the Ouroboros in the world of electronics would be the *linear feedback shift register (LFSR)*,[2] in which the output from a standard shift register is cunningly manipulated and fed back into its input in such a way as to cause the function to endlessly cycle through a sequence of patterns.

Many-to-One Implementations

LFSRs are simple to construct and are useful for a wide variety of applications, but are often sadly neglected by designers. One of the more common forms of LFSR is formed from a simple shift register with feedback from two or more points, or *taps*, in the register chain (Figure F-1).

The taps in this example are at bit 0 and bit 2, and can be referenced as [0,2]. All of the register elements share a common *clock* input, which is omitted from the symbol for reasons of clarity. The data input to the LFSR is generated by

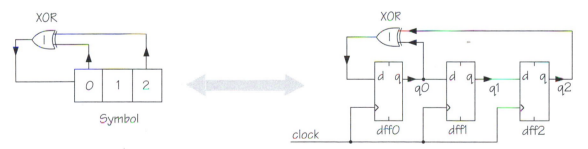

Figure F-1. LFSR with XOR feedback path

---

[1] Not to be confused with the *Amphisbaena*, a serpent in classical mythology having a head at each end and capable of moving in either direction.

[2] In conversation, LFSR is spelled out as "L-F-S-R".

XOR-ing or XNOR-ing the tap bits; the remaining bits function as a standard shift register. The sequence of values generated by an LFSR is determined by its feedback function (XOR versus XNOR) and tap selection (Figure F-2).

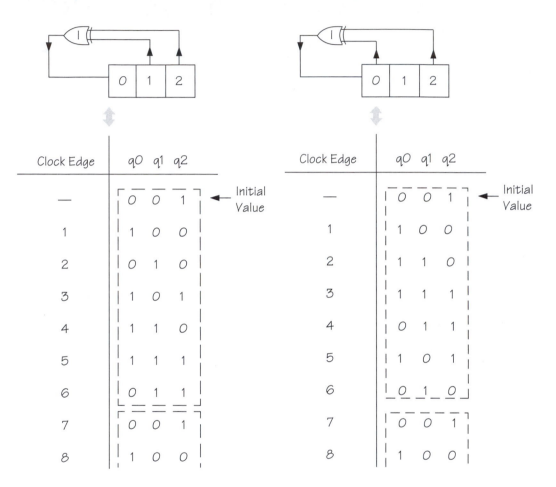

Figure F-2. Comparison of alternative tap selections

Both LFSRs start with the same initial value but, due to the different taps, their sequences rapidly diverge as clock pulses are applied. In some cases an LFSR will end up cycling round a loop comprising a limited number of values. However, both of the LFSRs shown in Figure F-2 are said to be of *maximal length* because they sequence through every possible value (excluding all of the bits being 0) before returning to their initial values.

A binary field with n bits can assume 2^n unique values, but a maximal-length LFSR with n register bits will only sequence through $(2^n - 1)$ values. This is because LFSRs with XOR feedback paths will not sequence through the value where all the bits are 0, while their XNOR equivalents will not sequence through the value where all the bits are 1 (Figure F-3).

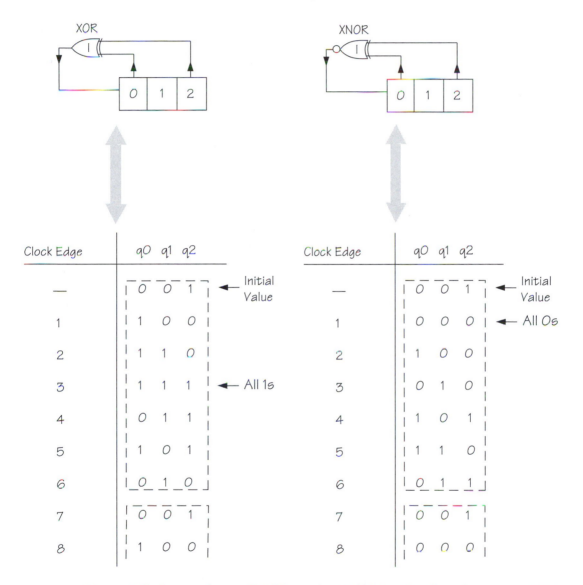

Figure F-3. Comparison of XOR versus XNOR feedback paths

More Taps Than You Know What to Do With

Each LFSR supports a number of tap combinations that will generate maximal-length sequences. The problem is weeding out the ones that do from the ones that don't, because badly chosen taps can result in the register entering a loop comprising only a limited number of states.

The author created a simple C program to determine the taps for maximal-length LFSRs with 2 to 32 bits. These values are presented for your delectation and delight in Figure F-4 (the * annotations indicate sequences whose length is a prime number).

The taps are identical for both XOR-based and XNOR-based LFSRs, although the resulting sequence will of course differ. As was previously noted, alternative tap combinations may also yield maximum-length LFSRs, although the resulting sequences will vary. For example, in the case of a 10-bit LFSR, there are two 2-tap combinations that result in a maximal-length sequence: [2,9] and [6,9]. There are also twenty 4-tap combinations, twenty-eight 6-tap combinations, and ten 8-tap combinations that satisfy the maximal-length criteria.

# of Bits	Length of Loop		Taps
2	3	*	[0,1]
3	7	*	[0,2]
4	15		[0,3]
5	31	*	[1,4]
6	63		[0,5]
7	127	*	[0,6]
8	255		[1,2,3,7]
9	511		[3,8]
10	1,023		[2,9]
11	2,047		[1,10]
12	4,095		[0,3,5,11]
13	8,191	*	[0,2,3,12]
14	16,383		[0,2,4,13]
15	32,767		[0,14]
16	65,535		[1,2,4,15]
17	131,071	*	[2,16]
18	262,143		[6,17]
19	524,287	*	[0,1,4,18]
20	1,048,575		[2,19]
21	2,097,151		[1,20]
22	4,194,303		[0,21]
23	8,388,607		[4,22]
24	16,777,215		[0,2,3,23]
25	33,554,431		[2,24]
26	67,108,863		[0,1,5,25]
27	134,217,727		[0,1,4,26]
28	268,435,455		[2,27]
29	536,870,911		[1,28]
30	1,073,741,823		[0,3,5,29]
31	2,147,483,647	*	[2,30]
32	4,294,967,295		[1,5,6,31]

Figure F-4. Taps for maximal length LFSRs with 2 to 32 bits

One-to-Many Implementations

Consider the case of an 8-bit LFSR, for which the minimum number of taps that will generate a maximal-length sequence is four. In the real world, XOR gates only have two inputs, so a four-input XOR function has to be created using three XOR gates arranged as two levels of logic. Even in those cases where an LFSR does support a minimum of two taps, you may actually wish to use a greater number of taps such as eight (which would result in three levels of XOR logic).

The problem is that increasing the levels of logic in the combinational feedback path can negatively impact the maximum clocking frequency of the function. One solution is to transpose the *many-to-one implementations* discussed above into their *one-to-many counterparts* (Figure F-5).

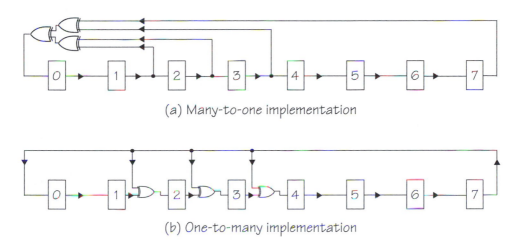

(a) Many-to-one implementation

(b) One-to-many implementation

Figure F-5: Many-to-one versus one-to-many implementations

The traditional many-to-one implementation for the eight-bit LFSR has taps at [7,3,2,1]. To convert this into its one-to-many counterpart, the most-significant tap (which is always the most significant bit) is fed back directly into the least significant bit, and is also individually XORed with the other original taps (bits [3,2,1] in this example). Note that although both styles result in maximal-length LFSRs, the actual sequences of values will differ between them. But the main point is that using the one-to-many style means that there is never more than one level of combinational logic in the feedback path, irrespective of the number of taps being employed.

Seeding an LFSR

One quirk with an XOR-based LFSR is that, if it happens to find itself in the all 0s value, it will happily continue to shift all 0s indefinitely (similarly for an XNOR-based LFSR and the all 1s value). This is of particular concern when power is first applied to the circuit. Each register bit can randomly initialize containing either a logic 0 or a logic 1, and the LFSR can therefore "wake up" containing its "forbidden" value. For this reason, it is necessary to initialize an LFSR with a *seed* value.

One method for loading a seed value is to use registers with reset or set inputs. A single control signal can be connected to the reset inputs on some of the registers and the set inputs on others. When this control signal is placed in its active state, the LFSR will load with a hard-wired seed value. However, in certain applications it is desirable to be able to vary the seed value. One technique for achieving this is to include a multiplexer at the input to the LFSR (Figure F-6).

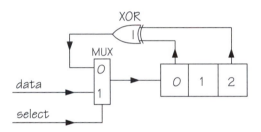
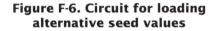

Figure F-6. Circuit for loading alternative seed values

When the multiplexer's *data* input is selected, the device functions as a standard shift register and any desired seed value may be loaded. After loading the seed value, the feedback path is selected and the device returns to its LFSR mode of operation.

FIFO Applications

The fact that an LFSR generates an unusual sequence of values is irrelevant in many applications. Consider a 4-bit × 16-word *first-in first-out (FIFO)* memory device (Figure F-7).

A brief summary of the FIFO's operation is as follows. The write and read pointers are essentially 4-bit registers whose outputs are processed by 4:16 decoders to select one of the sixteen words in the memory array. The reset input is used to initialize the device, primarily by clearing the write and read pointers such that they both point to the same memory word. The initialization also causes the empty output to be placed in its active state and the full output to be placed in its inactive state.

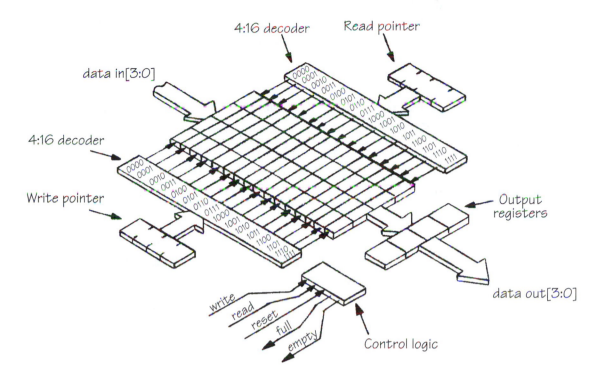

Figure F-7. First-in first-out (FIFO) memory

The write and read pointers chase each other around the memory array in an endless loop. An active edge on the *write* input causes any data on the *data in[3:0]* bus to be written into the word pointed to by the write pointer; the *empty* output is placed in its inactive state (because the device is no longer empty) and the write pointer is incremented to point to the next empty word.

Data can be written into the FIFO until all the words in the array contain values. When the write pointer catches up to the read pointer, the *full* output is placed in its active state (indicating that the device is full) and no more data can be written into the device.

An active edge on the *read* input causes the data in the word pointed to by the read pointer to be copied into the output register; the *full* output is placed in its inactive state and the read pointer is incremented to point to the next word containing data.[3]

---

[3] These discussions assume *write-and-increment* and *read-and-increment* techniques, but some FIFOs employ an *increment-and-write* and *increment-and-read* approach.

Data can be read out of the FIFO until the array is empty. When the read pointer catches up to the write pointer, the *empty* output is placed in its active state, and no more data can be read out of the device.

The write and read pointers for a 16-word FIFO are often implemented using 4-bit binary counters.[4] However, a moment's reflection reveals that there is no intrinsic advantage to a binary sequence for this particular application, and the sequence generated by a 4-bit LFSR would serve equally well. In fact, the two functions operate in a very similar manner as is illustrated by their block diagrams (Figure F-8).

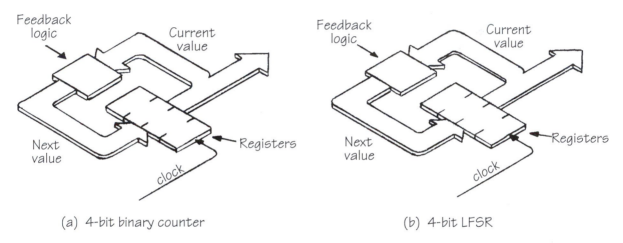

(a) 4-bit binary counter   (b) 4-bit LFSR

Figure F-8. Binary counter versus LFSR

However, the combinational logic for the 4-bit LFSR consists of a single, two-input XOR gate, while the combinational logic for the 4-bit binary counter requires a number of AND and OR gates. This means that the LFSR requires fewer tracks and is significantly more efficient in terms of silicon real estate. Additionally, the LFSR's feedback only passes through a single level of logic, while the binary counter's feedback passes through multiple levels of logic. This means that the new data value is available sooner for the LFSR, which can therefore be clocked at a higher frequency. These differentiations become even more pronounced for FIFOs with more words requiring pointers with more bits. Thus, LFSR's are the obvious choice for the discerning designer of FIFOs.

---

[4] The implementation of a 4-bit binary counter was discussed in Chapter 11.

Modifying LFSRs to Sequence 2n Values

The sole downside to the 4-bit LFSRs in the FIFO scenario above is that they will sequence through only 15 values ($2^4 - 1$), as compared to the binary counter's sequence of 16 values (2^4). Designers may not regard this to be a problem, especially in the case of larger FIFOs. However, if it is required for an LFSR to sequence through every possible value, there is a simple solution (Figure F-9).

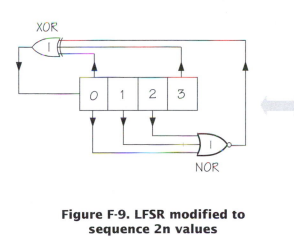

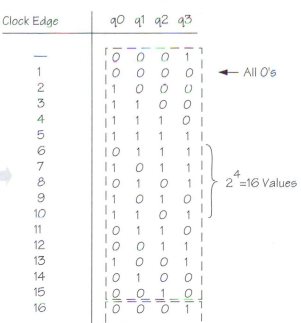

Clock Edge	q0	q1	q2	q3	
—	0	0	0	1	
1	0	0	0	0	← All 0's
2	1	0	0	0	
3	1	1	0	0	
4	1	1	1	0	
5	1	1	1	1	
6	0	1	1	1	
7	1	0	1	1	
8	0	1	0	1	$2^4 = 16$ Values
9	1	0	1	0	
10	1	1	0	1	
11	0	1	1	0	
12	0	0	1	1	
13	1	0	0	1	
14	0	1	0	0	
15	0	0	1	0	
16	0	0	0	1	

Figure F-9. LFSR modified to sequence 2n values

For the value where all of the bits are 0 to appear, the preceding value must have comprised a logic 1 in the most-significant bit (MSB)[5] and logic 0s in the remaining bit positions. In an *unmodified* LFSR, the next clock would result in a logic 1 in the least-significant bit (LSB) and logic 0s in the remaining bit positions. However, in the *modified* LFSR, the output from the NOR is a logic 0 for every case but two: the value preceding the one where all the bits are 0 and the value where all the bits are 0. These two values force the NOR's output to a logic 1, which inverts the usual output from the XOR. This in turn causes the sequence to first enter the all-0s value and then resume its normal course. (In the case of LFSRs with XNOR feedback paths, the NOR can be replaced with an AND, which causes the sequence to cycle through the value where all of the bits are 1.)

---

[5] As is often the case with any form of shift register, the MSB in these examples is taken to be on the right-hand side of the register, and the LSB is taken to be on the left-hand side (this is opposite to the way we usually do things).

Accessing the Previous Value

In some applications it is required to make use of a register's previous value. For example, in certain FIFO implementations, the "full" condition is detected when the write pointer is pointing to the location preceding the location pointed to by the read pointer.[6] This implies that a comparator must be used to compare the *current* value in the write pointer with the *previous* value in the read pointer. Similarly, the "empty" condition may be detected when the read pointer is pointing to the location preceding the location pointed to by the write pointer. This implies that a second comparator must be used to compare the *current* value in the read pointer with the *previous* value in the write pointer.

In the case of binary counters (assuming that, for some reason, we decided to use them for a FIFO application), there are two techniques by which the previous value in the sequence may be accessed. The first requires the provision of an additional set of so-called *shadow registers*. Every time the counter is incremented, its current contents are first copied into the shadow registers. Alternatively, a block of combinational logic can be used to decode the previous value from the current value. Unfortunately, both of these techniques involve a substantial overhead in terms of additional logic. By comparison, LFSRs inherently remember their previous value. All that is required is the addition of a single register bit appended to the MSB (Figure F-10).

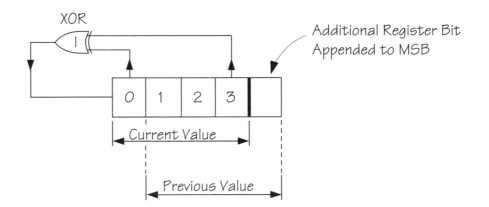

Figure F-10. Accessing an LFSR's previous value

---

[6] Try saying that quickly!

Encryption and Decryption Applications

The unusual sequence of values generated by an LFSR can be gainfully employed in the encryption (scrambling) and decryption (unscrambling) of data. A stream of data bits can be encrypted by XOR-ing them with the output from an LFSR (Figure F-11).

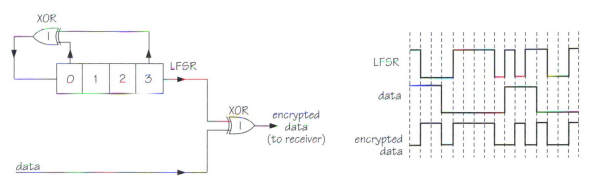

Figure F-11. Data encryption using an LFSR

The stream of encrypted data bits seen by a receiver can be decrypted by XOR-ing them with the output of an identical LFSR.[7]

Cyclic Redundancy Check Applications

A traditional application for LFSRs is in *cyclic redundancy check* (CRC) calculations, which can be used to detect errors in data communications. The stream of data bits being transmitted is used to modify the values fed back into an LFSR (Figure F-12).

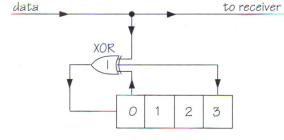

Figure F-12. Cyclic redundancy check (CRC) calculations

The final CRC value stored in the LFSR is known as a *checksum*, and is dependent on every bit in the data stream. After all of the data bits have been transmitted, the transmitter sends its checksum value to the receiver. The receiver contains an identical CRC

---

[7] This is obviously a very trivial form of encryption that's not very secure, but it's "cheap-and-cheerful" and may be useful in certain applications.

calculator and generates its own checksum value from the incoming data. Once all of the data bits have arrived, the receiver compares its internally generated checksum value with the checksum sent by the transmitter to determine whether any corruption occurred during the course of the transmission. This form of error detection is very efficient in terms of the small number of bits that have to be transmitted in addition to the data.

In the real world, a 4-bit CRC calculator would not be considered to provide sufficient confidence in the integrity of the transmitted data. This is due to the fact that a 4-bit LFSR can only represent 16 unique values, which means that there is a significant probability that multiple errors in the data stream could result in the two checksum values being identical. However, as the number of bits in a CRC calculator increases, the probability that multiple errors will cause identical checksum values approaches zero. For this reason, CRC calculators typically use a minimum of 16-bits providing 65,536 unique values.

There are a variety of standard communications protocols, each of which specifies the number of bits employed in their CRC calculations and the taps to be used. The taps are selected such that an error in a single data bit will cause the maximum possible disruption to the resulting checksum value. Thus, in addition to being referred to as *maximal-length*, these LFSRs may also be qualified as *maximal-displacement*.

In addition to checking data integrity in communications systems, CRCs find a wide variety of other uses: for example, the detection of computer viruses. For the purposes of this discussion, a computer virus may be defined as a self-replicating program released into a computer system for a number of purposes. These purposes range from the simply mischievous, such as displaying humorous or annoying messages, to the downright nefarious, such as corrupting data or destroying (or subverting) the operating system.

One mechanism by which a computer virus may both hide and propagate itself is to attach itself to an existing program. Whenever that program is executed, it first triggers the virus to replicate itself, yet a cursory check of the system shows only the expected files to be present. In order to combat this form of attack, a unique checksum can be generated for each program on the system, where the value of each checksum is based on the binary instructions forming the program with which it is associated. At some later date, an anti-virus program can be used to recalculate the checksum values for each program, and to compare them to the original values. A difference in the two values associated with a program may indicate that a virus has attached itself to that program.[8]

Data Compression Applications

The CRC calculators discussed above can also be used in a data compression role. One such application is found in the circuit board test strategy known as *functional test*. The board is plugged into a functional tester by means of its edge connector. The tester applies a pattern of signals to the board's inputs, allows sufficient time for any effects to propagate around the board, and then compares the actual values seen on the outputs with a set of expected values stored in the system. This process is repeated for a series of input patterns, which may number in the tens or hundreds of thousands (Figure F-13).

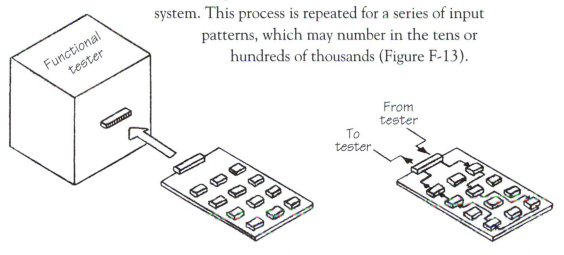

Figure F-13. Functional test

The illustration above is simplified for reasons of clarity. In practice, the edge connector may contain hundreds of pins while the board may contain thousands of components and tracks. If the board fails the preliminary tests, a more sophisticated form of analysis known as *guided probe* may be employed to identify the cause of the failure (Figure F-14).

---

[8] Unfortunately, the creators of computer viruses are quite sophisticated, and some viruses are armed with the ability to perform their own CRC calculations. When a virus of this type attaches itself to a program, it can pad itself with dummy binary values, which are selected so as to cause an anti-virus program to return an identical checksum value to the original.

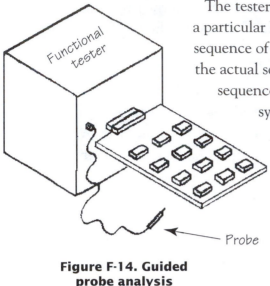

**Figure F-14. Guided
probe analysis**

The tester instructs the operator to place the probe at a particular location on the board, and then the entire sequence of test patterns is rerun. The tester compares the actual sequence of values seen by the probe with a sequence of expected values that are stored in the system. This process (placing the probe and running the tests) is repeated until the tester has isolated the faulty component or track.

A major consideration when supporting a guided probe strategy is the amount of expected data that must be stored. Consider a test sequence comprising 10,000 patterns driving a board containing 10,000 tracks. If the data were not compressed, the system would have to store 10,000 bits of expected data per track, which amounts to 100,000,000 bits of data for the board. Additionally, for each application of the guided probe, the tester would have to compare the 10,000 data bits observed by the probe with the 10,000 bits of expected data stored in the system. Thus, using data in an uncompressed form is an expensive option in terms of storage and processing requirements.

One solution to these problems is to employ LFSR-based CRC calculators. The sequence of expected values for each track can be passed through a 16-bit CRC calculator implemented in software. Similarly, the sequence of actual values seen by the guided probe can be passed through an identical CRC calculator implemented in hardware. In this case, the calculated checksum values are also known as *signatures*, and a guided probe process based on this technique is known as *signature analysis*. Irrespective of the number of test patterns used, the system has to store only two bytes of data for each track. Additionally, for each application of the guided probe, the tester has to compare only the two bytes of data gathered by the probe with two bytes of expected data stored in the system. Thus, compressing the data results in storage requirements that are orders of magnitude smaller and comparison times that are orders of magnitude faster than the uncompressed data approach.

Built-In Self-Test Applications

One test strategy which may be employed in complex integrated circuits is that of *built-in self-test (BIST)*. Devices using BIST contain special test generation and result gathering circuits (Figure F-15).

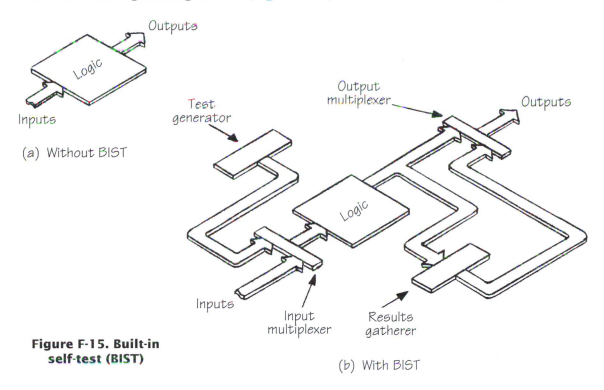

(a) Without BIST

(b) With BIST

Figure F-15. Built-in self-test (BIST)

A multiplexer is used to select between the standard inputs and those from the *test generator*. A second multiplexer selects between the standard outputs and those from the *results gatherer*. Both the test generator and results gatherer can be implemented using LFSRs (Figure F-16).[9]

The LFSR forming the test generator is used to create a sequence of test patterns, while the LFSR forming the results gatherer is used to capture the results. The results-gathering LFSR features modifications that allow it to accept parallel data. (Note that the two LFSRs are not obliged to contain the same number of bits, because the number of inputs to the logic being tested may be different to the number of outputs from the logic.)

---

[9] The standard inputs and outputs along with their multiplexers have been omitted from this illustration for reasons of clarity.

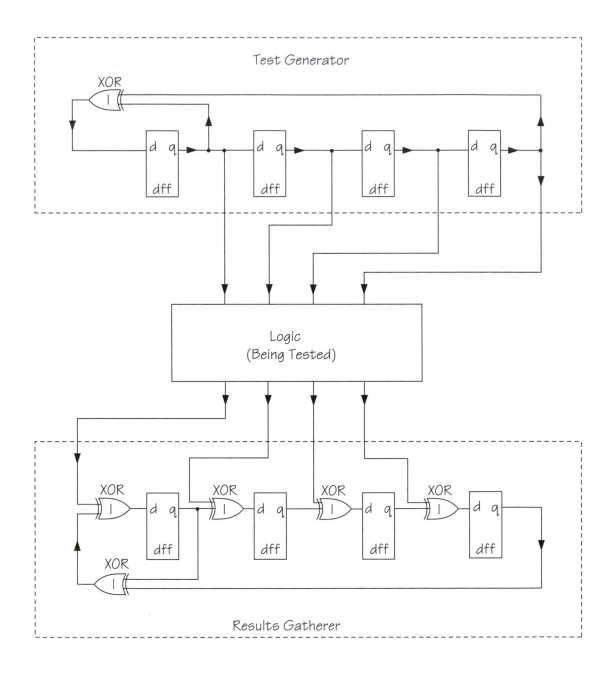

Figure F-16. BIST implemented using LFSRs

Pseudo-Random Number Applications

Many computer programs rely on an element of randomness. Computer games such as "Space Invaders" employ random events to increase the player's enjoyment. Graphics programs may exploit random numbers to generate intricate patterns. All forms of computer simulation may utilize random numbers to more accurately represent the real world. For example, digital simulations[10] may benefit from the portrayal of random stimulus such as external interrupts. Random stimulus can result in more realistic design verification, which can uncover problems that may not be revealed by more structured tests.

Random number generators can be constructed in both hardware and software. The majority of these generators are not truly random, but they give the appearance of being random and are therefore said to be *pseudo-random*. In fact, pseudo-random numbers have an advantage over truly random numbers, because the majority of computer applications typically require repeatability. For example, a designer repeating a digital simulation would expect to receive identical answers to those from the previous run. However, designers also need the ability to modify the seed value of the pseudo-random number generator so as to spawn different sequences of values as required.

There are a variety of methods available for generating pseudo-random numbers. A popular cheap-and-cheerful technique uses the remainder from a division operation as the next value in a pseudo-random sequence. For example, a C function which returns a pseudo-random number in the range 0 to 32767 could be written as follows:[11]

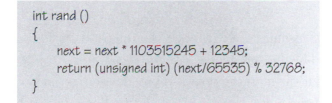

```
int rand ()
{
    next = next * 1103515245 + 12345;
    return (unsigned int) (next/65535) % 32768;
}
```

---

[10] Digital simulation is based on a program called a *logic simulator*, which is used to build a virtual representation of an electronic design in the computer's memory. The simulator then applies stimulus to the design's virtual inputs, simulates the effect of these signals as they propagate through the design, and checks the responses at the design's virtual outputs.

[11] The C Programming Language (Second Edition) by Brian W. Kernighan and Dennis M. Ritchie.

The variable *next* would be declared as global and initialized with some seed value. Also, an additional function would be provided to allow the programmer to load *next* with a new seed value if required. Every time this function is accessed it returns a new pseudo-random value. Unfortunately, pseudo-random implementations based on division do not always produce a "white spectrum". The series of generated values may be composed of a collection of sub-series, in which case the results will not be of maximal length. By comparison, the sequence of values generated by a software implementation of a maximal-length LFSR provides a reasonably good pseudo-random source, but is somewhat more expensive in terms of processing requirements.

A simple technique for checking the "randomness" of a pseudo-random number generator is as follows. Assume the existence of a function called rand() which returns a pseudo-random number as a 16-bit value. The most significant and least significant bytes may be considered to represent the y-axis and x-axis of a graph respectively. To test the randomness of the function, a simple C program segment may be created to plot a reasonable sample of random numbers:

```
int x_axis, y_axis, i, my_number;          /* Declare the variables.                   */
for (i = 1; i <= 1000; i++) {              /* Perform a loop 1,000 times.              */
    my_number = rand();                     /* Get a pseudo-random number.              */
    x_axis = my_number & 0x00FF;            /* Extract the least-significant byte.       */
    y_axis = (my_number & 0xFF00) >> 8;     /* Extract the most-significant byte.        */
    plot(x_axis, y_axis);                   /* Plot the value.                          */
}
```

This example assumes the existence of a function plot() that will cause the point represented by the variables x_axis and y_axis to be displayed on a computer screen. (Note that the masking operation used when extracting the most significant byte is included for reasons of clarity, but is in fact redundant.)

If the resulting points are uniformly distributed across the plotting area, then the pseudo-random generator is acceptable. But if the points are grouped around the diagonal axis or form clusters, then the generator is less than optimal (Figure F-17).

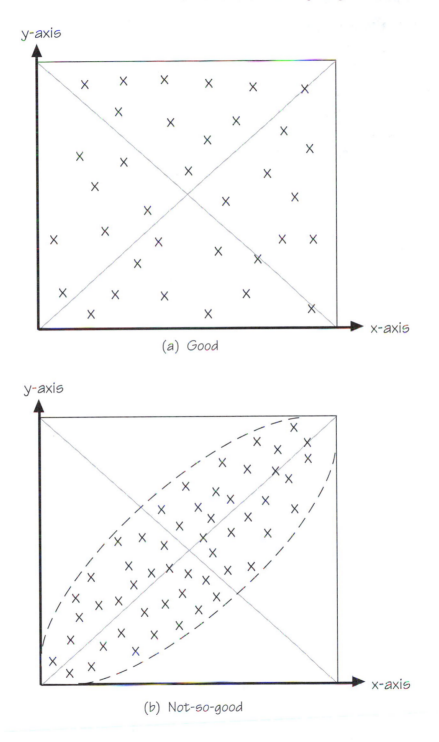

(a) Good

(b) Not-so-good

Figure F-17. Checking for "randomness"

Last But Not Least

LFSRs are simple to construct and are useful for a wide variety of applications, but be warned that choosing the optimal polynomial (which ultimately boils down to selecting the tap points) for a particular application is a task that is usually reserved for a master of the mystic arts, not to mention that the maths can be hairy enough to make a grown man break down and cry (and don't even get the author started on the subject of *cyclotomic polynomials*,[12] which are key to the tap-selection process).

---

[12] Mainly because he hasn't got the faintest clue what a *cyclotomic polynomial* is!

Pass-Transistor Logic

The term *pass-transistor logic* refers to techniques for connecting MOS transistors such that *data* signals pass between their *source* and *drain* terminals. These techniques minimize the number of transistors required to implement a function, but they are not recommended for the novice or the unwary because strange and unexpected effects can ensue. Pass-transistor logic is typically employed by designers of full-custom integrated circuits or ASIC cell libraries. The following examples introduce the concepts of pass-transistor logic, but they are not intended to indicate recommended design practices.

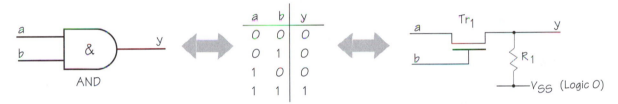

Figure G-1. Pass-transistor implementation of an AND gate

In the case of the AND (Figure G-1), the resistance of R_1 is assumed to be sufficiently high that its effect on output y is that of a very weak logic 0. When input b is set to logic 0, the NMOS transistor Tr_1 is turned OFF and y is "pulled down" to logic 0 by resistor R_1. When input b is set to logic 1, Tr_1 is turned ON and output y is connected to input a. In this case, a logic 0 on input a leaves y at logic 0, but a logic 1 on a will overdrive the effect of resistor R_1 and force y to a logic 1.

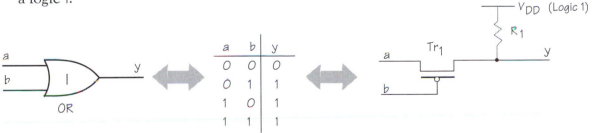

Figure G-2. Pass-transistor implementation of an OR gate

In the case of the OR (Figure G-2), a logic 1 applied to input *b* turns the PMOS transistor Tr_1 OFF, and output *y* is "pulled up" to logic 1 by resistor R_1. When input *b* is set to logic 0, Tr_1 is turned ON and output *y* is connected to input *a*. In this case, a logic 1 on input a leaves y at logic 1, but a logic 0 on *a* will overdrive the effect of resistor R_1 and force y to a logic 0.

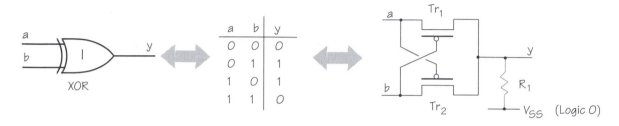

a	b	y
0	0	0
0	1	1
1	0	1
1	1	0

Figure G-3. Pass-transistor implementation of an XOR gate

In the case of the XOR (Figure G-3), logic 1s applied to inputs *a* and *b* turn Tr_1 and Tr_2 OFF, respectively, leaving output y to be "pulled down" to logic 0 by resistor R_1. When input *b* is set to logic 0, Tr_1 is turned ON and output y is connected to input *a*. In this case, a logic 0 on input *a* leaves y at logic 0, but a logic 1 on *a* will overdrive the effect of resistor R_1 and force y to a logic 1. Similarly, when input *a* is set to logic 0, Tr_2 is turned ON and output y is connected to input *b*. In this case, a logic 0 on input *b* leaves y at logic 0, but a logic 1 on *b* will overdrive the effect of resistor R_1 and force y to a logic 1 (phew!).

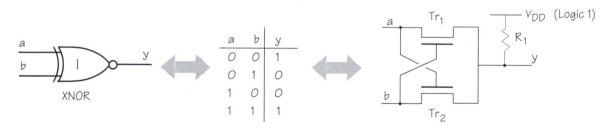

a	b	y
0	0	1
0	1	0
1	0	0
1	1	1

Figure G-4. Pass-transistor implementation of an XNOR gate

The case of the XNOR (Figure G-4) is simply the inverse of that for the XOR. Logic 0s applied to inputs *a* and *b* turn transistors Tr_1 and Tr_2 OFF, leaving output y to be "pulled up" to logic 1 by resistor R_1. When input *b* is set to logic 1, Tr_1 is turned ON and output y is connected to input *a*. In this case, a logic 1 on input *a* leaves *y* at logic 1, but a logic 0 on *a* will overdrive the effect of resistor R_1 and force *y* to a logic 0. Similarly, when input *a* is set to logic 1, Tr_2 is turned ON and output *y* is connected to input *b*. In this case, a logic 1 on input *b* leaves *y* at logic 1, but a logic 0 on *b* will overdrive the effect of resistor R_1 and force *y* to a logic 0.

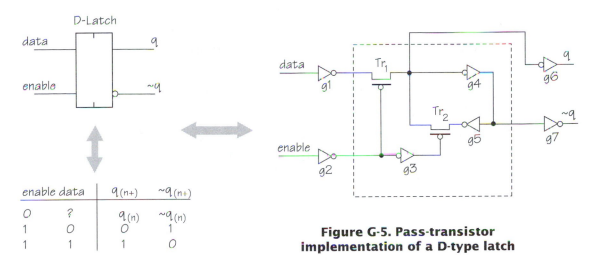

enable	data	$q_{(n+)}$	$\sim q_{(n+)}$
0	?	$q_{(n)}$	$\sim q_{(n)}$
1	0	0	1
1	1	1	0

Figure G-5. Pass-transistor implementation of a D-type latch

Before examining the pass-transistor implementation of a D-type latch with an active-high enable shown in Figure G-5, we should briefly review our definition of these functions. First, when the *enable* input is in its active state (a logic 1), the value on the *data* input is passed through the device to appear on the *q* and *~q* outputs.[1] Second, if the *data* input changes while *enable* remains active, the outputs will respond to reflect the new value. Third, when *enable* is driven to its inactive state, the outputs remember their previous values and no longer respond to any changes on the *data* input.

Now onto the pass-transistor implementation.[2] When *enable* is in its active state, the output from *g2* turns transistor Tr_1 ON, and the output from *g3* turns transistor Tr_2 OFF. Thus, the value on the *data* input passes through *g1*, Tr_1, and *g6* to appear on the *q* output, and through *g1*, Tr_1, *g4*, and *g7* to appear at

---

[1] As usual, the *data* appears in inverted form on the *~q* (complementary) output.

[2] Note the use of assertion-level logic symbols in the circuit diagram. These symbols were introduced in Appendix A.

the ~q output.[3] Any changes on *data* will be reflected on q and ~q after the delays associated with the gates and transistors have been satisfied. When *enable* is driven to its inactive state, the output from g2 turns Tr_1 OFF and the output from g3 turns Tr_2 ON. Disabling Tr_1 blocks the path from the *data* input, while enabling Tr_2 completes the self-sustaining loop formed by g4 and g5. It is this loop that acts as the memory for the device.

Compare this pass-transistor implementation with its standard counterpart introduced in Chapter 11. The standard implementation required one NOT, two ANDs, and two NORs totaling twenty-two transistors. By comparison, the pass-transistor implementation requires two discrete transistors and seven NOTs totaling only sixteen transistors. In fact, the heart of the pass transistor implementation consisting of Tr_1, Tr_2, g3, g4, and g5 requires only eight transistors. The remaining gates, g1, g2, g6, and g7, are used only to buffer the latch's input and outputs from the outside world.[4]

---

[3] Note the use of the tilde '~' character to indicate the complementary output ~q. The use of tildes was discussed in Appendix A.

[4] NOTs are used in preference to BUFs because NOTs require fewer transistors and are faster (these considerations were introduced in Chapter 6).

A No-Holds-Barred Seafood Gumbo

Well, my little avocado pears, festooned with slivers of spring onions and deluged with lashings of spicy Italian dressing . . . Having absorbed the myriad juicy snippets of information that are strewn so lavishly throughout *Bebop to the Boolean Boogie*, you are now uniquely qualified to regale an assembled throng with a discourse on almost any subject under the sun. So, this would be the perfect time to rest our weary brains and turn our attention to other facets of life's rich tapestry. For example, as soon as I've finished penning these last few words, I must return to practicing my down-hill mud-wrestling skills as, I have no doubt, so do you.

But first we need to recline, relax, lay back, unwind, and take the time to recharge our batteries—and what could be more appropriate than a steaming bowl of no-holds-barred seafood gumbo? So, here's the recipe for a fulsomely-flavored Epicurean taste-fest sensation sufficient to cause the most sophisticated of gastronomes to start salivating surreptitiously and to make a grown man break down and cry. This pert little beauty will pulsate promiscuously across your pallet and pummel it with a passion, titillate your taste buds and have them tap-dancing the tango on your tongue, reverberate and resonate resound-ingly throughout your nervous system, and warm the cockles of your heart. In short, this frisky little number will grab you by the short-and-curlies, swing you sybaritically around the room in a syncopated symphony of delight, and leave you groveling on your knees, gnashing your teeth, and gasping for more.

Beware! If you are under 21, male, or a politician, don't attempt to do anything on the culinary front without your mother's permission and supervision, because kitchens contain sharp things, hot things, and a wide variety of other poten-tially dangerous things.

The following ingredients are for the main body of the gumbo—you'll have to sort out any rice, bread, and side dishes for yourself:

2 cups of diced onions	2 bay leaves
1½ cups of diced green bell peppers	½ teaspoon dried thyme leaves
1 cup of diced celery	¼ teaspoon dried oregano
2 cups of halved button mushrooms	½ teaspoons of salt
3 large cloves of finely diced garlic	1½ teaspoons of white pepper
1 finely diced scotch bonnet or habañero pepper	½ teaspoon of black pepper
	½ teaspoon of cayenne pepper
10 thick-cut slices of bacon	2 teaspoons of Gumbo Filé
1 pound of Cajun-style sausage	
1 pound of peeled, medium-sized shrimp	5½ cups of chicken stock
½ pound of scallops	Lots of butter (more below)
¾ pound white fish cut into slices	¾ cup flour
1 small tin of anchovies	

If you can't get Cajun-style sausage, then Polish sausage will do nicely. Note that the teaspoon quantities in the list of ingredients do not refer to level measures, nor should you to attempt to set a new world record for the amount you can balance on one spoon—just try to aim for roughly the same sensuously rounded profile you get when you're casually spooning sugar into a cup of coffee. So, without further ado, let's gird up our loins and proceed to the fray:

1. First of all there's an art to cooking, and it starts by doing the washing up you've been putting off all day and putting all of the pots away.

3. Grill (broil) the bacon until it's crispy and crunchy; then put it on a plate to cool and set it to one side.

4. Prepare all of the vegetables, mushrooms, garlic, and scotch bonnet or habañero pepper (be careful with the latter—it's best to wear gloves here—

because these are ferociously hot and if you get the juice on your finger and then touch anywhere near your eyes you're going to be a very unhappy camper). Put them all in separate bowls, except for the scotch bonnet and garlic, which can go together.

5.   Chop the Cajun-style sausage into ¼-inch pieces (cut at about a 45 degree angle because this looks nicer); put them in a bowl and set it to one side.

6.   Mix the salt, thyme, oregano, Gumbo Filé, and the white, black, and cayenne peppers together in a cup and set it to one side (you'll need your hands free later).

7.   Wash up all the knives, chopping boards, and everything else you've used and put them all away, and then wipe down all of your working surfaces. Trust me—you'll feel better when everything is clean and tidy (have I ever lied to you before?). Take a five-minute break and quaff[1] a glass of wine—after all, who deserves it more than you?

8.   Put the chicken stock into a large chili pan and bring it to the boil. Then reduce the heat to a low, slow simmer and leave it on the back burner.

9.   Using a medium to medium-high heat, melt ¾ of a cup of butter in a large, heavy skillet until it starts to bubble. Gradually add the flour using a whisk and stir constantly until the resulting roux is a dark-ish, red-ish brown (the darker the better—try to be brave here and leave it longer than you expect). Remove the skillet from the heat, but keep on stirring until it's cooled down enough so that the mixture won't stick to it and burn.

10. Maintain the stock at a low simmer and add the mixture that you've just made, stirring it in one spoonful at a time, and waiting for each spoonful to dissolve before adding the next.

11. Clean the skillet, put it on a medium-high to high heat, and melt a chunk of butter. Stirring all the time, sauté the celery for one minute, add the bell peppers and sauté for 1½ minutes, add the onions and sauté for 1½ minutes, then add the scotch bonnets and garlic along with the mixture of herbs, salt, and pepper and sauté for one more minute. Finally, chuck the whole lot into the chili pan with the stock.

---

[1] Quaffing is like drinking, except that you tend to spill more down your chest.

12. Break the bacon into ½-inch pieces and toss them into the stock. Flake the anchovies with a fork and cast them into the stock. Hurl in the Cajun-style sausage and the bay leaves. Also, if you happen to have any lying around, add a couple of teaspoons of English Worcestershire sauce. Cover the chili pan and leave on a low simmer.

13. Return the skillet to a medium-high to high heat and melt another chunk of butter. Sauté the mushrooms until they're golden brown and squealing for more, then use them to swell the contents of the chili pan.

14. Simmer the whole mixture (stirring often) for at least one hour, which, by some strange quirk of fate, will give you all the time you need to wash the skillet and the dishes you used and put them away again. If you're ravenous you can proceed immediately to the next step. However, if you're wise, you'll remove the heat and leave your cunningly captivating creation to stand overnight (chilies, stews, curries, and gumbos always taste better if the ingredients have the time to formally introduce themselves). When you're ready to chow down, heat it back up again and proceed to the next step.

15. Add the shrimp, scallops, and fish. Bring to the boil then return to a simmer. Maintain the simmer until the seafood is cooked (I personally opt for around 15 minutes just to make sure) and you're ready to rock and roll.

This little beauty will put hairs on your chest, make them curl, and then take them off again. Seriously, this gumbo really is seductively, scintillatingly, scorchingly tasty—your friends will be singing your praises and dancing in the streets.

You can serve your gorgeous gourmet gumbo over steamed or boiled rice, with crusty French bread, or with whatever else your heart desires. The quantities given above will serve eight to ten manly-man sized portions with a little something left over for the following day. Of course, no meal would be complete without some wine—and the perfect complement to your scrumptious repast is . . . to be found in a very large bottle. Enjoy!

Abbreviations and Acronyms

Electronic engineers and computer buffs simply love abbreviations and acronyms. They are really happy if they can use one abbreviation to represent multiple things or different abbreviations to represent the same things, and they become close to ecstatic if they can generate a plethora of abbreviations from permutations of the same three or four letters. The situation is further confused, because the meanings behind abbreviations and acronyms may evolve over time. The list below contains the abbreviations and acronyms used in this book. For further reading, one could do a lot worse than *The Computer Glossary* by Alan Freedman.

2-D or 2D	Two Dimensional	**COB**	Chip-On-Board
3-D or 3D	Three Dimensional	**COC**	Chip-On-Chip
A/D	Analog-to-Digital	**COF**	Chip-On-Flex
ASIC	Application-Specific Integrated Circuit	**COT**	Customer-Owned Tooling
		CPLD	Complex PLD
ASSP	Application-Specific Standard Part	**CPU**	Central Processing Unit
		CRC	Cyclic Redundancy Check
BDD	Binary Decision Diagram	**CSIC**	Customer-Specific Integrated Circuit
BEDO	Burst EDO		
BGA	Ball Grid Array	**CSP**	Chip Scale Package
BiCMOS	Bipolar-CMOS	**CVD**	Chemical Vapor Deposition
BiNMOS	Bipolar-NMOS	**CVI**	Chemical Vapor Infiltration
BIST	Built-In Self-Test	**D/A**	Digital-to-Analog
BJT	Bipolar Junction Transistor	**DDR**	Double Data Rate
CAD	Computer-Aided Design	**DFT**	Design-for-Test
CAE	Computer-Aided Engineering	**DIL**	Dual-In-Line
CGA	Column Grid Array	**DRAM**	Dynamic RAM
CGH	Computer-Generated Hologram	**DSM**	Deep-Submicron
CMOS	Complementary Metal-Oxide Semiconductor	**DSP**	Digital Signal Processor
		DTL	Diode-Transistor Logic
CMP	Chemical Mechanical Polishing	**DIMM**	Dual In-line Memory Module
CNFET	Carbon Nanotube FET	**DWB**	Discrete Wired Board

E²PROM	Electrically-Erasable Programmable Read-Only Memory	**GAL**	Generic Array Logic[2]
EBE	Electron Beam Epitaxy	**GIGO**	Garbage-In Garbage-Out
ECC	Error-Correcting Code	**HDL**	Hardware Description Language
ECL	Emitter-Coupled Logic	**HTCC**	High-Temperature Cofired Ceramic
EDA	Electronic Design Automation	**IC**	Integrated Circuit
EDIF	Electronic Design Interchange Format	**ICR**	In-Circuit Reconfigurable
EDO	Extended Data Out	**IDM**	Integrated Device Manufacturer
EDSAC	Electronic Delay Storage Automatic Calculator	**I/O**	Input/Output
		IP	Intellectual Property
EDVAC	Electronic Discrete Variable Automatic Computer	**IR**	Infrared
		ISP	In-System Programmable
EEPROM	Electrically-Erasable Programmable Read-Only Memory	**JEDEC**	Joint Electronic Device Engineering Council
ENIAC	Electronic Numerical Integrator and Calculator	**LED**	Light-Emitting Diode
		LFSR	Linear Feedback Shift Register
EPROM	Erasable Programmable Read-Only Memory	**LIFO**	Last-In-First-Out
		LSB	Least Significant Bit
ESD	Electrostatic Discharge	**LSB**	Least Significant Byte
EUV	Extreme Ultraviolet	**LSI**	Large-Scale Integration
EUVL	Extreme Ultraviolet Lithography	**LTCC**	Low-Temperature Cofired Ceramic
FET	Field-Effect Transistor	**LTH**	Lead Through-Hole
FIFO	First-In First-Out	**MBE**	Molecular Beam Epitaxy
FPC	Flexible Printed Circuit	**MCM**	Multichip Module
FPD	Field-Programmable Device	**MCM-C**	Multichip Module – Ceramic substrate
FPGA	Field-Programmable Gate Array		
FPIC	Field-Programmable Interconnect Chip[1]	**MCM-D**	Multichip Module – Dielectric-covered substrate
FPID	Field-Programmable Interconnect Device	**MCM-L**	Multichip Module – Laminate substrate
FPM	Fast Page Mode	**MCM-S**	Multichip Module – Semiconductor substrate
FSM	Finite State Machine		

---

[1] FPIC is a trademark of Aptix Corporation.

[2] GAL is a registered trademark of Lattice Semiconductor Corporation.

MEMS	MicroElectroMechanical Systems	**RAM**	Random Access Memory
MOS	Metal-Oxide Semiconductor	**RDRAM**	Rambus DRAM
MOSFET	Metal-Oxide Semiconductor Field-Effect Transistor	**RIMM**	*See the Glossary for more details*
MRAM	Magnetic Random Access Memory	**ROM**	Read-Only Memory
MSB	Most Significant Bit	**RTL**	Register Transfer Level
MSB	Most Significant Byte	**RTL**	Resistor-Transistor Logic
MSI	Medium-Scale Integration	**RWM**	Read-Write Memory
MTJ	Magnetic Tunnel Junction	**SDRAM**	Synchronous DRAM
NMOS	N-channel MOS	**SIMM**	Single In-line Memory Module
NPN	N-type – P-type – N-type	**SIPO**	Serial-In Parallel-Out
NRE	Non-Recurring Engineering	**SISO**	Serial-In Serial-Out
nvRAM	Non-Volatile RAM	**SMD**	Surface Mount Device
OMEMS	Optical MEMS	**SMOBC**	Solder Mask Over Bare Copper
OTP	One Time Programmable	**SMT**	Surface Mount Technology
PAL	Programmable Array Logic[3]	**SoC**	System-on-Chip
PCB	Printed Circuit Board	**SOP**	Small Outline Package
PGA	Pad Grid Array	**SPLD**	Simple PLD
PGA	Pin Grid Array	**SRAM**	Static RAM
PHB	Photochemical Hole Burning	**SSI**	Small-Scale Integration
PISO	Parallel-In Serial-Out	**TAB**	Tape Automated Bonding
PLA	Programmable Logic Array	**TTL**	Transistor-Transistor Logic
PLD	Programmable Logic Device	**μC**	Microcontroller
PMOS	P-channel MOS	**UDSM**	Ultra-Deep-Submicron
PNP	P-type – N-type – P-type	**ULA**	Uncommitted Logic Array
PROM	Programmable Read-Only Memory	**ULSI**	Ultra-Large-Scale Integration
PTH	Plated Through-Hole	**UNIVAC**	Universal Automatic Computer
PWB	Printed Wire Board	**μP**	Microprocessor
QDR	Quad Data Rate	**UV**	Ultraviolet
QFP	Quad Flat Pack	**VHDL**	VHSIC HDL
		VHSIC	Very High Speed Integrated Circuit
		VLSI	Very-Large-Scale Integration

---

[3] PAL is a registered trademark of Monolithic Memories Inc.

Glossary

Absolute Scale of Temperature – A scale of temperature that was invented by the British mathematician and physicist William Thomas, first Baron of Kelvin. Under the *absolute*, or *Kelvin*, scale of temperature, 0K (corresponding to –273°C) is the coldest possible temperature and is known as *absolute zero*.

Absolute Zero – *see Absolute Scale of Temperature*.

Active-High – Once again, beware, here be dragons. In this book an *active-high* signal is defined as a signal whose active state is considered to be a logic 1. This definition, which is used by the engineers in the trenches, allows all forms of logic, including *assertion-level* logic, to be represented without any confusion, regardless of whether *positive-* or *negative-logic* implementations are employed. Unfortunately, some academics (and even textbooks) define an active-high signal as being one whose asserted (TRUE or logic 1) state is at a higher voltage level than its unasserted (FALSE or logic 0) state. However, this definition serves only to cause confusion when combined with negative-logic implementations. Thus, when using the term active-high in discussions with other people, you are strongly advised to make sure that you're all talking about the same thing before you start.

Active-Low – Beware, here be dragons. In this book an *active-low* signal is defined as a signal whose active state is considered to be a logic 0. This definition, which is used by the engineers in the trenches, allows all forms of logic, including *assertion-level* logic, to be represented without any confusion, regardless of whether *positive-* or *negative-logic* implementations are employed.

Unfortunately, some academics (and even textbooks) define an active-low signal as being one whose asserted (TRUE or logic 1) state is at a lower voltage level than its unasserted (FALSE or logic 0) state. However, this definition serves only to cause confusion when combined with negative-logic implementations. Thus, when using the term active-low in discussions with other people, you are strongly advised to make sure that you're all talking about the same thing before you start.

Active Substrate – A hybrid or multichip module substrate formed from a semiconductor. Termed *active* because components such as transistors can be fabricated directly into the substrate.

Active Trimming – The process of trimming components such as resistors while the circuit is under power. Such components are fabricated directly onto the substrate of a hybrid or multichip module, and the trimming is usually performed using a laser beam.

Actuator – A transducer that converts an electronic signal into a physical equivalent. For example, a loudspeaker is an actuator that converts electronic signals into corresponding sounds.

Adaptive Hardware – Refers to devices that allow new design variations to be "compiled" in real-time, which may be thought of as dynamically creating subroutines in hardware *See also Virtual Hardware and Cache Logic*.

Additive Process – A process in which conducting material is added to specific areas of a substrate. Groups of tracks, individual tracks, or portions of tracks can be built up to precise thicknesses by iterating the process multiple times with selective masking.

Address Bus – A unidirectional set of signals used by a computer to point to memory locations in which it is interested.

A/D (Analog-to-Digital) – The process of converting an analog value into its digital equivalent.

Analog – A continuous value that most closely resembles the real world and can be as precise as the measuring technique allows.

Analog Circuit – A collection of components used to process or generate analog signals.

Analog-to-Digital – *see* A/D.

Analogue – The way they spell "analog" in England.

Anisotropic Adhesives – Special adhesives that contain minute particles of conductive material. These adhesives find particular application with the flipped-chip techniques used to mount bare die on the substrates of hybrids, multichip modules, or circuit boards. The conducting particles are only brought in contact with each other at the sites where the raised pads on the die are pressed down over their corresponding pads on the substrate, thereby forming good electrical connections between the pads.

Anti-Fuse Technology – A programmable logic device technology in which conducting paths (*anti-fuses*) are "grown" by applying signals of relatively high voltage and current to the device's inputs.

Anti-Pad – The area of copper etched away around a via or a plated through-hole on a power or ground plane, thereby preventing an electrical connection being made to that plane.

Application-Specific Integrated Circuit – *see* ASIC.

Application-Specific Standard Part – *see* ASSP.

ASIC (Application-Specific Integrated Circuit) – A device whose function is determined by a designer for a particular application or group of applications.

ASIC Cell – A logic function in the cell library defined by the manufacturer of an application-specific integrated circuit.

Assertion-Level Logic – A technique used to more precisely represent the function of logic gates with active-low inputs.

Associative Rules – Algebraic rules that state that the order in which pairs of variables are associated together will not affect the result of an operation: for example,[1] $(a \& b) \& c \equiv a \& (b \& c)$.

ASSP (Application-Specific Standard Part) – Refers to complex integrated circuits created by a device manufacturer using ASIC technologies, where these components are to be sold as standard parts to anybody who wants to buy them.

Asynchronous – A signal whose data is acknowledged or acted upon immediately, regardless of any clock signal.

Alto – Unit qualifier (symbol = a) representing one millionth of one millionth of one millionth, or 10^{-18}. For example, 3 as stands for 3×10^{-18} seconds.

Backplane – The medium used to interconnect a number of circuit boards. Typically refers to a special, heavy-duty printed or discrete wired circuit board.

Ball Grid Array – *see* BGA.

Bare Die – An unpackaged integrated circuit.

---

[1] Note that the symbol $\equiv$ indicates "is equivalent to" or "is the same as."

Barrier Layer – *see Overglassing.*

Base – Refers to the number of digits in a numbering system. For example, the decimal numbering system is said to be base-10. (May also be referred to as the "radix.")

Basic Cell – A predefined group of unconnected components that is replicated across the surface of a gate array.

BDD (Binary Decision Diagram) – A method of representing Boolean equations as decision trees.

Bebop – A form of music characterized by fast tempos and agitated rhythms that became highly popular in the decade following World War II.

BEDO (Burst EDO) – An asynchronous form of DRAM-based computer memory that was popular for a while in the latter half of the 1990s. Along with other asynchronous memory techniques, BEDO was eventually superceded by SDRAM technologies *See also FPM, EDO, and SDRAM.*

BGA (Ball Grid Array) – A packaging technology similar to a *pad grid array (PGA)*, in which a device's external connections are arranged as an array of conducting pads on the base of the package. However, in the case of a ball grid array, small balls of solder are attached to the conducting pads.

BiCMOS (Bipolar-CMOS) – A technology in which the function of each logic gate is implemented using low-power CMOS, while the output stage is implemented using high-drive bipolar transistors.

Binary – Base-2 numbering system (the type of numbers that computers use internally).

Binary Decision Diagram – *see BDD.*

Binary Digit – A numeral in the binary scale of notation. A binary digit (typi-

cally abbreviated to "bit") can adopt one of two values: 0 or 1.

Binary Encoding – A form of state assignment for state machines that requires the minimum number of state variables.

Binary Logic – Digital logic gates based on two distinct voltage levels. The two voltages are used to represent the binary values 0 and 1, and their logical equivalents FALSE and TRUE.

BiNMOS (Bipolar-NMOS) – A relatively new low-voltage integrated circuit technology in which complex combinations of bipolar and NMOS transistors are used for sophisticated output stages providing both high speed and low static power dissipation.

Bipolar Junction Transistor – *see BJT.*

Bi-quinary – A system that utilizes two bases, base-2 and base-5, to represent decimal numbers. Each decimal digit is represented by the sum of two parts, one of which has the value of decimal zero or five, and the other the values of zero through four. The abacus is one practical example of the use of a bi-quinary system.

BIST (Built-in Self-test) – A test strategy in which additional logic is built into a component, thereby allowing it to test itself.

Bit – Abbreviation *of binary digit.* A binary digit can adopt one of two values: 0 or 1.

BJTs (Bipolar Junction Transistors) – A family of transistors.

Blind Via – A via that is only visible from one side of the substrate.

Bobble – A small circle used on the inputs to a logic gate symbol to indicate an active low input or control, or on the outputs to indicate a negation (inversion) or complementary signal. Some engineers prefer to use the term "bubble."

Boolean Algebra – A mathematical way of representing logical expressions.

Bootstrapping – A sequence of initialization operations performed by a computer when it is first powered up.

Bounce Pad – A special pattern etched onto the power or ground plane of a microwire circuit board. The bounce pad is used in conjunction with a laser beam to create blind vias. The laser beam evaporates the epoxy forming the outer layers of the board and continues down to the bounce pad which reflects, or bounces, it back up, thereby terminating the via.

Braze – To unite or fuse two pieces of metal by heating them in conjunction with a hard solder with a high melting point.

Buckyballs – Prior to the mid-1980s, the only major forms of pure carbon known to us were graphite and diamond. In 1985, however, a third form consisting of spheres formed from 60 carbon atoms (C_{60}) was discovered. Officially known as Buckministerfullerine[2]—named after the American architect R. Buckminister Fuller who designed geodesic domes with the same fundamental symmetry—these spheres are more commonly known as "buckyballs." In 2000, scientists with the U.S. Department of Energy's Lawrence Berkeley National Laboratory (Berkeley Lab) and the University of California at Berkeley reported that they had managed to fashion a transistor from a single buckyball.

Built-In Self-Test – *see BIST.*

Bulk Storage – Refers to some form of media, typically magnetic, such as tape or a disk drive, which can be used to store large quantities of information relatively inexpensively.

Buried Via – A via used to link conducting layers internal to a substrate. Such a via is not visible from either side of the substrate.

Burst EDO – *see BEDO.*

Bus – A set of signals performing a common function and carrying similar data. Typically represented using vector notation: for example, data[7:0].

Bundle – A set of signals related in some way that makes it appropriate to group them together for ease of representation or manipulation. May contain both scalar and vector elements: for example, {a,b,c,d[5:0]}.

Byte – A group of eight binary digits, or bits.

Cache Logic[3] – A proprietary name intended to reflect designs realized in virtual hardware. Derived from its similarity to the concept of *Cache Memory.*

Cache Memory – A small, high-speed, memory (usually SRAM) used to buffer the central processing unit from any slower, lower cost memory devices such as DRAM. The high-speed cache memory is used to store active instructions and data,[4] while the bulk of the instructions and data reside in the slower memory.

Canonical Form – In a mathematical context this term is taken to mean a generic or basic representation. Canonical forms provide the means to compare two expressions without falling into the trap of trying to compare "apples" with "oranges."

---

[2] Science magazine voted Buckministerfullerine the "Molecule of the Year" in 1991.

[3] Cache Logic is a trademark of Atmel Corporation.

[4] In this context, "active" refers to data or instructions that a program is currently using, or which the operating system believes that the program will want to use in the immediate future.

Capacitance – A measure of the ability of two adjacent conductors separated by an insulator to hold a charge when a voltage differential is applied between them. Capacitance is measured in units of Farads.

Carbon Nanotube FET – *see CNFET*.

Catalyst – A substance that initiates a chemical reaction under different conditions (such as lower temperatures) than would otherwise be possible. The catalyst itself remains unchanged at the end of the reaction.

Cell – *see ASIC Cell, Basic Cell, Cell Library, and Memory Cell*.

Cell Library – The collective name for the set of logic functions defined by the manufacturer of an application-specific integrated circuit. The designer decides which types of cells should be realized and connected together to make the device perform its desired function.

Central Processing Unit – *see CPU*.

Ceramic – An inorganic, nonmetallic material, such as alumina, beryllia, steatite, or forsterite, which is fired at a high temperature and is often used in electronics as a substrate or to create component packages.

CGA (Column Grid Array) – A packaging technology similar to a *pad grid array (PGA)*, in which a device's external connections are arranged as an array of conducting pads on the base of the package. However, in the case of a column grid array, small columns of solder are attached to the conducting pads.

CGH (Computer-Generated Hologram) – In the context of this book, this refers to a slice of quartz or similar material into which three-dimensional patterns are cut using a laser. The angles of the patterns cut into the quartz are precisely calculated for use in the optical communication strategy known as holographic interconnect. All of these calculations are performed by a computer, and the laser used to cut the three-dimensional patterns into the quartz is also controlled by a computer. Thus, the slice of quartz is referred to as a *computer-generated hologram*.

Channel – **(1)** The area between two arrays of basic cells in a channeled gate array. **(2)** The gap between the source and drain regions in a MOS transistor.

Channeled Gate Array – An *application-specific integrated circuit (ASIC)* organized as arrays of basic cells. The areas between the arrays are known as channels.

Channel-Less Gate Array – An *application-specific integrated circuit (ASIC)* organized as a single large array of basic cells. May also be referred to as a "sea of cells" or a "sea of gates" device.

Checksum – The final cyclic-redundancy check value stored in a linear feedback shift register (or software equivalent). Also known as a "signature" in the guided-probe variant of a functional test.

Chemically-Amplified Resist – In the case of a chemically-amplified resist, the application of a relatively small quantity of ultraviolet light stimulates the formation of chemicals in the resist, where these chemicals accelerate the degrading process. This reduces the amount of ultraviolet light required to degrade the resist and allows the creation of finer features with improved accuracy.

Chemical Mechanical Polishing – *see CMP*.

Chemical Vapor Deposition – *see CVD*.

Chemical Vapor Infiltration – *see CVI*.

Chip – Popular name for an integrated circuit.

Chip-on-Board – *see* COB.

Chip-on-Chip – *see* COC.

Chip-on-Flex – *see* COF.

Chip Scale Package – *see* CSP.

Circuit Board – The generic name for a wide variety of interconnection techniques, which include rigid, flexible, and rigid-flex boards in single-sided, double-sided, multilayer, and discrete wired configurations.

CMOS (Complementary Metal Oxide Semiconductor) – Logic gates constructed using both NMOS and PMOS transistors connected in a complementary manner.

CMP (Chemical Mechanical Polishing) – A process used to re-planarize a wafer— smoothing the surface by polishing out the "bumps" caused by adding a metalization (tracking) layer.

CNFET (Carbon Nanotube FET) – A new type of transistor announced by IBM in the summer of 2002. This device is based on a field-effect transistor featuring a carbon nanotube measuring 1.4 nanometers in diameter as the channel (the rest of the transistor is formed using conventional silicon and metal processing technologies). It will be a number of years before these devices become commercially available, but they are anticipated to significantly outperform their silicon-only counterparts.

Coaxial Cable – A conductor in the form of a central wire surrounded first by a dielectric (insulating) layer, and then by a conducting tube which serves to shield the central wire from external interference.

COB (Chip-on-Board) – A process in which unpackaged integrated circuits are physically and electrically attached to a circuit board, and are then encapsulated with a "glob" of protective material such as epoxy.

COC (Chip-on-Chip) – A process in which unpackaged integrated circuits are mounted on top of each other. Each die is very thin and it is possible to have over a hundred die forming a 3D cube.

Coefficient of Thermal Expansion – Defines the amount that a material expands and contracts due to changes in temperature. If materials with different coefficients of thermal expansion are bonded together, changes in temperature will cause shear forces at the interface between them.

COF (Chip-on-Flex) – Similar to *chip-on-board*, except that the unpackaged integrated circuits are attached to a flexible printed circuit.

Cofired Ceramic – A substrate formed from multiple layers of "green" ceramic that are bonded together and fired at the same time.

Column Grid Array – *see* CGA.

Combinatorial – *see* Combinational.

Combinational – A digital function whose output value is directly related to the current combination of values on its inputs.

Commutative Rules – Algebraic rules that state that the order in which variables are specified will not affect the result of an operation: for example,[5] $a \,\&\, b \equiv b \,\&\, a$.

---

[5] Note that the symbol $\equiv$ indicates "is equivalent to" or "is the same as."

Comparator (digital) – A logic function that compares two binary values and outputs the results in terms of binary signals representing *less-than* and/or *equal-to* and/or *greater than*.

Compiled Cell Technology – A technique used to create portions of a standard cell *application-specific integrated circuit (ASIC)*. The masks used to create components and interconnections are directly generated from Boolean representations using a program called a *silicon compiler*. May also be used to create data-path and memory functions.

Complementary Output – Refers to a function with two outputs carrying complementary logical values. One output is referred to as the *true output* and the other as the *complementary output*.

Complementary Rules – Rules in Boolean Algebra derived from the combination of a single variable with the inverse of itself.

Complex Programmable Logic Device – *see CPLD.*

Computer-Generated Hologram – *see CGH.*

Computer Virus – There are many different types of computer viruses but, for the purposes of this book, a computer virus may be defined as a self-replicating program released into a computer system for a number of purposes. These purposes range from the simply mischievous, such as displaying humorous or annoying messages, to the downright nefarious, such as corrupting data or destroying (or subverting) the operating system.

Conditioning – *see Signal Conditioning.*

Conductive Ink Technology – A technique in which tracks are screen-printed directly onto the surface of a circuit board using conductive ink.

Conjunction – Propositions combined with an AND operator: for example, "You

have a parrot on your head AND you have a fish in your ear." The result of a conjunction is true if all the propositions comprising that conjunction are true.

Configurable Hardware – A product whose function may be customized once or a very few times. *See also Adaptive Computing, Reconfigurable Hardware, Remotely Reconfigurable Hardware, Dynamically Reconfigurable Hardware, and Virtual Hardware.*

CPLD (Complex PLD) – A device that contains a number of PLA or PAL functions sharing a common programmable interconnection matrix.

CPU (Central Processing Unit) – The "brain" of a computer where all of the decision-making and number crunching is performed.

CRC (Cyclic Redundancy Check) – A calculation used to detect errors in data communications, typically performed using a linear feedback shift register. Similar calculations may be used for a variety of other purposes such as data compression.

Crossover – Used in discrete wire technology to designate the intersection of two or more insulated wires.

CSIC (Customer-Specific Integrated Circuit) – An alternative and possibly more accurate name for an ASIC, but this term is rarely used in the industry and shows little indication of finding favor with the masses.

CSP (Chip Scale Package) – An integrated circuit packaging technique in which the package is only fractionally larger than the silicon die.

Cure – To harden a material using heat, ultraviolet light, or some other process.

Cured – Refers to a material that has been hardened using heat, ultraviolet light or some other process.

Customer-Specific Integrated Circuit –
see CSIC.

CVD (Chemical Vapor Deposition) –
A process for growing thin films on a
substrate, in which a gas containing the
required molecules is converted into
plasma by heating it to extremely high
temperatures using microwaves. The
plasma carries atoms to the surface of the
substrate where they are attracted to the
crystalline structure of the substrate. This
underlying structure acts as a template.
The new atoms continue to develop the
structure to build up a layer on the
substrate's surface.

CVI (Chemical Vapor Infiltration) – A
process similar to *chemical vapor deposition
(CVD)* but, in this case, the process
commences by placing a crystalline
powder of the required substance in a
mold. Additionally, thin posts, or
columns, can be preformed in the mold,
and the powder can be deposited around
them. When exposed to the same plasma
as used in the CVD technique, the
powder coalesces into a polycrystalline
mass. After the CVI process has been
performed, the posts can be dissolved,
leaving holes through the crystal for use
in creating vias. CVI processes can
produce layers twice the thickness of
those obtained using CVD techniques
at a fraction of the cost.

Cyclic Redundancy Check – *see* CRC.

D/A (Digital-to-Analog) – The process of
converting a digital value into its analog
equivalent.

Data Bus – A bi-directional set of signals
used by a computer to convey information from a memory location to the
central processing unit and vice versa.
More generally, a set of signals used to
convey data between digital functions.

Data-Path Function – A well-defined
function such as an adder, counter, or
multiplier used to process digital data.

Daughter Board – If a backplane contains
active components such as integrated
circuits, then it is usually referred to as a
motherboard. In this case, the boards
plugged into it are referred to as daughter
boards.

DDR (Double Data Rate) – A modern form
of SDRAM-based memory. The original
SDRAM specification was based on using
one of the clock edges only (say the rising
edge) to read/write data out-of/into the
memory. DDR refers to memory designed
in such a way that data can be read/
written on both edges of the clock. This
effectively doubles the amount of data
that can be pushed through the system
without increasing the clock frequency
(like many things this sounds simple if
you say it fast, but making this work is
trickier than it may at first appear).
See also QDR *and* SDRAM.

Decimal – Base-10 numbering system.

Decoder (digital) – A logic function that
uses a binary value, or address, to select
between a number of outputs and to
assert the selected output by placing it
in its active state.

Deep Sub-Micron – *see* DSM.

Delamination – Occurs when a composite
material formed from a number of layers
is stressed, thermally or otherwise, such
that the layers begin to separate.

DeMorgan Transformation – The transformation of a Boolean expression into an
alternate, and often more convenient,
form.

Die – (1) An unpackaged integrated circuit.
In this case, the plural of die is also die
(in much the same way that "a shoal of
herring" is the plural of "herring"). **(2)** A
piece of metal with a design engraved or
embossed on it for stamping onto another
material, upon which the design appears
in relief.

Dielectric Layer – (1) An insulating layer used to separate two signal layers. **(2)** An insulating layer used to modify the electrical characteristics of an MCM-D substrate.

Die Separation – The process of separating individual die from the wafer by marking the wafer with a diamond scribe and fracturing it along the scribed lines.

Die Stacking – A technique used in specialist applications in which several bare die are stacked on top of each other to form a sandwich. The die are connected together and then packaged as a single entity.

Diffusion Layer – The surface layer of a piece of semiconductor into which impurities are diffused to form P-type and N-type material. In addition to forming components, the diffusion layer may also be used to create embedded traces.

Digital – A value represented as being in one of a finite number of discrete states called *quanta*. The accuracy of a digital value is dependent on the number of quanta used to represent it.

Digital Circuit – A collection of logic gates used to process or generate digital signals.

Digital Signal Processor – *see DSP.*

Digital-to-Analog – *see D/A.*

DIMM (Dual In-line Memory Module) – A single memory integrated circuit can only contain a limited amount of data, so a number are gathered together onto a small circuit board called a *memory module*. Each memory module has a line of gold-plated pads on both sides of one edge of the board. These pads plug into a corresponding connector on the main computer board. In the case of a *dual in-line memory module* (DIMM), the pads on opposite sides of the board are electrically isolated from each other and form two separate contacts. *See also SIMM and RIMM.*

Diode – A two-terminal device that conducts electricity in only one direction; in the other direction it behaves like an open switch. These days the term diode is almost invariably taken to refer to a semiconductor device, although alternative implementations such as vacuum tubes are available.

Diode-Transistor Logic – *see DTL.*

Discrete Device – Typically taken to refer to an electronic component such as a resistor, capacitor, diode, or transistor that is presented in an individual package. More rarely, the term may be used in connection with a simple integrated circuit containing a small number of primitive gates.

Discrete Wire Board – *see DWB.*

Discrete Wire Technology – The technology used to fabricate discrete wire boards.

Disjunction – Propositions combined with an OR operator; for example, *"You have a parrot on your head OR you have a fish in your ear."* The result of a disjunction is true if at least one of the propositions comprising that disjunction is true.

Distributive Rules – Two very important rules in Boolean Algebra. The first states that the AND operator distributes over the OR operator: for example,[6] $a \,\&\, (b \mid c) \equiv (a \,\&\, b) \mid (a \,\&\, c)$. The second states that the OR operator distributes over the AND operator: for example, $a \mid (b \,\&\, c) \equiv (a \mid b) \,\&\, (a \mid c)$.

Doping – The process of inserting selected impurities into a semiconductor to create P-type or N-type material.

Double Data Rate – *see DDR.*

---

[6] Note that the symbol $\equiv$ indicates "is equivalent to" or "is the same as."

Double-Sided – A printed circuit board with tracks on both sides.

DRAM (Dynamic RAM) – A memory device in which each cell is formed from a transistor-capacitor pair. Called *dynamic* because the capacitor loses its charge over time, and each cell must be periodically recharged if it is to retain its data.

DSM (Deep Submicron) – Typically taken to refer to integrated circuits containing structures that are smaller than 0.5 microns (one half of one millionth of a meter).

DSP (Digital Signal Processor) – A special form of microprocessor that has been designed to perform a specific processing task on a specific type of digital data much faster and more efficiently than can be achieved using a general-purpose microprocessor.

DTL (Diode-Transistor Logic) – Logic gates implemented using particular configurations of diodes and bipolar junction transistors. For the majority of today's designers, diode-transistor logic is of historical interest only.

Dual In-line Memory Module – *see DIMM*.

Duo-decimal – Base-12 numbering system.

DWB (Discrete Wire Board) – A form of circuit board in which a special computer-controlled wiring machine ultrasonically bonds extremely fine insulated wires into the surface layer of the board. This discipline has enjoyed only limited recognition, but some believe that it is poised to emerge as the technology-of-choice for high-speed designers.

Dynamic Flex – A type of *flexible printed circuit* which is used in applications that are required to undergo constant flexing such as ribbon cables in printers.

Dynamically Reconfigurable Hardware – A product whose function may be customized on-the-fly while remaining resident in the system. *See also Configur-able Hardware, Reconfigurable Hardware, Remotely Reconfigurable Hardware, and Virtual Hardware*.

Dynamic RAM – *see DRAM*.

EBE (Electron Beam Epitaxy) – A technique for creating thin films on substrates in precise patterns. The substrate is first coated with a layer of dopant material before being placed in a high vacuum. A guided beam of electrons is fired at the substrate, causing the dopant to be driven into it, effectively allowing molecular-thin layers to be "painted" onto the substrate where required.

ECC (Error-Correcting Code) – Computer systems are very complicated and there's always the chance that an error will occur when reading or writing to the memory (a stray pulse of "noise" may flip a logic 0 to a logic 1 while your back is turned). Thus, serious computers use ECC memory, which includes extra bits and special circuitry that tests the accuracy of data as it passes in and out of memory and corrects any (simple) errors.

ECL (Emitter-Coupled Logic) – Logic gates implemented using particular configurations of bipolar junction transistors.

Edge-Sensitive – An input that only affects a function when it transitions from one logic value to another.

EDO (Extended Data Out) – An asynchronous form of DRAM-based computer memory that was popular for a while in the latter half of the 1990s. Along with other asynchronous memory techniques, EDO was eventually superceded by SDRAM technologies. *See also FPM, BEDO, and SDRAM*.

EEPROM or E²PROM (Electrically-Erasable Programmable Read-Only Memory) – A memory device whose contents can be electrically programmed by the designer. Additionally, the contents can be electrically erased, allowing the device to be reprogrammed.

Electrically-Erasable Programmable Read-Only Memory – *see EEPROM.*

Electromigration – **(1)** A process in which structures on an integrated circuit's substrate (particularly structures in deep sub-micron technologies) are eroded by the flow of electrons in much the same way as a river erodes land. **(2)** The process of forming transistor-like regions in a semiconductor using an intense magnetic field.

Electron Beam Epitaxy – *see EBE.*

Electron Beam Lithography – An integrated circuit fabrication process in which fine beams of electrons are used to draw extremely high-resolution patterns directly into the resist without the use of a mask.

Electrostatic Discharge – *see ESD.*

Emitter-Coupled Logic – *see ECL.*

Enzyme – One of numerous complex proteins that are produced by living cells and catalyze biochemical reactions at body temperatures.

EPROM (Erasable Programmable Read-Only Memory) – A memory device whose contents can be electrically programmed by the designer. Additionally, the contents can be erased by exposing the die to ultraviolet light through a quartz window mounted in the top of the component's package.

Equivalent Gate – A concept in which each type of logic function is assigned an equivalent gate value for the purposes of comparing functions and devices. However, the definition of an equivalent gate varies depending on whom you're talking to.

Equivalent Integrated Circuit – A concept used to compare the component density supported by diverse interconnection technologies such as circuit boards, hybrids, and multichip modules.

Erasable Programmable Read-Only Memory – *see EPROM.*

Error-Correcting Code – *see ECC.*

ESD (Electro-Static Discharge) – This refers to a charged person, or object, discharging static electricity, which can be generated in the process of moving around. Although the current associated with such a static charge is extremely low, the electric potential can be in the millions of volts and can severely damage electronic components. CMOS devices are particularly prone to damage from static electricity.

Etching – The process of selectively removing any material not protected by a resist using an appropriate solvent or acid. In some cases the unwanted material is removed using an electrolytic process.

Eutectic Bond – A bond formed when two pieces of metal, or metal-coated materials, are pressed together and vibrated at ultrasonic frequencies.

Extended Data Out – *see EDO.*

Face-Surface Mirror – A mirror in which the reflective coating is located on the face of the mirror, and not behind it as would be the case in the mirror you look into when you shave or apply your makeup (or both depending on how liberated you are).

Falling-Edge – *see Negative-Edge.*

Fan-In Via – *see Fan-Out Via.*

Fan-Out Via – In the case of surface mount devices, each component pad is usually connected by a short length of track to a via which forms a link to other conducting layers, and this via is known as a fan-out via. Some engineers attempt to differentiate vias that fall inside the device's footprint (under the body of the device) from vias that fall outside the device's footprint by referring to the former as *fan-in vias*, but this is not a widely used term.

Fast Page Mode – *see FPM.*

Femto – Unit qualifier (symbol = f) representing one thousandth of one millionth of one millionth, or 10^{-15}. For example, 3 fs stands for 3×10^{-15} seconds.

FET – A transistor whose control, or gate, signal creates an electro-magnetic field, which turns the transistor ON or OFF.

Field-Effect Transistor – *see FET.*

Field-Programmable Device – *see FPD.*

Field-Programmable Gate Array – *see FPGA.*

Field-Programmable Interconnect Chip – *see FPIC*[7].

Field-Programmable Interconnect Device – *see FPID.*

FIFO (First-In First-Out) – A memory device in which data is read out in the same order that it was written in.

Finite State Machine – *see FSM.*

Firmware – Refers to programs, or sequences of instructions, that are hard-coded into non-volatile memory devices.

First-In First Out – *see FIFO.*

Flash – *see Gold Flash.*

FLASH Memory – An evolutionary technology that combines the best features of the EPROM and E^2PROM technologies. The name FLASH is derived from the technology's fast reprogramming time compared to EPROM.

Flex – *see FPC.*

Flexible Printed Circuit – *see FPC.*

Flipped-Chip – A generic name for processes in which unpackaged integrated circuits are "flipped over" and mounted directly onto a substrate with their component sides facing the substrate.

Flipped-TAB – A combination of *flipped-chip* and *tape-automated bonding* (*TAB*).

Footprint – The area occupied by a device mounted on a substrate.

FPC (Flexible Printed Circuit) – A specialist circuit board technology, often abbreviated to "flex," in which tracks are printed onto flexible materials. There are a number of flavors of flex, including *static flex*, *dynamic flex*, and *rigid flex*.

FPD (Field Programmable Device) – A generic name that encompasses SPLDs, CPLDs, and FPGAs.

FPGA (Field-Programmable Gate Array) – A programmable logic device which is more versatile than traditional programmable devices such as PALs and PLAs, but less versatile than an *application-specific integrated circuit* (*ASIC*). Some field-programmable gate arrays use anti-fuses such as those found in programmable logic devices, while others are based on SRAM equivalents.

FPIC (Field-Programmable Interconnect Chip)[8] – An alternate, proprietary name for a *field-programmable interconnect device* (*FPID*).

FPID (Field-Programmable Interconnect Device) – A device used to connect logic devices together, and which can be dynamically reconfigured in the same way as standard SRAM-based FPGAs. Because each FPID may have around 1,000 pins, only a few such devices are typically required on a circuit board.

FPM (Fast Page Mode) – An asynchronous form of DRAM-based computer memory, which was popular for a while in the latter half of the 1990s. Along with other asynchronous memory techniques, FPM was eventually superceded by SDRAM technologies. *See also EDO, BEDO, and SDRAM.*

---

[7] FPIC is a trademark of Aptix Corporation.

[8] FPIC is a trademark of Aptix Corporation.

FR4 – The most commonly used insulating base material for circuit boards. FR4 is made from woven glass fibers that are bonded together with an epoxy. The board is cured using a combination of temperature and pressure, which causes the glass fibers to melt and bond together, thereby giving the board strength and rigidity. The first two characters stand for "Flame Retardant" and you can count the number of people who know what the "4" stands for on the fingers of one hand. FR4 is technically a form of fiberglass, and some people do refer to these composites as *fiberglass boards* or *fiberglass substrates*, but not often.

Free-Space Optical Interconnect – A form of optical interconnect in which laser diode transmitters communicate directly with phototransistor receivers without employing optical fibers or optical waveguides.

FSM (Finite State Machine) – The actual implementation (in hardware or software) of a function that can be considered to consist of a finite set of states through which it sequences.

Full Custom – An application-specific integrated circuit in which the design engineers have complete control over every mask layer used to fabricate the device. The ASIC vendor does not provide a cell library or prefabricate any components on the substrate.

Functional Latency – Refers to the fact that, at any given time, only a portion of the logic functions in a device or system are typically active (doing anything useful).

Functional Test – A test strategy in which signals are applied to a circuit's inputs, and the resulting signals—which are observed on the circuit's outputs—are compared to known good values.

Fuse – *see Fusible-Link Technology*.

Fusible Link Technology – A programmable logic device technology that employs links called *fuses*. Individual fuses can be removed by applying pulses of relatively high voltage and current to the device's inputs.

Fuzz-Button – A small ball of fibrous gold used in one technique for attaching components such as multichip modules to circuit boards. Fuzz-buttons are inserted between the pads on the base of the package and their corresponding pads on the board. When the package is forced against the board, the fuzz-buttons compress to form good electrical connections. One of the main advantages of the fuzz-button approach is that it allows broken devices to be quickly removed and replaced. Even though fuzz-button technology would appear to be inherently unreliable, it is used in such devices as missiles, so one can only assume that it is fairly robust.

GaAs (Gallium Arsenide) – A 3:5 valence high-speed semiconductor formed from a mixture of gallium and arsenic. GaAs transistors can switch approximately eight times faster than their silicon equivalents. However, GaAs is hard to work with, which results in GaAs transistors being more expensive than their silicon cousins.

GAL (Generic Array Logic) – A variation on a PAL device from a company called Lattice Semiconductor Corporation.[9]

Gallium Arsenide – *see GaAs*.

Garbage-In Garbage-Out – *see GIGO*.

Gate Array – An application-specific integrated circuit in which the manufacturer prefabricates devices containing arrays of unconnected components

---

9 GAL is a registered trademark of Lattice Semiconductor Corporation.

(transistors and resistors) organized in groups called *basic cells*. The designer specifies the function of the device in terms of cells from the cell library and the connections between them, and the manufacturer then generates the masks used to create the metalization layers.

Generic array logic – *see* GAL.

Geometry – Refers to the size of structures created on an integrated circuit. The structures typically referenced are the width of the tracks and the length of the transistor's channels; the dimensions of other features are derived as ratios of these structures.

Giga – Unit qualifier (symbol = G) representing one thousand million, or 10^9. For example, 3 GHz stands for 3×10^9 hertz.

GIGO (Garbage-In Garbage-Out) – An electronic engineer's joke, also familiar to the writers of computer programs.

Glue Logic – Simple logic gates used to interface more complex functions.

Gold Flash – An extremely thin layer of gold with a thickness measured on the molecular level, which is either electroplated or chemically plated[10] onto a surface.

Gray Code – A sequence of binary values in which each pair of adjacent values differs by only a single bit: for example, 00, 01, 11, 10.

Green Ceramic – Unfired, malleable ceramic.

Ground Plane – A conducting layer in, or on, a substrate providing a grounding, or reference, point for components. There may be several ground planes separated by insulating layers.

Guard Condition – A Boolean expression associated with a state transition in a state diagram or state table. The expression must be satisfied for that state transition to be executed.

Guided Probe – A form of functional test in which the operator is guided in the probing of a circuit to isolate a faulty component or track.

Guided Wave – A form of optical interconnect, in which optical waveguides are fabricated directly on the substrate of a multichip module. These waveguides can be created using variations on standard opto-lithographic thin-film processes.

Hard Macro (Macro Cell) – A logic function defined by the manufacturer of an *application-specific integrated circuit* (ASIC). The function is described in terms of the simple functions provided in the cell library and the connections between them. The manufacturer also defines how the cells forming the macro will be assigned to basic cells and the routing of tracks between the basic cells.

Hardware – Generally understood to refer to any of the physical portions constituting an electronic system, including components, circuit boards, power supplies, cabinets, and monitors.

Hardware Description Language – *see* HDL.

HDL (Hardware Description Language) – Today's digital integrated circuits can end up containing millions of logic gates, and it simply isn't possible to capture and manage designs of this complexity at the schematic (circuit diagram) level. Thus, as opposed to using schematics, the functionality of a high-end integrated circuit is now captured in textual form using a (HDL). The two most popular HDLs are Verilog and VHDL.

Hertz – *see* Hz.

---

[10] That is, using an electroless plating process.

Heterojunction – The interface between two regions of dissimilar semiconductor materials. The interface of a heterojunction has naturally occurring electric fields that can be used to accelerate electrons, and transistors created using heterojunctions can switch much faster than their homojunction counterparts of the same size.

Hexadecimal – Base-16 numbering system. Each hexadecimal digit can be directly mapped onto four binary digits, or bits.

High Impedance State – The state of a signal that is not currently being driven by anything. A high-impedance state is indicated by the 'Z' character.

Hologram – A three-dimensional image (from the Greek *holos*, meaning "whole" and *gram*, meaning "message").

Holographic Interconnect – A form of optical interconnect based on a thin slice of quartz, into which three-dimensional images are cut using a laser beam. Thus, the quartz is referred to as a *computer-generated hologram*, and this interconnection strategy is referred to as holographic.

Holography – The art of creating three-dimensional images known as holograms.

Homojunction – An interface between two regions of semiconductor having the same basic composition but opposing types of doping. Homojunctions dominate current processes because they are easier to fabricate than their *heterojunction* cousins.

Hybrid – An electronic sub-system in which a number of integrated circuits (packaged and/or unpackaged) and discrete components are attached directly to a common substrate. Connections between the components are formed on the surface of the substrate, and some components such as resistors and inductors may also be fabricated directly onto the substrate.

Hydrogen Bond – The electrons in a water molecule are not distributed equally, because the oxygen atom is a bigger, more robust fellow which grabs more than its fair share. The end result is that the oxygen atom has an overall negative charge, while the two hydrogen atoms are left feeling somewhat on the positive side. This unequal distribution of charge means that the hydrogen atoms are attracted to anything with a negative bias: for example, the oxygen atom of another water molecule. The resulting bond is known as a hydrogen bond.

Hz (hertz) – Unit of frequency. One hertz equals one cycle—or one oscillation—per second.

IC (Integrated Circuit) – A device in which components such as resistors, diodes, and transistors are formed on the surface of a single piece of semiconductor.

ICR (In-Circuit Reconfigurable)- An SRAM-based or similar component that can be dynamically reprogrammed on-the-fly while remaining resident in the system.

Idempotent Rules – Rules derived from the combination of a single Boolean variable with itself.

IDM (Integrated Device Manufacturer) – A company that focuses on designing and manufacturing integrated circuits as opposed to complete electronic systems.

Impedance – The resistance to the flow of current caused by resistive, capacitive, or inductive devices (or undesired parasitic elements) in a circuit.

In-Circuit Reconfigurable – *see ICR*.

Inclusion – A defect in a crystalline structure.

Inductance – A property of a conductor that allows it to store energy in a magnetic field which is induced by a current flowing through it. The base unit of inductance is the *Henry*.

Inert Gas – *see Noble Gas.*

In-System Programmable – *see ISP.*

Integrated Circuit – *see IC.*

Integrated Device Manufacturer – *see IDM.*

Intellectual Property – *see IP.*

Invar – An alloy similar to bronze.

Involution Rule – A rule that states that an even number of Boolean inversions cancel each other out.

Ion – A particle formed when an electron is added to, or subtracted from, a neutral atom or group of atoms.

Ion Implantation – A process in which beams of ions are directed at a semiconductor to alter its type and conductivity in certain regions.

IP (Intellectual Property) – When a team of electronics engineers is tasked with designing a complex integrated circuit, rather than "reinvent the wheel," they may decide to purchase the plans for one or more functional blocks that have already been created by someone else. The plans for these functional blocks are known as *intellectual property (IP)*. IP blocks can range all the way up to sophisticated communications functions and microprocessors. The more complex functions—like microprocessors—may be referred to as "cores."

ISP (In-System Programmable) – An E^2-based, FLASH-based, or similar component that can be reprogrammed while remaining resident on the circuit board.

JEDEC (Joint Electronic Device Engineering Council) – A council that creates, approves, arbitrates, and oversees industry standards for electronic devices. In programmable logic, the term JEDEC refers to a textual file containing information used to program a device. The file format is a JEDEC-approved standard and is commonly referred to as a *JEDEC file.*

Jelly Bean Device – An integrated circuit containing a small number of simple logic functions.

Joint Electronic Device Engineering Council – *see JEDEC.*

Jumper – A small piece of wire used to link two tracks on a circuit board.

Karnaugh Map – A graphical technique for representing a logical function. Karnaugh maps are often useful for the purposes of minimization.

Kelvin Scale of Temperature – A scale of temperature that was invented by the British mathematician and physicist William Thomas, first Baron of Kelvin. Under the *Kelvin*, or *absolute*, scale of temperature, 0K (corresponding to $-273°C$) is the coldest possible temperature and is known as *absolute zero*.

Kilo – Unit qualifier (symbol = K) representing one thousand, or 10^3. For example, 3 KHz stands for 3×10^3 hertz.

Kipper – A fish cured by smoking and salting.[11]

Latch-Up Condition – A condition in which a circuit draws uncontrolled amounts of current, and certain voltages are forced, or "latched-up," to some level. Particularly relevant in the case of CMOS devices, which can latch-up if their operating conditions are violated.

Laminate – A material constructed from thin layers or sheets. Often used in the context of circuit boards.

Large-Scale Integration – *see LSI.*

---

[11] Particularly tasty as a breakfast dish.

Laser Diode – A special semiconductor diode that emits a beam of coherent light.

Last-In First-Out – *see LIFO.*

Lateral Thermal Conductivity – Good lateral thermal conductivity means that the heat generated by components mounted on a substrate can be conducted horizontally across the substrate and out through its leads.

Lead – **(1)** A metallic element (chemical symbol Pb). **(2)** A metal conductor used to provide a connection from the inside of a device package to the outside world for soldering or other mounting techniques. Leads are also commonly called *pins*.

Lead Frame – A metallic frame containing leads (pins) and a base to which an unpackaged integrated circuit is attached. Following encapsulation, the outer part of the frame is cut away and the leads are bent into the required shapes.

Lead Through-Hole – *see LTH.*

Least-Significant Bit – *see LSB.*

Least-Significant Byte – *see LSB.*

LED (Light-Emitting Diode) – A semiconductor diode that behaves in a similar manner to a normal diode except that, when turned on, it emits light in the visible or infrared (IR) regions of the electromagnetic spectrum.

Level-Sensitive – An input whose effect on a function depends only on its current logic value or level, and is not directly related to it transitioning from one logic value to another.

LFSR (Linear Feedback Shift Register) – A shift register whose data input is generated as an XOR or XNOR of two or more elements in the register chain

LIFO (Last-in First-Out) – A memory device in which data is read out in the reverse order to which it was written in.

Light-Emitting Diode – *see LED.*

Line – Used to refer to the width of a track: for example, *"This circuit board track has a line-width of 5 mils (five thousandths of an inch)."*

Linear Feedback Shift Register – *see LFSR.*

Literal – A variable (either true or inverted) in a Boolean equation.

Low-Fired Cofired – Similar in principle to standard cofired ceramic substrate techniques. However, low-fired cofired uses modern ceramic materials with compositions that allow them to be fired at temperatures as low as 850°C. Firing at these temperatures in an inert atmosphere such as nitrogen allows non-refractory metals such as copper to be used to create tracks.

Logic Function – A mathematical function that performs a digital operation on digital data and returns a digital value.

Logic Gate – The physical implementation of a logic function.

Logic Synthesis – A process in which a program is used to automatically convert a high-level textual representation of a design (specified using a *hardware description language (HDL)* at the *register transfer level (RTL)* of abstraction) into equivalent Boolean Equations (like the ones introduced in Chapter 9). The synthesis tool automatically performs simplifications and minimizations, and eventually outputs a gate-level netlist.

LSB – **(1) (Least Significant Bit)** The binary digit, or bit, in a binary number that represents the least significant value (typically the right-hand bit). **(2) (Least Significant Byte)** – The byte in a multi-byte word that represents the least significant values (typically the right-hand byte).

LSI (Large Scale Integration) – Refers to the number of logic gates in a device. By one convention, large-scale integration represents a device containing 100 to 999 gates.

LTH (Lead Through-Hole) – A technique for populating circuit boards in which component leads are inserted into plated through-holes. Often abbreviated to "through-hole" or "thru-hole." When all of the components have been inserted, they are soldered to the board, usually using a wave soldering technique.

Macro Cell – *see Hard Macro*.

Macro Function – *see Soft Macro*.

Magnetic Random Access Memory – *see MRAM*.

Magnetic Tunnel Junction – *see MTJ*.

Mask – *see Optical Mask*.

Mask Programmable – A device such as a read-only memory, which is programmed during its construction using a unique set of masks.

Maximal Displacement – A linear feedback shift register whose taps are selected such that changing a single bit in the input data stream will cause the maximum possible disruption to the register's contents.

Maximal Length – A linear feedback shift register that sequences through $(2^n - 1)$ states before returning to its original value.

Max's Algorithm – A never-before-seen-in-public, soon-to-be-famous, mind bogglingly convoluted algorithm for automatically extracting Reed-Müller expressions.

Maxterm – The logical OR of the inverted variables associated with an input combination to a logical function.

MBE (Molecular Beam Epitaxy) – A technique for creating thin films on substrates in precise patterns, in which the substrate is placed in a high vacuum, and a guided beam of ionized molecules is fired at it, effectively allowing molecular-thin layers to be "painted" onto the substrate where required.

MCM (Multichip Module) – A generic name for a group of advanced interconnection and packaging technologies featuring unpackaged integrated circuits mounted directly onto a common substrate.

Medium-Scale Integration – *see MSI*.

Mega – Unit qualifier (symbol = M) representing one million, or 10^6. For example, 3 MHz stands for 3×10^6 hertz.

Memory Cell – A unit of memory used to store a single binary digit, or bit, of data.

Memory Word – A number of memory cells logically and physically grouped together. All the cells in a word are typically written to, or read from, at the same time.

Metalization Layer – A layer of conducting material on an integrated circuit that is selectively deposited or etched to form connections between logic gates. There may be several metalization layers separated by dielectric (insulating) layers.

Metal-Oxide Semiconductor – *see MOS*.

Meta-Stable – A condition where the outputs of a logic function are oscillating uncontrollably between undefined values.

Micro – Unit qualifier (symbol $= \mu$) representing one millionth, or 10^{-6}. For example, 3 μS stands for 3×10^{-6} seconds.

Microcontroller (μC) – A microprocessor augmented with special-purpose inputs, outputs, and control logic like counter-timers.

Microprocessor (μP) – A general-purpose computer implemented on a single integrated circuit (or sometimes on a group of related chips called a *chipset*).

Microwire[12] – A trade name for one incarnation of *discrete wire technology*. Microwire augments the main attributes of multiwire with laser-drilled blind vias, allowing these boards to support the maximum number of tracks and components.

Milli – Unit qualifier (symbol = m) representing one thousandth, or 10^{-3}. For example, 3 ms stands for 3×10^{-3} seconds.

Minimization – The process of reducing the complexity of a Boolean expression.

Minterm – The logical AND of the variables associated with an input combination to a logical function.

Mixed-signal – Typically refers to an integrated circuit that contains both analog and digital elements.

Mod – *see Modulus*.

Modulus – Refers to the number of states that a function such as a counter will pass through before returning to its original value. For example, a function that counts from 0000_2 to 1111_2 (0 to 15 in decimal) has a modulus of 16 and would be called a modulo-16 or mod-16 counter.

Modulo – *see Modulus*.

Molecular Beam Epitaxy – *see MBE*.

Moore's Law – In 1965, Gordon Moore (who was to co-found Intel Corporation in 1968) noted that new generations of memory devices were released approximately every 18 months, and that each new generation of devices contained roughly twice the capacity of its predecessor. This observation subsequently became known as *Moore's Law*, and it

has been applied to a wide variety of electronics trends.

MOS (Metal-Oxide Semiconductor) – A family of transistors.

Most-Significant Bit – *see MSB*.

Most-Significant Byte – *see MSB*.

Motherboard – A backplane containing active components such as integrated circuits.

MRAM (Magnetic RAM) – A form of memory expected to come online circa 2003 to 2004 that has the potential to combine the high speed of SRAM, the storage capacity of DRAM, and the non-volatility of FLASH, while consuming very little power.

MSB – **(1) (Most Significant Bit)** The binary digit, or bit, in a binary number that represents the most significant value (typically the left-hand bit). **(2) (Most Significant Byte)** The byte in a multi-byte word that represents the most significant values (typically the left-hand byte).

MSI Medium Scale Integration) – Refers to the number of logic gates in a device. By one convention, medium-scale integration represents a device containing 13 to 99 gates.

MTJ (Magnetic Tunnel Junction) – A sandwich of two ferromagnetic layers separated by a thin insulating layer. An MRAM memory cell is created by the intersection of two wires (say a "row" line and a "column" line) with an MJT sandwiched between them.

Multichip Module – *see MCM*.

Multilayer – A printed circuit board constructed from a number of very thin single-sided and/or double-sided boards,

---

[12] Microwave is a registered trademark of Advanced Interconnection Technology (AIT), Islip, New York, USA.

which are bonded together using a combination of temperature and pressure.

Multiplexer (digital) – A logic function that uses a binary value, or address, to select between a number of inputs and conveys the data from the selected input to the output.

Multiwire[13] – A trade name for the primary incarnation of discrete wire technology.

Nano – Unit qualifier (symbol = n) representing one thousandth of one millionth, or 10^{-9}. For example, 3 ns stands for 3×10^{-9} seconds.

Nanobot – A molecular-sized robot. *See also Nanotechnology*.

Nanophase Materials – A form of matter that was only (relatively) recently discovered, in which small clusters of atoms form the building blocks of a larger structure. These structures differ from those of naturally occurring crystals, in which individual atoms arrange themselves into a lattice.

Nanotechnology – This is an elusive term that is used by different research-and-development teams to refer to whatever it is that they're working on at the time. However, regardless of their area of interest, nanotechnology always refers to something extremely small. Perhaps the "purest" form of nanotechnology is that of molecular-sized units that assemble themselves into larger products.

Nanotubes – A structure that may be visualized as taking a thin sheet of carbon and rolling it into a tube. Nanotubes can be formed with walls that are only one atom thick. The resulting tube has a diameter of 1 nano (one thousandth of one millionth of a meter). Nanotubes are an almost ideal material: they are stronger than steel, have excellent thermal stability, and they are also tremendous conductors of heat and electricity. In addition to acting as wires, nanotubes can be persuaded to act as transistors—this means that we now have the potential to replace silicon transistors with molecular-sized equivalents at a level where standard semiconductors cease to function.

Negative-Edge – Beware, here be dragons! In this book a *negative-edge* is defined as a transition from a logic 1 to a logic 0. This definition is therefore consistent with the other definitions used throughout the book, namely those for *active-high, active-low, assertion-level logic, positive-logic,* and *negative-logic*. It should be noted, however, that some would define a negative-edge as *"a transition from a higher to a lower voltage level, passing through a threshold voltage level."* However, this definition serves only to cause confusion when combined with negative-logic implementations.

Negative Ion – An atom or group of atoms with an extra electron.

Negative Logic – A convention dictating the relationship between logical values and the physical voltages used to represent them. Under this convention, the more negative potential is considered to represent TRUE and the more positive potential is considered to represent FALSE.

Negative Resist – A process in which ultraviolet radiation passing through the transparent areas of a mask causes the resist to be cured. The uncured areas are then removed using an appropriate solvent.

Negative-True – *see Negative Logic*.

Nibble – *see Nybble*.

---

[13] Multiwire is a registered trademark of Advanced Interconnection Technology (AIT), Islip, New York, USA

NMOS (N-channel MOS) – Refers to the order in which the semiconductor is doped in a MOS device. That is, which structures are constructed as N-type versus P-type material.

Noble Gas – Gases whose outermost electron shells are completely filled with electrons. Such gases are extremely stable and it is difficult to coerce them to form compounds with other elements. There are six noble gasses: helium,[14] neon, argon, krypton, xenon, and radon. This group of elements was originally known as the *inert gasses*, but in the early 1960s it was found to be possible to combine krypton, xenon, and radon with fluorine to create compounds. Although helium, neon, and argon continue to resist, there is an increasing trend to refer to this group of gasses as *noble* rather than *inert*.

Noble Metal – Metals such as gold, silver, and platinum that are extremely inactive and are unaffected by air,[15] heat, moisture, and most solvents.

Noise – The miscellaneous rubbish that gets added to a signal on its journey through a circuit. Noise can be caused by capacitive or inductive coupling, or from externally generated electromagnetic interference.

Non-recurring Engineering – *see NRE*.

Non-Volatile – A memory device that does not lose its data when power is removed from the system.

Non-Volatile RAM – A device which is generally formed from an SRAM die mounted in a package with a very small battery, or as a mixture of SRAM and E^2PROM cells fabricated on the same die.

NPN (N-type – P-type – N-type) – Refers to the order in which the semiconductor is doped in a bipolar junction transistor.

NRE (Non-Recurring Engineering) – This typically refers to the costs associated with developing an ASIC or ASSP. The NRE depends on a number of factors, including the complexity of the design, the style of packaging, and who does what in the design flow (that is, how the various tasks are divided between the system design house and the ASIC vendor).

N-type – A piece of semiconductor doped with impurities that make it amenable to donating electrons.

Nybble – A group of four binary digits, or bits.

Octal – Base-8 numbering system. Each octal digit can be directly mapped onto three binary digits, or bits.

Ohm – Unit of resistance. The Greek letter omega, Ω, is often used to represent ohms; for example, 1 MΩ indicates one million ohms.

One-Hot Encoding – A form of state assignment for state machines in which each state is represented by an individual state variable.

One-Time Programmable – A device such as a PAL, PLA, or PROM that can be programmed a single time and whose contents cannot be subsequently erased.

Operating System – The collective name for the set of master programs that control the core operation and the base-level user-interface of a computer.

---

[14] The second most common element in the universe (after hydrogen).

[15] Before you start penning letters of protest, silver can be attacked by sulfur and sulfides, which occur naturally in the atmosphere. Thus, the tarnishing seen on your mother's silver candlesticks is actually silver sulfide and not silver oxide.

Optical Interconnect – The generic name for interconnection strategies based on opto-electronic systems, including *fiber-optics*, *free-space*, *guided-wave*, and *holographic* techniques.

Optical Lithography – A process in which radiation at optical wavelengths (usually in the *ultraviolet* (UV) and *extreme ultraviolet* (EUV) ranges) is passed through a mask, and the resulting patterns are projected onto a layer of resist coating the substrate material.

Optical Mask – A sheet of material carrying patterns that are either transparent or opaque to the wavelengths used in an optical-lithographic process. Such a mask can carry millions of fine lines and geometric shapes.

Opto-Electronic – Refers to a system which combines optical and electronic components.

Organic Resist – A material that is used to coat a substrate and is then selectively cured to form an impervious layer. These materials are called *organic* because they are based on carbon compounds as are living creatures.

Organic Solvent – A solvent for organic materials such as those used to form organic resists.

Organic Substrate – Substrate materials such as FR4, in which woven glass fibers are bonded together with an epoxy. These materials are called *organic* because epoxies are based on carbon compounds as are living creatures.

Overglassing – One of the final stages in the integrated circuit fabrication process in which the entire surface of the wafer is coated with a layer of silicon dioxide or silicon nitride. This layer may also be referred to as the *barrier layer* or the *passivation layer*. An additional lithographic step is required to pattern holes in this layer to allow connections to be made to the pads.

Padcap – A special flavor of circuit board used for high-reliability military applications, also known as *pads-only-outer-layers*. Distinguished by the fact that the outer surfaces of the board have pads but no tracks. Signal layers are only created on the inner planes, and tracks are connected to the surface pads by vias.

Pad – An area of metalization on a substrate used for probing or to connect to a via, plated through-hole, or an external interconnect.

Pad Grid Array – *see* PGA.

Pad Stack – Refers to any *pads*, *anti-pads*, and *thermal relief pads* associated with a via or a plated through-hole as it passes through the layers forming the substrate.

PAL (Programmable Array Logic)[16] – A programmable logic device in which the AND array is programmable but the OR array is predefined. See *also PLA, PLD, and PROM*.

Parallel-In Serial-Out – *see* PISO.

Parasitic Effects – The effects caused by undesired resistive, capacitive, or inductive elements inherent in the material or topology of a track or component.

Passivation Layer – *see Overglassing*.

Passive Trimming – A process in which a laser beam is used to trim components such as thick-film and thin-film resistors on an otherwise unpopulated and unpowered hybrid or multichip module substrate. Probes are placed at each end of a component to monitor its value while the laser cuts away (evaporates) some of the material forming the component.

---

[16] PAL is a registered trademark of Monolithic Memories, Inc.

Pass-Transistor Logic – A technique for connecting MOS transistors such that data signals pass between their source and drain terminals. Pass-transistor logic minimizes the number of transistors required to implement a function, and is typically employed by designers of cell libraries or full-custom integrated circuits.

PC100 – A popular form of SDRAM-based computer memory that runs at 100 MHz (the data bus is 64-bits wide, although 128-bit wide versions (using dual 64-bit cards in parallel) have been used in high-end machines).

PC133 – A popular form of SDRAM-based computer memory that runs at 133 MHz (the data bus is 64-bits wide, although 128-bit wide versions (using dual 64-bit cards in parallel) have been used in high-end machines).

PCB (Printed Circuit Board) – A type of circuit board that has conducting tracks superimposed, or "printed," on one or both sides, and may also contain internal signal layers and power and ground planes. An alternative name, *printed wire board (PWB)*, is commonly used in America.

Peta – Unit qualifier (symbol = P) representing one thousand million million, or 10^{15}. For example, 3 PHz stands for 3×10^{15} hertz.

PGA (1) (Pad Grid Array) – A packaging technology in which a device's external connections are arranged as an array of conducting pads on the base of the package. **(2) (Pin Grid Array)** – A packaging technology in which a device's external connections are arranged as an array of conducting leads, or pins, on the base of the package.

PHB (Photochemical Hole Burning) – An optical memory technique, in which a laser in the visible waveband is directed at a microscopic point on the surface of a slice of glass that has been doped with organic dyes or rare-earth elements. The laser excites electrons in the glass such that they change the absorption characteristics of that area of the glass and leave a *band*, or *hole*, in the absorption spectrum.

Photochemical Hole Burning – *see PHB.*

Phototransistor – A special transistor that converts an optical input in the form of light into an equivalent electronic signal in the form of a voltage or current.

Pico – Unit qualifier (symbol = p) representing one millionth of one millionth, or 10^{-12}. For example, 3 ps stands for 3×10^{-12} seconds.

Pin – *see Lead.*

Pin Grid Array – *see PGA.*

PISO (Parallel-In Serial Out) – Refers to a shift register in which the data is loaded in parallel and read out serially.

PLA (Programmable Logic Array) – The most user-configurable of the traditional programmable logic devices, because both the AND and OR arrays are programmable. *See also PAL, PLD, and PROM.*

Place Value – Refers to a numbering system in which the value of a particular digit depends both on the digit itself and its position in the number.

Plasma – A gaseous state in which the atoms or molecules are dissociated to form ions.

Plated Through-Hole – *see PTH.*

PLD (Programmable Logic Device) – The generic name for a device constructed in such a way that the designer can configure, or "program" it to perform a specific function. *See also PAL, PLA, and PROM.*

PMOS (P-channel MOS) – Refers to the order in which the semiconductor is doped in a MOS device. That is, which structures are constructed as P-type versus N-type material.

PNP (P-type – N-type – P-type) – Refers to the order in which the semiconductor is doped in a bipolar junction transistor.

Polysilicon Layer – An internal layer in an integrated circuit used to create the gate electrodes of MOS transistors. In addition to forming gate electrodes, the polysilicon layer can also be used to interconnect components. There may be several polysilicon layers separated by dielectric (insulating) layers.

Populating – The act of attaching components to a substrate.

Positive-Edge – Beware, here be dragons! In this book a *positive-edge* is defined as a transition from a logic 0 to a logic 1. This definition is therefore consistent with the other definitions used throughout the book, namely those for *active-high*, *active-low*, *assertion-level logic*, *positive-logic*, and *negative-logic*. It should be noted, however, that some would define a positive-edge as "*a transition from a lower to a higher voltage level, passing through a threshold voltage level.*" However, this definition serves only to cause confusion when combined with negative-logic implementations.

Positive Ion – An atom or group of atoms lacking an electron.

Positive Logic – A convention that dictates the relationship between logical values and the physical voltages used to represent them. Under this convention, the more positive potential is considered to represent TRUE and the more negative potential is considered to represent FALSE.

Positive Resist – A process in which radiation passing through the transparent areas of a mask causes previously cured resist to be degraded. The degraded areas

are then removed using an appropriate solvent.

Positive-True – *see Positive Logic.*

Power Plane – A conducting layer in or on the substrate providing power to the components. There may be several power planes separated by insulating layers.

Precedence of Operators – Determines the order in which operations are performed. For example, in standard arithmetic the multiplication operator has a higher precedence than the addition operator. Thus, in the equation $6 + 2 \times 3$, the multiplication is performed before the addition and the result is 12. Similarly, in Boolean Algebra, the AND operator has a higher precedence than the OR operator.

Prepreg – Non-conducting semi-cured layers of FR4 used to separate conducting layers in a multilayer circuit board.

Primitives – Simple logic functions such as BUF, NOT, AND, NAND, OR, NOR, XOR, and XNOR. These may also be referred to as *primitive logic gates.*

Printed Circuit Board – *see PCB.*

Printed Wire Board – *see PWB.*

Product-of-Sums – A Boolean equation in which all of the *maxterms* corresponding to the lines in the truth table for which the output is a logic *0* are combined using AND operators.

Product Term – A set of literals linked by an AND operator.

Programmable Array Logic – *see PAL.*

Programmable Logic Array – *see PLA.*

Programmable Logic Device – *see PLD.*

Programmable Read-Only Memory – *see PROM.*

PROM (Programmable Read-Only Memory) – A programmable logic device in which the OR array is programmable but the AND array is predefined. Usually considered to be a memory device whose contents can be electrically programmed (once) by the designer. *See also PAL, PLA, and PLD.*

Proposition – A statement that is either true or false with no ambiguity. For example, the proposition *"I just tipped a bucket of burning oil into your lap"* is either true or false, but there's no ambiguity about it.

Protein – A complex organic molecule formed from chains of amino acids, which are themselves formed from combinations of certain atoms, namely: carbon, hydrogen, nitrogen, oxygen, usually sulfur, and occasionally phosphorous or iron. Additionally, the chain of amino acids *"folds in on itself,"* thereby forming an extremely complex three-dimensional shape. Organic molecules have a number of useful properties, not the least of which is that their structures are intrinsically *"self healing"* and reject contamination. Also, in addition to being extremely small, many organic molecules have excellent electronic properties. Unlike metallic conductors, they transfer energy by moving electron excitations from place to place rather than relocating entire electrons. This can result in switching speeds that are orders of magnitude faster than their semiconductor equivalents.

Protein Memory – A form of memory based on organic proteins. *See also Protein.*

Protein Switch – A form of switch based on organic proteins. *See also Protein.*

Pseudo-Random – An artificial sequence of values that give the appearance of being random, but which are also repeatable.

PTH (Plated Through-Hole) – **(1)** A hole in a double-sided or multilayer board that is used to accommodate a through-hole component lead and is plated with copper. **(2)** An alternative name for the *lead through-hole (LTH)* technique for populating circuit boards in which component leads are inserted into plated through-holes.

P-type – A piece of semiconductor doped with impurities that make it amenable to accepting electrons.

PWB (Printed Wire Board) – A type of circuit board that has conducting tracks superimposed, or "printed," on one or both sides, and may also contain internal signal layers and power and ground planes. An alternative name, *printed circuit board (PCB)*, is predominantly used in Europe and Asia.

QDR (Quad Data Rate) – A modern form of SDRAM-based memory. The original SDRAM specification was based on using one of the clock edges only (say the rising edge) to read/write data out-of/into the memory. QDR refers to memory that has separate data in and data out busses both designed in such a way that data can be read/written on both edges of the clock. This effectively quadruples the amount of data that can be pushed through the system without increasing the clock frequency (like many things this sounds simple if you say it fast, but making this work is trickier than it may at first appear). *See also DDR and SDRAM.*

QFP (Quad Flat Pack) – The most commonly used package in surface mount technology to achieve a high lead count in a small area. Leads are presented on all four sides of a thin square package.

Quad Date Rate – *see QDR.*

Quad Flat Pack – *see QFP.*

Quantization – Part of the process by which an analog signal is converted into a series of digital values. First of all the analog signal is sampled at specific times.

For each sample, the complete range of values that the analog signal can assume is divided into a set of discrete bands or quanta. Quantization refers to the process of determining which band the current sample falls into. *See also Sampling.*

Quinary – Base-5 numbering system.

Radix – Refers to the number of digits in a numbering system. For example, the decimal numbering system is said to be radix-10. May also be referred to as the "base."

RAM (Random Access Memory) – A data-storage device from which data can be read out and new data can be written in. Unless otherwise indicated, the term RAM is typically taken to refer to a semiconductor device in the form of an integrated circuit.

Rambus[17] – An alternative computer memory concept that began to gain attention toward the end of the 1990s (this is based on core DRAM concepts presented in a cunning manner). By around 2000, personal computers equipped with memory modules formed from Rambus DRAM (RDRAM) devices were available. These memory modules were 16-bits (2-bytes) wide, and were accessed on both edges of a 400 MHz clock, thereby providing a peak data throughput of $2 \times 2 \times 400 = 1,600$ megabytes per second (these are officially quoted as 800 MHz devices, but that's just marketing spin trying to make clocking on both edges of a 400 MHz clock sound more impressive).
In the summer of 2002, a new flavor of Rambus was announced with a 32-bit wide bus using both edges of a 533 MHz clock (the official announcements quote "*1,066 MHz Rambus DRAM*," but this is the same marketing spin as noted above). Whatever, this will provide a peak

bandwidth of 4,200 megabytes per second.

Random-Access Memory – *see RAM*.

RDRAM (Rambus DRAM) – A form of DRAM computer memory based on Rambus technology. *See also Rambus.*

Read-Only Memory – *see ROM*.

Read-Write Memory – *see RWM*.

Real Estate – Refers to the amount of area available on a substrate.

Reconfigurable Hardware – A product whose function may be customized many times. *See also Configurable Hardware, Remotely Reconfigurable Hardware, Dynamically Reconfigurable Hardware, and Virtual Hardware.*

Reed-Müller Logic – Logic functions implemented using only XOR and XNOR gates.

Reflow Oven – An oven employing infra-red (IR) radiation or hot air.

Reflow Soldering – A *surface mount technology* process in which the substrate and attached components are passed through a reflow oven to melt the solder paste.

Refractory Metal – Metals such as *tungsten*, *titanium*, and *molybdenum* (try saying this last one quickly), which are capable of withstanding extremely high, or refractory, temperatures.

Register Transfer Level – *see RTL*.

Remotely Reconfigurable Hardware – A product whose function may be customized remotely, by telephone or radio, while remaining resident in the system. *See also Configurable Hardware, Reconfigurable Hardware, Dynamically Reconfigurable Hardware, and Virtual Hardware.*

---

[17] Rambus is a trademark of Rambus Inc.

Resist – A material that is used to coat a substrate and is then selectively cured to form an impervious layer.

Resistor-Transistor Logic – *see RTL.*

Rigid Flex – Hybrid constructions which combine standard rigid circuit boards with flexible printed circuits (flex), thereby reducing the component count, weight, and susceptibility to vibration of the circuit, and greatly increasing its reliability.

RIMM – This really doesn't stand for anything per se, but it is the trademarked name for a *Rambus* memory module. RIMMs are similar in concept to DIMMs, but have a different pin count and configuration. *See also DIMM and SIMM.*

Rising-Edge – *see Positive-Edge.*

ROM (Read-Only Memory) – A data storage device from which data can be read out, but new data cannot be written in. Unless otherwise indicated, the term ROM is typically taken to refer to a semiconductor device in the form of an integrated circuit.

RTL – (1) (Register Transfer Level) – A relatively high level of abstraction at which design engineers capture the functionality of digital integrated circuit designs described using a *hardware description language (HDL)*. **(2) (Resistor-Transistor Logic)** – Logic gates implemented using particular configurations of resistors and bipolar junction transistors. For the majority of today's designers, resistor-transistor logic is of historical interest only.

RWM (Read-Write Memory) – Alternative (and possibly more appropriate) name for a *random access memory (RAM)*.

Sampling – Part of the process by which an analog signal is converted into a series of digital values. Sampling refers to observing the value of the analog signal at specific times. *See also Quantization.*

Scalar Notation – A notation in which each signal is assigned a unique name; for example, a3, a2, a1, and a0. *See also Vector Notation.*

Scaling – A technique for making transistors switch faster by reducing their size. This strategy is known as *scaling*, because all of the transistor's features are typically reduced by the same proportion.

Schematic – Common name for a circuit diagram.

Scrubbing – The process of vibrating two pieces of metal (or metal-coated materials) at ultrasonic frequencies to create a friction weld.

SDRAM (Synchronous DRAM) – Until the latter half of the 1990s, DRAM-based computer memories were asynchronous, which means they weren't synchronized to the system clock. Over time, the industry migrated to *synchronous DRAM (SDRAM)*, which is synchronized to the system clock and makes everyone's lives much easier. Note that SDRAM is based on core DRAM concepts – it's all in the way you tweak the chips and connect them together.

Sea-of-Cells – Popular name for a channelless gate array.

Sea-of-Gates – Popular name for a channelless gate array.

Seed Value – An initial value loaded into a linear feedback shift register or random number generator.

Semiconductor – A special class of material that can exhibit both conducting and insulating properties.

Sensor – A transducer that detects a physical quantity and converts it into a form suitable for processing. For example, a microphone is a sensor that detects sound and converts it into a corresponding voltage or current.

Sequential – A function whose output value depends not only on its current input values, but also on previous input values. That is, the output value depends on a sequence of input values.

Serial-In Parallel-Out – *see SIPO*.

Serial-In Serial-Out – *see SISO*.

Sexagesimal – Base-60 numbering system.

Side-Emitting Laser Diode – A laser diode constricted at the edge of an integrated circuit's substrate such that, when power is applied, the resulting laser beam is emitted horizontally: that is, parallel to the surface of the substrate.

Signal Conditioning – Amplifying, filtering, or otherwise processing a (typically analog) signal.

Signal Layer – A layer carrying tracks in a circuit board, hybrid, or multichip module. *See also Wiring Layer*.

Signature – Refers to the *checksum* value from a *cyclic-redundancy-check* (CRC) when used in the guided-probe form of functional test.

Signature Analysis – A guided-probe functional-test technique based on *signatures*.

Sign Bit – The most significant binary digit, or bit, of a signed binary number (if set to a logic 1, this bit represents a negative quantity).

Signed Binary Number – A binary number in which the most-significant bit is used to represent a negative quantity. Thus, a signed binary number can be used to represent both positive and negative values.

Sign-Magnitude – Negative numbers in standard arithmetic are typically represented in sign-magnitude form by prefixing the value with a minus sign: for example, –27. For reasons of efficiency, computers rarely employ the sign-magnitude form. Instead, they use *signed binary numbers* to represent negative values.

Silicon Bumping – The process of depositing additional metalization on a die's pads to raise them fractionally above the level of the *barrier layer*.

Silicon Chip – Although a variety of semiconductor materials are available, the most commonly used is silicon, and integrated circuits are popularly known as *silicon chips*, or simply *chips*.

Silicon Compiler – The program used in compiled cell technology to generate the masks used to create components and interconnections. May also be used to create datapath functions and memory functions.

SIMM (Single In-line Memory Module) – A single memory integrated circuit can only contain a limited amount of data, so a number are gathered together onto a small circuit board called a *memory module*. Each memory module has a line of gold-plated pads on both sides of one edge of the board. These pads plug into a corresponding connector on the main computer board. A *single in-line memory module (SIMM)* has the same electrical signal on corresponding pads on the front and back of the board (that is, the pads on opposite sides of the board are "tied together"). *See also DIMM and RIMM*.

Simple PLD – *see SPLD*.

Single In-line Memory Module – *see SIMM*.

Single-Sided – A printed circuit board with tracks on one side only.

Sintering – A process in which ultra-fine metal powders weld together at temperatures much lower than those required for larger pieces of the same materials.

SIPO (Serial-In Parallel-Out) – Refers to a shift register in which the data is loaded in serially and read out in parallel.

SISO (Serial-In Serial-Out) – Refers to a shift register in which the data is both loaded in and read out serially.

Skin Effect – In the case of high frequency signals, electrons are only conducted on the outer surface, or skin, of a conductor. This phenomenon is known as the *skin effect*.

Small-Scale Integration – *see SSI.*

SMD (Surface Mount Device) – A component whose packaging is designed for use with *surface mount technology*.

SMT (Surface Mount Technology) – A technique for populating hybrids, multi-chip modules, and circuit boards, in which packaged components are mounted directly onto the surface of the substrate. A layer of solder paste is screen-printed onto the pads, and the components are attached by pushing their leads into the paste. When all of the components have been attached, the solder paste is melted using either *reflow soldering* or *vapor-phase soldering*.

SMOBC (Solder Mask Over Bare Copper) – A technique in which the *solder mask* is applied in advance of the *tin-lead plating*. This prevents solder from leaking under the mask when the tin-lead alloy melts during the process of attaching components to the board.

SoC (System-on-Chip) – In the past, an electronic system was typically composed of a number of integrated circuits, each with its own particular function (say a microprocessor, a communications function, some memory devices, etc.). For many of today's high-end applications, however, electronics engineers are combining all of these functions on a single device, which may be referred to as a *system-on-chip* (*SoC*).

Soft Macro (Macro Function) – A logic function defined by the manufacturer of an *application-specific integrated circuit* (*ASIC*) or by a third-party IP provider. The function is described in terms of the simple functions provided in the cell library and connections between them. The assignment of cells to basic cells and the routing of the tracks is determined at the same time, and using the same tools, as for the other cells specified by the designer.

Software – Refers to intangible programs, or sequences of instructions, that are executed by *hardware*.

Solder – An alloy of tin and lead with a comparatively low melting point used to join less fusible metals. Typical solder contains 60% tin and 40% lead; increasing the proportion of lead results in a softer solder with a lower melting point, while decreasing the proportion of lead results in a harder solder with a higher melting point. Note that the solder used in a brazing process is of a different type, being a hard solder with a comparatively high melting point composed of an alloy of copper and zinc (brass).

Solder Bump Bonding – A flip-chip technique (also known as *Solder Bumping*) for attaching a die to a package substrate. A minute ball of solder is attached to each pad on the die, and the die is flipped over and attached to the package substrate. Each pad on the die has a corresponding pad on the package substrate, and the package-die combo is heated so as to melt the solder balls and form good electrical connections between the die and the substrate. *See also Solder Bumping.*

Solder Bumping – A flipped chip technique (also known as *Solder Bump Bonding*) in which spheres of solder are formed on the die's pads. The die is flipped and the solder bumps are brought into contact with corresponding pads on the substrate. When all the chips have been mounted on the substrate, the solder bumps are melted using reflow soldering or vapor-phase soldering. *See also Solder Bump Bonding.*

Solder Mask – A layer applied to the surface of a printed circuit board that prevents solder from sticking to any metalization except where holes are patterned into the mask.

Solder Mask Over Bare Copper – *see* SMOBC.

Space – Used to refer to the width of the gap between adjacent tracks.

SPLD (Simple PLD) – Originally all PLDs contained a modest number of equivalent logic gates and were fairly simple. As more *complex PLDs (CPLDs)* arrived on the scene, however, it became common to refer to their simpler cousins as *simple PLDs (SPLDs)*.

SRAM (Static RAM) – A memory device in which each cell is formed from four or six transistors configured as a latch or a flip-flop. The term *static* is used because, once a value has been loaded into an SRAM cell, it will remain unchanged until it is explicitly altered or until power is removed from the device.

SSI (Small Scale Integration) – Refers to the number of logic gates in a device. By one convention, small-scale integration represents a device containing 1 to 12 gates.

Standard Cell – A form of *application-specific integrated circuit (ASIC)*, which, unlike a gate array, does not use the concept of a *basic cell* and does not have any prefabricated components. The ASIC vendor creates custom masks for every stage of the device's fabrication, allowing each logic function to be created using the minimum number of transistors.

State Assignment – The process by which the states in a state machine are assigned to the binary patterns that are to be stored in the state variables.

State Diagram – A graphical representation of the operation of a state machine.

State Machine – *see* FSM.

Statement – A sentence that asserts or denies an attribute about an object or group of objects. For example, *"Your face resembles a cabbage."*

State Table – A tabular representation of the operation of a state machine. Similar to a truth table, but also includes the *current state* as an input and the *next state* as an output.

State Transition – An arc connecting two states in a state diagram.

State Variable – One of a set of registers whose values represent the current state occupied by a state machine.

Static Flex – A type of *flexible printed circuit* that can be manipulated into permanent three-dimensional shapes for applications such as calculators and high-tech cameras, which require efficient use of volume and not just area.

Static RAM – *see* SRAM.

Steady State – A condition in which nothing is changing or happening.

Subatomic Erosion – *see Electromigration.*

Substrate – Generic name for the base layer of an integrated circuit, hybrid, multichip module, or circuit board. Substrates may be formed from a wide variety of materials, including semiconductors, ceramics, FR4 (fiberglass), glass, sapphire, or diamond depending on the application. Note that the term "substrate" has traditionally not been widely used in the circuit board world, at least not by the people who manufacture the boards. However, there is an increasing tendency to refer to a circuit board as a substrate by the people who populate the boards. The main reason for this is that circuit boards are often used as substrates in hybrids and multichip modules, and there is a trend towards a standard terminology across all forms of interconnection technology.

Subtractive Process – A process in which a substrate is first covered with conducting material, and then any unwanted material is subsequently removed, or subtracted.

Sum-of-Products – A Boolean equation in which all of the *minterms* corresponding to the lines in the truth table for which the output is a logic 1 are combined using OR operators.

Superconductor – A material with zero resistance to the flow of electric current.

Surface-Emitting Laser Diode – A laser diode constricted on an integrated circuit's substrate such that, when power is applied, the resulting laser beam is emitted directly away from the surface of the substrate.

Surface Mount Device – *see SMD*.

Surface Mount Technology – *see SMT*.

Symbolic Logic – A mathematical form in which propositions and their relationships may be represented symbolically using Boolean equations, truth tables, Karnaugh Maps, or similar techniques.

Synchronous – **(1)** A signal whose data is not acknowledged or acted upon until the next active edge of a clock signal. **(2)** A system whose operation is synchronized by a clock signal.

Synchronous DRAM – *see SDRAM*.

System-on-Chip – *see SoC*.

TAB (Tape-Automated Bonding) – A process in which transparent flexible tape has tracks created on its surface. The pads on unpackaged integrated circuits are attached to corresponding pads on the tape, which is then stored in a reel. Silver-loaded epoxy is screen printed on the substrate at the site where the device is to be located and onto the pads to which the device's leads are to be connected. The reel of TAB tape is fed through an automatic machine, which pushes the device and the TAB leads into the epoxy. When the silver-loaded epoxy is cured using *reflow soldering* or *vapor-phase soldering*, it forms electrical connections between the TAB leads and the pads on the substrate.

Tap – A register output which is used to generate the next data input to a linear feedback shift register.

Tape Automated Bonding – *see TAB*.

Tera – Unit qualifier (symbol = T) representing one million million, or 10^{12}. For example, 3 THz stands for 3×10^{12} hertz.

Tertiary – Base-3 numbering system.

Tertiary Digit – A numeral in the tertiary scale of notation. A tertiary digit can adopt one of three states: 0, 1, or 2. Often abbreviated to "trit."

Tertiary Logic – An experimental technology in which logic gates are based on three distinct voltage levels. The three voltages are used to represent the tertiary digits 0, 1, and 2, and their logical equivalents FALSE, TRUE, and MAYBE.

Thermal Relief Pad – A special pattern etched around a via or a plated through-hole to connect it into a power or ground plane. A thermal relief pad is necessary to prevent too much heat being absorbed into the power or ground plane when the board is being soldered.

Thermal Tracking – Typically used to refer to the problems associated with optical interconnection systems whose alignment may be disturbed by changes in temperature.

Thick-Film Process – A process used in the manufacture of hybrids and multichip modules in which signal and dielectric (insulating) layers are screen-printed onto the substrate.

Thin-Film Process – A process used in the manufacture of hybrids and multichip modules in which signal layers and dielectric (insulating) layers are created using opto-lithographic techniques.

Through-Hole – *see Lead Through-Hole, Plated Through-Hole, and Through-Hole Via*.

Through-Hole Via – A via that passes all the way through the substrate.

Thru-Hole – A commonly used abbreviation of "through-hole."

Time-of-Flight – The time taken for a signal to propagate from one logic gate, integrated circuit output, or opto-electronic component to another.

Tin-Lead Plating – An electroless plating process in which exposed areas of copper on a circuit board are coated with a layer of tin-lead alloy. The alloy is used to prevent the copper from oxidizing and provides protection against contamination.

Tinning – *see Tin-Lead Plating*.

Toggle – Refers to the contents or outputs of a logic function switching to the inverse of their previous logic values.

Trace – *see Track*.

Track – A conducting connection between electronic components. May also be called a *trace* or a *signal*. In the case of integrated circuits, such interconnections are often referred to collectively as *metalization*.

Transducer – A device that converts input energy of one form into output energy of another.

Transistor – A three-terminal semiconductor device that, in the digital world, can be considered to operate like a switch.

Tri-State Function – A function whose output can adopt three states: 0, 1, and Z (high-impedance). The function does not drive any value in the Z state and, in many respects, may be considered to be disconnected from the rest of the circuit.

Trit – Abbreviation of *tertiary digit*. A tertiary digit can adopt one of three values: 0, 1, or 2.

Truth Table – A convenient way to represent the operation of a digital circuit as columns of input values and their corresponding output responses.

TTL (Transistor-Transistor Logic) – Logic gates implemented using particular configurations of bipolar junction transistors.

Transistor-Transistor Logic – *see TTL*.

μC – *see Microcontroller*.

μP – *see Microprocessor*.

ULA (Uncommitted Logic Array) – One of the original names used to refer to gate array devices. This term has largely fallen into disuse.

Ultra-Large-Scale Integration – *see ULSI*.

ULSI (Ultra-Large-Scale Integration) – Refers to the number of logic gates in a device. By one convention, ultra-large-scale integration represents a device containing a million or more gates.

Uncommitted Logic Array – *see ULA*.

Unsigned Binary Number – A binary number in which all the bits are used to represent positive quantities. Thus, an unsigned binary number can only be used to represent positive values.

Vapor-Phase Soldering – A surface mount process in which a substrate carrying components attached by solder paste is lowered into the vapor-cloud of a tank containing boiling hydrocarbons. This melts the solder paste thereby forming good electrical connections. However, vapor-phase soldering is becoming increasingly less popular due to environmental concerns.

Vaporware – Refers to either hardware or software that exists only in the minds of the people who are trying to sell it to you.

Vector Notation – A notation in which a single name is used to reference a group of signals, and individual signals within the group are referenced by means of an index: for example, a[3], a[2], a[1], and a[0]. This concept of a vector is commonly used in the context of electronics tools such as schematic capture packages, logic simulators, and graphical waveform displays.
Some people prefer to use the phrase *binary group notation*, but the *phrase vector notation* is commonly used by practicing electronic engineers. (Note that this type of vector notation is in no way related to the algebraic concept of vector notation for Cartesian 2-space or 3-space.)

Very-Large-Scale Integration – *see VLSI*.

VHDL – A *hardware description language* (*HDL*), which came out of the American Department of Defense (DoD), and which has evolved into an open standard. VHDL is an acronym for *VHSIC HDL* (where VHSIC is itself an acronym for *very high speed integrated circuit*).

Via – A hole filled or lined with a conducting material, which is used to link two or more conducting layers in a substrate.

Virtual Hardware – An extension of dynamically configurable hardware based on a new generation of FPGAs that were introduced around the beginning of 1994. In addition to supporting the dynamic reconfiguration of selected portions of the internal logic, these devices also feature the following: no disruption to the device's inputs and outputs; no disruption to the system-level clocking; the continued operation of any portions of the device that are not undergoing reconfiguration; and no disruption to the contents of internal registers during reconfiguration, even in the area being reconfigured. *See also Configurable Hardware, Reconfigurable Hardware, Remotely Reconfigurable Hardware, and Dynamically Reconfigurable Hardware*.

Virtual Memory – A trick used by a computer's operating system to pretend that it has access to more memory than is actually available. For example, a program running on the computer may require 500 megabytes to store its data, but the computer may have only 128 megabytes of memory available. To get around this problem, whenever the program attempts to access a memory location that does not physically exist, the operating system performs a slight-of-hand and exchanges some of the contents in the memory with data on the hard disk.

Virus – *see Computer Virus*.

VLSI (Very Large Scale Integration) – Refers to the number of logic gates in a device. By one convention, very-large-scale integration represents a device containing 1,000 to 999,999 gates.

Volatile – Refers to a memory device that loses any data it contains when power is removed from the system: for example, random-access memory in the form of SRAM or DRAM.

Wafer – A paper-thin slice cut from a cylindrical crystal of pure semiconductor.

Wafer Probing – The process of testing individual integrated circuits while they still form part of a wafer. An automated tester places probes on the device's pads, applies power to the power pads, injects a series of signals into the input pads, and monitors the corresponding signals returned from the output pads.

Waveguide – A transparent path bounded by non-transparent, reflective areas, which is fabricated directly onto the surface of a substrate. Used in the optical interconnection strategy known as *guided-wave*.

Wave Soldering – A process used to solder circuit boards populated with through-hole components. A machine creates a wave[18] of hot, liquid solder which travels across the surface of the tank. The populated circuit boards are passed over the wave-soldering machine on a conveyer belt. The velocity of the conveyer belt is carefully controlled and synchronized such that the solder wave brushes across the bottom of the board only once.

Wire Bonding – The process of connecting the pads on an unpackaged integrated circuit to corresponding pads on a substrate using wires that are finer than a human hair. Wire bonding may also be used to connect the pads on an unpackaged integrated circuit, hybrid, or multichip module to the leads of the component package.

Wiring Layer – A layer carrying wires in a discrete wired board. *See also Signal Layer*.

Word – A group of signals or logic functions performing a common task and carrying or storing similar data: for example, a value on a computer's data bus could be referred to as a "data word" or " a word of data."

X Architecture – An initiative proposed by a group of companies in 2001 to use diagonal tracks to connect functions on silicon chips (as opposed to traditional North-South and East-West tracking layers). Initial evaluations apparently show that this diagonal interconnect strategy can increase chip performance by 10% and reduce power consumption by 20%. However, it may take some time for design tools and processes (and popular acceptance) to catch up and start using this technique.

X-Ray Lithography – Similar in principle to optical lithography, but capable of constructing much finer features due to the shorter wavelengths involved. However, X-ray lithography requires an intense source of X-rays, is more difficult to use, and is considerably more expensive than optical lithography.

Yield – The number of devices that work as planned, specified as a percentage of the total number actually fabricated.

---

[18] A large ripple, actually.

Index

f = footnote, **s** = sidebar, **p** = pronunciation, **CD** = CD-ROM

f = footnote, **s** = sidebar, **p** = pronunciation, **CD** = CD-ROM

f = footnote, **s** = sidebar, **p** = pronunciation, **CD** = CD-ROM

f = footnote, **s** = sidebar, **p** = pronunciation, **CD** = CD-ROM

f = footnote, **s** = sidebar, **p** = pronunciation, **CD** = CD-ROM

f = footnote, **s** = sidebar, **p** = pronunciation, **CD** = CD-ROM

f = footnote, **s** = sidebar, **p** = pronunciation, **CD** = CD-ROM

f = footnote, **s** = sidebar, **p** = pronunciation, **CD** = CD-ROM

I

f = footnote, **s** = sidebar, **p** = pronunciation, **CD** = CD-ROM

f = footnote, **s** = sidebar, **p** = pronunciation, **CD** = CD-ROM

f = footnote, s = sidebar, p = pronunciation, **CD** = CD-ROM

f = footnote, **s** = sidebar, **p** = pronunciation, **CD** = CD-ROM

f = footnote, **s** = sidebar, **p** = pronunciation, **CD** = CD-ROM

f = footnote, **s** = sidebar, **p** = pronunciation, **CD** = CD-ROM

f = footnote, **s** = sidebar, **p** = pronunciation, **CD** = CD-ROM

f = footnote, **s** = sidebar, **p** = pronunciation, **CD** = CD-ROM

f = footnote, **s** = sidebar, **p** = pronunciation, **CD** = CD-ROM

f = footnote, **s** = sidebar, **p** = pronunciation, **CD** = CD-ROM

f = footnote, **s** = sidebar, **p** = pronunciation, **CD** = CD-ROM

LIMITED WARRANTY AND DISCLAIMER OF LIABILITY